The 3D Studio MAX® R2 Quick Reference

Michele Bousquet

Autodesk®

Press

I(T)P® International Thomson Publishing

Albany • Bonn • Boston • Cincinnati • Detroit • London • Madrid
Melbourne • Mexico City • New York • Pacific Grove • Paris • San Francisco
Singapore • Tokyo • Toronto • Washington

Trademarks

3D Studio MAX™ is a trademark of Autodesk, Inc. AutoCAD® and the AutoCAD logo are registered trademarks of Autodesk, Inc. Microsoft and Windows 95 are registered trademarks of the Microsoft Corporation. Windows NT is a trademark of Microsoft Corp. Online Companion is a trademark of International Thomson Publishing company. All other product names are acknowledged as trademarks of their respective owners.

Autodesk Press Staff:

Publisher: Alar Elken
Aquisitions Editor: Sandy Clark
Production Coordinator: Jennifer Gaines
Art and Design Coordinator: Mary Beth Vought
Editorial Assistant: Christopher Leonard

COPYRIGHT © 1998
Delmar Publishers Inc.
Autodesk Press imprint

an International Thomson Publishing company
The ITP logo is a trademark under license.
Printed in the United States of America

For more information, contact:
Autodesk Press
3 Columbia Circle, Box 15-015
Albany, New York USA 12212-5015

International Thomson Publishing Europe
Berkshire House 168-173
High Holborn
London, WC1V 7AA
United Kingdom

Thomas Nelson Australia
102 Dodds Street
South Melbourne, Victoria 3205
Australia
Nelson Canada
1120 Birchmont Road
Scarborough, Ontario
Canada, M1K 5G4

International Thomson Publishing Southern Africa
Building 18, Constantia Park
240 Old Pretoria Road
P.O. Box 2459
Halfway House, 1685 South Africa

International Thomson Editores
Campos Eliseos 385, Piso 7
Colonia Polanco
11560 Mexico D. F. Mexico

International Thomson Publishing GmbH
Konigswinterer Strasse 418
53227 Bonn Germany

International Thomson Publishing France
Tour Maine-Montparnasse
33, Avenue du Maine
75755 Paris Cedex 15, France

International Thomson Publishing -Japan
Hirakawacho Kyowa Building, 3F
2-2-1 Hirakawa-cho Chiyoda-ku
Tokyo 102 Japan

International Thomson Publishing Asia
221 Henderson Road
#05-10 Henderson Building
Singapore 0315

1 2 3 4 5 6 7 8 9 10 XXX 03 02 01 00 99 98 97

Library of Congress Cataloging-in-Publication Data

Bousquet, Michele, 1962-
    The 3D Studio MAX 2.0 Quick Reference / Michele Bousquet,
       p.   cm.
    ISBN 0-7668-0152-7
    1. Computer animation. 2. 3D Studio. 3. Computer graphics.
    I. Title.
    TR897.8.B676   1998
    006.6'93--dc21   97-43425
                       CIP

Table of Contents

Menus   1

Menus (cont'd)

Toolbar   167

Toolbar (cont'd)

Panels   251

Panels (cont'd)

Mat Editor 571

Appendix   681

How to Use This Book

This book is divided into five sections. Each section lists entries in alphabetical order. Sections of the book are indicated by tabs at the edges of pages.

Menus

The *Menus* section describes each menu option. To find an entry, look under the last menu option in the sequence. For example, to find information on the *File/Preferences* menu option, look under **Preferences** in the menu section. Two exceptions are *Edit/Select by/Name* and *Edit/Select by/Color*, which are described under the **Select By Color/Name** entry.

Toolbar

The *Toolbar* section describes the buttons and pulldowns across the top and bottom of the screen. To find out a button's name, move the cursor over the button and wait a moment until the button name appears below the button. A picture of the button appears next to the entry name.

When a button is a flyout, an airplane ✈ appears across from the button name. When a button is accessed from a flyout, an airplane appears along with a picture of the button used to access the flyout.

Panels

The *Panels* section describes each option available on the panel at the right or left of viewports. For detailed information on how entries are listed, see the beginning of the *Panels* section.

Mat Editor

The *Mat Editor* section describes the options on the Material Editor. Entries include all buttons on the Material Editor toolbar and all material and map types. A few additional entries, such as **About the Material Editor** and **Choosing Maps**, provide general information on common tasks in the Material Editor. These entries are referenced where necessary from other entries in the *Mat Editor* section.

Appendix

The *Appendix* includes important information on rendering concepts. Several of these concepts are referenced from entries concerned with rendering, such as the **Render** entry in the *Menus* section.

Format Notes

Each entry has a section called General Usage. This section is designed to show you at least one way the entry can be used.

Where appropriate, Usage Notes are included to give you important information on the entry. A Sample Usage section is sometimes included with a simple tutorial to illustrate how an entry can be used.

When a button is referenced, a picture of the button appears the first time it is referenced on that page, but not for subsequent references on the same page.

In this book, the following references appear in boldface.

- Entries in any section
- Button names
- Keyboard keys
- Options on a dialog or panel
- Sections on a dialog or panel

The following references are italicized.

- Menu options
- Names of sections in this book

The following references are not boldfaced or italicized.

- Dialog names
- Rollout names

Examples of formatting in this book:

- The Time Configuration dialog can be accessed by clicking **Time Configuration** . The **Length** value under the **Animation** section sets the length, in frames, of the current animation.

- To clone an object, select the object and choose *Edit/Clone* from the menu. You can also clone an object by holding down the **<Shift>** key while transforming the object.

- The settings under the Rotational Joints and Sliding Joints rollouts set limits for inverse kinematics. See the **IK** entry in the *Panels* section for more information on inverse kinematics.

Acknowledgments

With every book, there are a few special people who are vital to the end result.

I would like to proffer my thanks to all those on the CompuServe Kinetix Forum who answered my questions thoroughly and promptly, especially David Marks and Cheryl Fromholzer of Kinetix, both of whom suffered through my endless niggling queries about this or that parameter. An extra special thanks to Dennis Phinney at Kinetix who came through for me in a time of need.

Thanks also to Larry Minton for his tireless technical editing, and to Michael George for his help on the NURBS and VRML tools. And of course, thanks to Monique Gillotti for sparing no red ink in practicing her unsurpassed proofreading skills.

A big thanks to my production team at Autodesk Press, Sandy Clark, Chris Leonard, Jennifer Gaines, Mary Beth Vought and Mary Beth Ray, for their sanity and patience through the howling demands of authorship.

Of course I have to thank my buddies at KR Graphics in Andover, Massachusetts, Kenny and John Citron, who helped with all the last-minute tasks and got this book out to pre-press. The jokes are still terrible and the coffee hasn't improved, but the company is still great.

Every author of a computer book should have a technician close by to keep the machine ticking, a live-in musician to soothe the nerves with lilting melodies, and a supportive significant other to say the right things when times are rough. I have the good fortune to have all three rolled into one. My special thanks to David Drew for helping me with this book in more ways than I can count.

Menus

This section describes the menus across the top of the 3D Studio MAX screen. Menu options are listed in alphabetical order. To find an entry, look under the last menu option in the sequence. For example, a description of the *Tools/Display Floater* menu option is under the **Display Floater** entry in this section of the book.

Two exceptions are *Edit/Select by/Name* and *Edit/Select by/Color*, which are described under the **Select By Color/Name** entry.

In 3D Studio MAX, you can also right-click on an object or screen item to access a pop-up menu. For example, a pop-up menu appears when you right-click on a viewport label. For information on commonly used pop-up menus, see the **Pop-up Menus** entry in this section of the book.

About 3D Studio MAX

Displays information about 3D Studio MAX.

General Usage

Choose *About 3D Studio MAX* from the *Help* menu. A dialog appears.

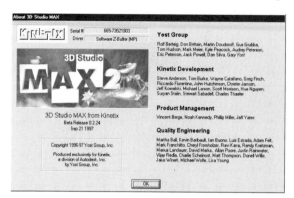

Your serial number and driver are shown at the top left of the screen. Credits are at the right. After a few moments, the credits will begin to roll, showing a long list of people who worked on 3D Studio MAX.

Click **OK** when done viewing the information on this dialog.

Dialog Options

Serial #   Displays the serial number for 3D Studio MAX.

Driver   Indicates the driver being used for graphics display. This driver was selected when MAX was installed, and cannot be changed here. To change the driver, choose the *File/Preferences* menu option and access the **Viewports** tab.

OK   Exits the dialog.

Activate Grid Object

Activates a grid object for use as the current grid.

General Usage

Create a grid object and select the object. Choose *Views/Grids/Activate Grid Object*. The selected grid becomes the current grid.

Usage Notes

By default, 3D Studio MAX provides a 3D home grid for creating objects and positioning them in 3D space. The home grid is the grid you see in each viewport.

There may be times when you want to create objects on a slope to match the rotation of other objects, or you want to work with a grid with its 0,0,0 origin at a particular location. In these cases, you can create your own grid and use it as the creation grid.

To create a grid object, go to the **Create** panel 🖉. Click **Helpers** 🔲. Click **Grid**, and click and drag in any viewport to create the grid object.

The grid can then be rotated or positioned anywhere in the scene. Select the grid and choose *Views/Grids/Activate Grid Object* to make the new grid the current grid. You can also align the new grid to the current viewport with *Views/Grids/Align to View*.

To switch back to the home grid, choose *Views/Grids/Activate Home Grid* from the menu.

Activate Home Grid

Sets the home grid as the active grid.

General Usage

This option is available when a grid object has previously been selected for use as the active grid.

Choose *Views/Grids/Activate Home Grid* to set the home grid as the active grid.

Displays help for additional topics not covered by the main help.

General Usage

Choose *Additional Help* from the *Help* menu. Highlight a topic on the list and click **Display Help**. Help for the chosen topic appears.

Usage Notes

Help for most topics can be found by choosing *Help/Online Reference* from the menu. The topics listed under *Help/Additional Help* require a large body of help information, so they have been given separate help entries.

Align

Aligns one or more objects with another object.

General Usage

Select one or more objects to be aligned with another object. Choose *Align* from the *Tools* menu.

Click on the object to which you wish to align the selected objects. The Align Selection dialog appears.

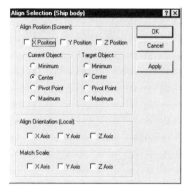

For information on the Align option, see the **Align** entry in the *Toolbar* section of this book.

Aligns one object with another using the normals of the target object.

General Usage

Select the object to be aligned. Choose *Align Normals* from the *Tools* menu. Click and drag over the selected object until the desired normal appears as a blue arrow. Release the mouse.

Click and drag over the object to which you want to align the selected object until the desired normal appears as a green arrow. Release the mouse.

The selected object jumps to align with the target object, with the two selected faces aligned. The Normal Align dialog appears.

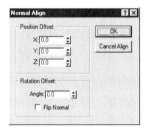

For information on the Align Normals option, see the **Normal Align** entry in the *Toolbar* section of this book.

Align To View

Aligns the active grid to the current viewport.

General Usage

Select a grid object. Choose *Views/Grid/Activate Grid Object* to activate the grid. Choose *Views/Grids/Align to View*. The grid is aligned to the viewport.

Sample Usage

Under the **Create** panel , click on **Helpers** and click **Grid**. Create a grid in any viewport. In an orthographic view, use **Arc Rotate** to rotate the view, changing it to a User view.

Choose *Views/Grid/Activate Grid Object* to activate the grid. Select the grid object in the User View. Choose *Views/Grids/Align to View* from the menu. The grid is aligned to the User view.

Creates a .ZIP or other archive file containing the scene and all bitmaps used in the scene.

General Usage

Load the scene to be archived. Choose *File/Archive* from the menu. A file selector appears. Enter or choose an archive filename and click **Save**. The archive is created.

Usage Notes

The default archive program used with this option is PKZIP. If the archive operation fails, it means 3D Studio MAX could not find PKZIP in the 3dsmax2 directory or any of the configured paths for your system.

You can set the archive program to another archiving program, or set the path for PKZIP, with *File/Preferences* under the **Files** tab.

Archiving is useful for transferring your scene to another computer for rendering or viewing. Rather than trying to remember the names of all the bitmaps you have used in your scene, you can use the Archive command to automatically create an archive file with all the necessary bitmaps. The archive file can then be copied to another computer and decompressed for a render-ready scene.

Array

Creates an array (a series of objects) from one or more objects.

General Usage

Select one or more objects to array. Choose *Tools/Array* from the menu. The Array dialog appears.

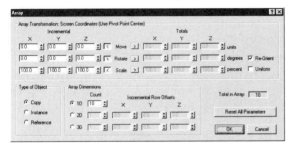

This dialog allows you to array objects in any direction with any transform.

For information on the Array tool, see the **Array** entry in the *Toolbar* section of this book.

Attaches one or more objects to an existing group.

General Usage

Select the object or objects that you wish to attach to a group. Choose *Attach* from the *Group* menu. Click on the group to which you wish to attach the object or objects. The objects are attached to the group.

In order to use *Group/Attach*, you must first have grouped other objects in the scene with *Group/Group*. Objects can be detached from a group with *Group/Detach*.

Grouping is often a better alternative to attaching, linking or performing a boolean union. Grouped objects can be readily transformed together. For example, a group can be easily rotated around a common center. With grouped objects, there is less danger of accidentally choosing the wrong object for transformation, as there is with linked objects. In addition, grouped objects don't lose their mapping coordinates as boolean objects do.

Background Image

Displays a map, image or animation in the current viewport as a background image.

General Usage

Activate the viewport in which you wish to see a background image. Choose *Background Image* from the *Views* menu. The Viewport Background dialog appears.

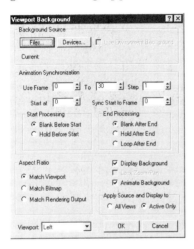

Click **Files** or **Devices** to choose the source for the background. Turn on **Display Background** to make the background appear in the current viewport. Set the aspect ratio to match the viewport, bitmap or rendering output.

Click **OK** to activate the settings. After a few moments, the background appears in the viewport. A different background can be assigned to each viewport.

To set up a bitmap image as the background, click **Files**. An animated file can be specified, such as an *.avi* file or a series of images in an *.ifl* file. To set up an environment map, see **Use Environment Map** below.

Note that setting up a background to appear in a viewport will not cause the background to appear in the rendered image. The background for rendering is set up with the *Rendering/Environment* menu option. For more information, see the **Environment** listing in the *Menus* section of this book.

Menus

Dialog Options

Files   Click on **Files** to select the bitmap image to display as the viewport background.

Devices   Click **Devices** to import a background from a plug-in device such as a video capture card. You must install the device, the driver and the plug-in software before you can use this option.

Use Environment Map   When this checkbox is on, an environment map previously assigned as the rendering background appears in the viewport. In order to enable this checkbox, you must first assign an environment map as the background. To do this, first set up the environment map under any map setting in the Material Editor. Then choose *Rendering/Environment* from the menu, click the **Environment Map** button (labeled **None**) and select the map from the Material/Map Browser. When you choose *Views/Background Image* from the menu, this checkbox will be enabled.

When you turn this checkbox on, all options except **Display Background**, **All Views** and **Active Only** are disabled.

To speed up display and redraw of the environment background, choose *File/Preferences* from the menu and turn on the **Low-Res Environment Background** checkbox under the **Viewports** tab.

Use Frame   The first input frame (background image, animation or device input) to be used as a background for the animated scene. You can further customize the frames to be used with the **Sync Start to Frame** number below.

To   The last input frame to be used as a background.

Step   Frame increment for background playback. For example, a **Step** value of 1 plays every frame, 2 plays every other frame and 5 plays every 5th frame.

Start at   Scene frame number at which you want the first background image frame to appear. To set up the action for frames before this frame, choose an option from the **Start Processing** section of the dialog.

Menus

Sync Start to Frame	The scene frame at which to start playing the background animation. The default is the **Use Frame**. Set the **Start Sync to Frame** number greater than or equal to the **Use Frame** number. For example, if the **Use Frame** number is 10, the **Sync Start to Frame** number is 15, and the **Start at** number is 40, then frame 15 of the background animation will be used as a background for frame 40.
Blank Before Start	Displays a blank background on frames before the **Start at** frame.
Hold Before Start	Displays the first used frame of the background animation on frames before the **Start at** frame.
Blank After End	Displays a blank background after the range of background frame has played.
Hold After End	Displays the last used frame of the background animation after the range of background frames has played.
Loop After End	Loops playback of the background animation.
Match Viewport	Scales the background image to fit in the viewport.
Match Bitmap	The background is displayed with the bitmap's aspect ratio. A border may appear at the top, bottom or sides of the viewport.
Match Rendering Output	Scales the background image to the rendering aspect ratio. A border may appear at the top, bottom or sides of the viewport.
Display Background	When this option is on, the background displays in the viewport.
Lock Zoom/Pan	This checkbox is enabled when **Match Bitmap** or **Match Rendering Output** is chosen as the display aspect ratio. When this checkbox is on and you zoom or pan in the viewport, the background zooms or pans along with the scene.

Animate Background   If the background image file is an animation file such as an AVI file or an IFL listing, turn on this option to use animation frames.

The appropriate frame of the animation will appear in the viewport when the time slider is moved. If you want to see an animated background while playing animation, choose *File/Preferences* to access the Preference Settings dialog, and turn on the **Update Background While Playing** checkbox under the **Viewports** tab.

OK   Accepts settings and exits the dialog.

Cancel   Cancels settings and exits the dialog.

Clone

Makes copies of the selected object or objects.

General Usage

Select an object or objects. Choose *Edit/Clone* from the menu. The Clone Options dialog appears. Select a clone option under Object.

You can also clone an object by holding down the **<Shift>** key while transforming the object. For example, if **Select and Move** is on, you can hold down the **<Shift>** key and click and drag on an object to move and copy it at the same time. The dialog above appears when you use **<Shift>** to clone an object.

Dialog Options

Object   Choose the type of clone you want to make. See **Sample Usage** below for a brief exercise that illustrates the differences between these three clone types.

Copy makes an independent copy of the object.

Instance makes a two-way dependent clone of the object. When the original object or instance is modified, the other object is automatically modified as well. Modifiers are shared by the original object and instance, but transforms are not. For example, a **Bend** modifier applied to one object will be automatically applied to the other, but scaling one of the objects with **Select and Uniform Scale** will not transform the other object.

Reference makes a one-way dependent clone. Modifiers applied to the original object are automatically applied to the reference, but modifiers applied to the reference are not applied to the original object. However, if modifiers were applied to the original object before the reference was made, changing the parameters of the previously applied modifier on either object will be applied to the other object.

Controller   Choose the type of clone used to control linked objects. These options are only available when two or more linked objects are selected for copying.

Choosing **Instance** instances the relationship between any linked children that are being cloned, and their parent object. Any change to one child's position in relationship to the parent will be instanced to the corresponding child clone. The Controller **Instance** instances child relationships regardless of whether **Copy, Instance** or **Reference** was chosen under **Object**.

Sample Usage

The easiest way to understand the difference between copies, instances and references is to try an example using the three types of clones.

1. In the Top viewport, create a sphere with a radius of about 50 units.

2. Click **Select and Move** ⊕. Hold down the **<Shift>** key and move the sphere to the right of itself in the Top viewport. The Clone Options dialog appears. Make sure **Copy** is on under the Objects section, and click **OK**.

3. Make another clone of the sphere next to the first clone, but this time make it an **Instance**.

4. Select the original object. Under the **Modify** panel, click **Taper**. Change the **Amount** of the taper to 2. The original object and the instance are both modified, while the copy is not.

5. Make another clone of the original object. Make this clone a **Reference**.

6. Select the original object. Under the **Modify** panel, click **Twist** and change the **Angle** to 60. The modifier is shared with the original object, which in turn shares it with the instance.

7. Select the reference object. Under the **Modify** panel, click **Noise**. Change **Scale** to 10, and change **X**, **Y** and **Z** under the Strength section to 10. The **Noise** modifier is applied only to the reference object.

8. Under the Modifier Stack rollout, pull down the reference object's modifier stack. Note that a line appears between the **Twist** and **Noise** modifiers. This line marks the difference between instanced and independent modifiers. Access the **Twist** modifier. Change the **Angle** parameter. The twist changes on the original object and the instance as well.

9. Apply another modifier such as **Skew**. The modifier is applied to the original object and instance as well. Because the modifier was applied while the stack was below the line, it is an instanced modifier and is shared with the original object, which in turn shares it with the instance.

Close

Closes an open group.

General Usage

Select one or more objects in the open group, or select the group bounding box. Choose *Group/Close* from the menu.

Sample Usage

Select one or more objects in the scene. Create a group with *Group/Group*. Open the group with *Group/Open*. Select and move one or more of the objects. Choose *Group/Close* from the menu to close the group again.

Sets up default subdirectory paths for all file operations including file export, plug-ins and bitmaps.

General Usage

Choose *File/Configure Paths* from the menu. The Configure Paths dialog appears.

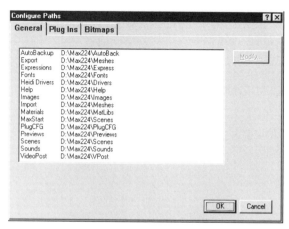

The dialog has three tabs for different types of paths.

General   Sets up default paths for most 3D Studio MAX file operations.

Plug Ins   Tells MAX where your plug-ins can be found.

Bitmaps   Tells MAX where to look for bitmaps when rendering. The order in which the paths are listed determines the order in which MAX will search subdirectories for a needed bitmap.

Dialog Options

Modify   To modify an existing path, highlight the path and click **Modify**. Choose a new path for the item. You can also choose a network drive as a path.

Delete   To delete a path from the list, highlight the path and click **Delete**. This option is available only under the **Plug Ins** and **Bitmaps** tabs.

Add   To add a path, highlight the path and click **Add**. Choose a new path for the item. When adding a path under the **Plug Ins** tab, enter a name for the new path in the **Description** field. This option is available only under the **Plug Ins** and **Bitmaps** tabs.

Move Up   MAX uses the directory listing under the **Bitmaps** tab to
Move Down search for bitmaps when rendering. Directories are searched in the order listed. If several directories are listed, the search can take a long time. To keep search time to a minimum, organize the directory listing so 3D Studio MAX is most likely to find the bitmap in the first few directory entries.

To change a directory's position on the list, highlight the entry. Click **Move Up** or **Move Down** to move the directory up or down one level on the list.

Connect to Support and Information

Connects to the Kinetix website.

General Usage

With your Internet software, connect to your service provider. In 3D Studio MAX, choose *Help/Connect to Support and Information*. MAX connects to the Kinetix website.

Usage Notes

Online help for 3D Studio MAX is updated regularly, and updated help files can be downloaded from the Kinetix website. Once downloaded, the help files can be accessed by choosing *Help/Online Reference* from the menu. See the Kinetix website for instructions on downloading help files.

The Kinetix website can be reached through any web browser at **www.ktx.com**.

Crossing

Changes the selection mode to crossing.

General Usage

When crossing mode is active, any selection region drawn will select all objects partially or completely inside the region.

The converse of crossing mode is window mode. When window mode is active, only objects completely inside a selection region will be selected.

To activate crossing mode, choose *Edit/Region/Crossing* from the menu. This also causes the **Window Selection** button on the prompt line to change to the **Crossing Selection** button .

Deactivates the **Show Map in Viewport** ⧈ function for all maps in the scene.

General Usage

Choose *Deactivate All Maps* from the *Views* menu. A message appears reminding you that this operation will turn off the **Show Map in Viewport**

button ⧈ for all materials being used in the scene.

This option cannot be undone. Once you have deactivated all maps, the only way to reactivate them is to load each material into the Material Editor and turn on the **Show Map in Viewport** button ⧈ again for each one.

Delete

Deletes selected objects and sub-objects from the scene.

General Usage

Select one or more objects or sub-objects to delete. Choose *Edit/Delete* from the menu.

You can also use the **<Delete>** key on the keyboard to delete objects or sub-objects.

Delete Track View

Deletes a closed Track View window.

General Usage

Choose *Delete Track View* from the *Track View* menu. A list of closed Track View windows appears.

To delete a Track View window, highlight it on the list and click **OK**.

Only closed Track View windows appear on the list. Minimized or open Track View windows do not appear on the list.

Detach

Detaches an object from its group.

General Usage

Select a group. Choose *Group/Open* from the menu. This opens the current group.

Select an object from the group. Choose *Group/Detach* from the menu. The object is detached from the group.

Usage Notes

Before an object can be detached from a group, it must be grouped or attached to an existing group.

To try this command sequence, create two or more objects in the scene. Select both objects. Choose *Group/Group* from the menu and accept the default group name. The objects are now grouped, which means you can transform or modify them as one object.

Choose *Group/Open* from the menu. This opens the group, allowing you to manipulate individual objects in the group. Select an object in the group. Choose *Group/Detach*. The object is detached from the group.

Displays a floating dialog for controlling object display and for hiding and freezing objects.

General Usage

Choose *Display Floater* from the *Tools* menu. The Display Floater appears.

This dialog allows you to hide, unhide, freeze and unfreeze objects. Click the **Object Level** tab for further hiding and freezing options.

Hiding objects removes them from the screen, but does not delete them from the scene. You can unhide hidden objects at any time.

Freezing objects leaves the objects on the screen, but prohibits them from being selected or altered. Objects turn dark gray when frozen.

Hide/Freeze

Hide   Click **Selected** to hide selected objects. Click **Unselected** to hide unselected objects. Click **By Name** to display a list where you can pick objects to hide. Click **By Hit** to click on objects to be hidden. When hiding objects by hit, every object you select will be hidden until you select another tool.

Freeze   Click **Selected** to freeze selected objects. Click **Unselected** to freeze unselected objects. Click **By Name** to display a list where you can pick objects to freeze. Click **By Hit** to click on objects to be frozen. When freezing objects by hit, every object you select will be frozen until you select another tool.

Unhide   Click **All** to unhide all hidden objects. Click **By Name** to select objects to unhide from a list.

Unfreeze   Click **All** to unfreeze all frozen objects. Click **By Name** to select objects to unfreeze from a list. Click **By Hit** to click on objects to unfreeze.

Object Level

On the **Object Level** tab, you can hide objects by category.

All	Turns on all the checkboxes in the Hide by Category section.
None	Turns off all the checkboxes in the Hide by Category section.
Invert	Inverts the selection of checkboxes.

The options under the Display Properties section pertain to currently selected objects.

Display as Box Displays the selected objects as bounding boxes.

Backface Cull Eliminates the display of faces behind other faces in wireframe display.

Edges Only Displays edges only, rather than the crossline separating triangular faces.

Vertex Ticks Displays tick marks at each vertex.

Trajectory Displays the animation trajectories of selected objects.

Edit Named Selections

Edits, deletes, combines or adds selection sets.

General Usage

Choose *Edit Named Selections* from the *Edit* menu. The Edit Named Selections dialog appears. This dialog can be used to edit named selection sets.

Dialog Options

Named Selections	Lists all named selection sets.
Objects	When a selection set is highlighted under **Named Selections**, the objects in the selection set appear on this list.
Combine	Combines two or more selection sets. Select the selection sets from the list, then click **Combine**. A dialog appears where you can enter the new selection set name. The new selection set is added to the list of **Named Selections**.
Delete	Deletes the selected selection set.
Subtract (A-B)	When two selection sets are selected, subtracts the contents of the second selected set from the first.
Subtract (B-A)	When two selection sets are selected, subtracts the contents of the second selected set from the first.
Intersection	When two selection sets are selected, this option creates a new selection set from the objects contained in both selection sets. Select two selection sets and click **Intersection**. A dialog appears for the name of the new selection set being formed. The new selection set is added to the list of **Named Selections**.

Add   Adds objects to the currently selected selection set. Select a selection set and click **Add**. A dialog appears where you can select objects to add to the selection set. Click the **Add** button to add the objects to the selection set. You can choose more than one selection set at a time and add the selected objects to all the selections sets at once.

Remove   Removes objects from the current selection set. Highlight the object from the Objects list and click **Remove** to remove the object from the selection set.

OK   Accepts all changes to named selection sets.

Cancel   Cancels all changes to named selection sets.

Usage Notes

Selection sets are created directly on the screen. To create a selection set, select objects for the selection set. Type a selection set name in the **Named Selection Sets** entry box on the Toolbar and press **<Enter>**. You can also choose selection sets from the **Named Selection Sets** pulldown menu. Objects can belong to more than one selection set.

Sets up the background and environment elements such as fog, fire and volume lights.

General Usage

Choose *Rendering/Environment* from the menu. The Environment dialog appears.

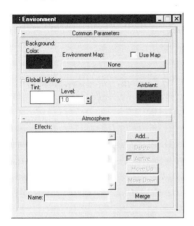

The top of the dialog sets up the background for the scene. To add fog, volume lights or combustion, click the **Add** button under the Atmosphere rollout. When an atmosphere option is selected, a rollout for the selected option appears in the lower section of the dialog.

Background

Color
The color for the background. Click the color swatch to access the Color Selector. This color is used for the background only if the **Use Map** checkbox is not checked.

Environment Map
If you want to use a map for the background, click the button labeled **None**. The Material/Map Browser appears. Under the **Browse From** section, pick the type of maps to display and choose the map from the list.

If the map has not yet been set up, turn on **New** under **Browse From** and select the map type from the list. To set up the map, exit the Material/Map Browser and access the **Material Editor** . Click **Get Material** to access the Material/Map Browser again and turn on **Scene** under **Browse From**. Select the environment map from the list. The map appears in the current Material Editor slot, where you can edit it as usual with the Material Editor.

You can also assign a map by dragging the map from the Material Editor to the **Environment Map** button. In this case, you will be prompted to choose whether to use a copy or instance of the map for the environment map.

Use Map   When this option is turned on, MAX uses the specified environment map as the background. When **Use Map** is turned off, MAX uses the **Color** for the background. This checkbox is automatically turned on when you click the **Environment Map** button and choose a map or map type.

Tint   Tints all lights in the scene. If the color swatch is any color but white, all lights in the scene except ambient light will be tinted with the **Tint** color.

Level   Sets a global multiplier for all lights except ambient. Each light has a Multiplier value on its Create and Modify panels for multiplying the light's intensity. Raising or lowering the **Level** allows you to increase or decrease the intensity of all lights in the scene at once.

Ambient   Sets the ambient light color for the scene. Ambient light is light that permeates the entire scene, such as the gray light at dawn or dusk. Lightening the ambient light color is not a solution to a poorly lit scene. As too much ambient light results in low-contrast images, this color should be kept very dark.

Usage Notes

You can set up a background to appear in viewports while you're drawing on the screen. For information on this procedure, see the **Background Image** entry in the *Menus* section of this book.

Atmosphere

Under the Atmosphere rollout, you can add effects such as fog, volume lights and combustion. For further information on atmospheric effects, see the heading for each type of effect below.

Effects   Effects added to the scene are listed in this window.

Name   Once you select an effect, you can assign it a custom name in this entry box.

Add   Click **Add** to add an effect to the scene. The Add Atmospheric Effect dialog appears.

Menus

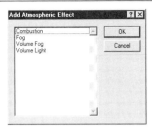

Combustion creates fire and smoke effects. **Fog** and **Volume Fog** create a foggy effect in the scene. A **Volume Light** makes a light appear to have physical volume.

Atmospheric effects do not appear in viewports. The scene must be rendered in order to see the effects. Atmospheric effects render only in Perspective or camera views.

Delete   Deletes the effect from the list.

Active   Indicates whether the effect is currently active. You can use this checkbox to turn off an effect when working with the scene to speed up rendering time.

Move Up
Move Down   Effects are rendered in the order listed. The rendered scene may look different when the order of effects is changed. You can rearrange the list by moving effects up or down. To move an effect, highlight the effect and click **Move Up** or **Move Down** to move it up or down on the list.

Merge   Merges atmospheric effects and their related objects from other files to the current scene.

Combustion

Combustion creates fire, smoke and explosions. The size of the effect is controlled by the size of the object to which the **Combustion** effect is assigned. For all effects, the time length of the effect and the colors used can be set. For fire, you can control the width of the flames and how fast they climb. For explosions, each phase of the explosion can be accurately timed.

To access the Combustion Parameters rollout, click **Add** and choose **Combustion** from the Add Atmospheric Effect dialog. Under the **Atmosphere** section of the Environment dialog, highlight **Combustion**. The Combustion Parameters rollout appears.

Menus

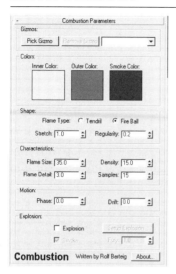

The **Gizmos** section is where you choose the object to enclose the **Combustion** effect. Only an atmospheric gizmo can be picked for **Combustion**. The **Combustion** effect is always contained inside the boundaries of the gizmo.

To create an atmospheric gizmo, go to the **Create** panel and click

Helpers . Choose **Atmospheric Apparatus** from the pulldown list. Create a **BoxGizmo**, **CylGizmo** or **SphereGizmo**. The shape of the gizmo determines the overall shape of the **Combustion** effect. Gizmos can also be scaled to further customize the shape of the effect. For more information on creating an atmospheric gizmo, see the **BoxGizmo**, **CylGizmo** or **SphereGizmo** entries in the *Panels* section of this book.

Pick Gizmo   In order for the **Combustion** effect to appear in your rendering, the effect must be assigned to an atmospheric apparatus. Click **Pick Gizmo** and pick one or more atmospheric gizmos from the scene. You can also press the **<H>** key to pick objects from a list.

If you pick more than one gizmo, the same **Combustion** settings will be applied to each one. To create multiple **Combustion** effects with different settings, click the **Add** button under Atmosphere to add more **Combustion** effects to the list, and choose a different gizmo for each one. Set the values for each individual gizmo.

Remove Gizmo   To remove a gizmo from the selection list, use the pulldown list to select the gizmo and click **Remove Gizmo**.

Under Colors, you can choose three colors for the **Combustion** effect.

Inner Color   The color for the densest part of the combustion effect.

Outer Color   The color for the less dense areas of the combustion effect.

Smoke Color   The smoke color for explosion effects. When both the **Smoke** and **Explosion** checkboxes under the Explosion section are checked, the **Inner Color** and **Outer Color** morph to the smoke color after the explosion. If either **Smoke** or **Explosion** are unchecked, this color is not used.

The **Shape** section is for setting the overall shape of the effect.

Flame Type   **Tendril** flames are oriented along the local Z axis of the gizmo. **Tendril** flames are suited to campfire type effects. **Fireball** flames are round and are more suited to explosions.

Stretch   Stretches flames along the local Z axis of the gizmo. Values over 1.0 stretch flames, while values under 1.0 compress flames.

Regardless of how high the **Stretch** value is, flames still reside inside the gizmo. To make flames longer in physical size, scale the gizmo along its local Z axis with **Select and Non-uniform Scale** .

Regularity   A value ranging from 0.0 to 1.0 which determines how fully the flame fills the gizmo. A value of 1.0 nearly fills the gizmo, with some degradation of flames near the gizmo boundary. A value of 0.0 centers the effect at the center of the gizmo with a large degree of falloff as the flames reach the outer boundary.

Under **Characteristics** you can set attributes for the **Combustion** effect. Settings are interdependent on the gizmo size and on each other.

Flame Size   The relative size of individual flames, or in other words, the relative size of the blotches of the **Inner Color** that make up the flames. Large sizes work well for fireballs, while smaller sizes are recommended for tendrils. If the **Flame Size** is very small, the **Samples** value may need to be increased in order to see the flames in the rendering.

Flame Detail   A value ranging from 0.0 to 10.0 which controls the fuzziness or sharpness of flames. Lower values produce smooth, fuzzy flames. With higher values, the transition from the **Inner Color** to the **Outer Color** is much briefer, making sharp, patterned flames. Lower values render faster than higher values.

Density   Controls the transparency and brightness of flames. Lower values make the flames seem more transparent and use more of the **Outer Color.** Higher values make the flames more opaque and use more of the **Inner Color**. The apparent density is also affected by the size of the gizmo. The larger the gizmo, the more dense and opaque the flames will appear to be.

The **Density** value can be animated. If **Explosion** is checked, **Density** will automatically animate from 0.0 at the start of the animation, reaching the entered value at the peak of the explosion. If **Density** is animated and **Explosion** is on, the combustion effect takes into account the animated **Density** value and the current phase of the explosion to determine the **Density** for that frame. In any case, if the animated **Density** is 0, no flames will appear in the scene.

Samples   Sets the sampling rate, or the number of times the flame is calculated. Higher values produce more accurate flames, but take longer to render. You may need to raise this value if the **Flame Size** is small or **Flame Detail** is high.

The settings under the **Motion** section determine how the **Combustion** effect will behave over a series of frames.

Phase   The **Phase** value must be animated in order to produce a burning fire or an explosion. If **Phase** is not animated, flames will render, but will not animate.

The **Phase** value has different meanings depending on whether the **Explosion** checkbox is checked. When **Explosion** is unchecked, different **Phase** values will produce varying flame effects. If the **Phase** value changes steadily, the tendrils or fireball will churn steadily. As the **Phase** value changes more rapidly, the churning effect will increase. If the change in the **Phase** value is steady, the effect is a steady burning fire.

Menus

If **Explosion** is checked, the **Phase** value determines which part of the explosion is taking place at any given time. The **Phase** value corresponds to different parts of an explosion as follows.

Phase value	Explosion behavior
0	Density is 0, no flames visible.
1-99	The density of flames increases toward the **Density** value entered, flames become larger and morph from **Inner Color** to **Outer Color**.
100	Explosion reaches peak density and size.
101-200	If **Smoke** is checked, **Inner Color** and **Outer Color** morph toward the **Smoke Color**. If not, the flames churn, retaining the **Inner Color** and **Outer Color**.
201-299	Smoke or explosion dissipates.
300	All smoke and flames have disappeared.
Over 300	No combustion effect.

In order to have an explosion take place, the **Phase** value must be animated with values between 0 and 300. The simplest way to set up an explosion is to turn on the **Animate** button on the Toolbar, move to a later frame such as 100, and change the **Phase** value to 300. This will create an explosion from frames 0 to 100.

Intermediate **Phase** values can be set at different frames during the explosion to make each phase of the explosion occur faster or slower.

You may find it helpful to view or edit the phase curve in Track View. To do this, click **Track View** . Expand the **Environment** track and **Combustion** track and locate the **Phase** track. Click **Function Curves** on the Track View window and highlight the **Phase** track label. The **Phase** function curve appears. This curve shows you the change in **Phase** over time.

Menus

Drift | Determines the rate at which flames move up the local Z axis of the apparatus. Higher values make the flames move faster. The **Drift** value is the number of units the flame will move over the course of the animation. For the smoothest effects, enter a multiple of the apparatus height.

The options under the **Explosion** section are for setting up an explosion effect within the gizmo.

Explosion | Turn on to create an explosion effect. The rate of explosion is determined by the animated **Phase** value.

Smoke | Turn on to create smoke in the explosion. When this checkbox and the Explosion checkbox are turned on, the **Inner Color** and **Outer Color** will morph to the **Smoke Color** as the **Phase** value increases from 100 to 200.

Setup Explosion | Click this button to enter the start and end frames for the explosion.

Fury | Varies the churning effect of the current explosion phase.

About | Click this button to see the version number and information about the author of **Combustion**.

Usage Notes

When working with combustion effects, the most important setting is the **Phase** value. This value must be animated in order to see any change at all over a series of frames. This is true no matter which settings you choose in other parts of the dialog.

Fog

The **Fog** effect creates fog over objects in the scene. Fog renders only in camera and Perspective views.

Standard fog is placed in relationship to the current camera view, so the fog is always a specific distance from the camera. Layered fog creates a flat slab of fog parallel to the Top viewport with a specific height and location.

Both types of fog are used to make objects in the distance appear to be partially or fully hidden by fog. If your animated sequence calls for a camera that flies through chunks of fog, use **Volume Fog**.

To access the Fog Parameters rollout, click **Add** and choose **Fog** from the Add Atmospheric Effect dialog. Under the **Atmosphere** section of the Environment dialog, highlight **Fog**. The Fog Parameters rollout appears.

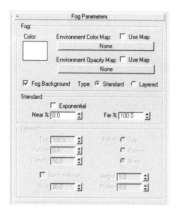

The **Fog** section of the Fog Parameters rollout sets up the colors, maps, and type of fog. For the fog color, a map can be used instead of a plain color.

Color Sets the color of the fog. Click on the color swatch to access the Color Selector. This color is used for the fog effect only if the **Use Map** checkbox is off.

Environment Color Map Allows you to use a map for the fog color. To assign a map for the fog color, click the button labeled **None**. The Material/Map Browser appears. Under the **Browse From** section, pick the type of maps to display and choose the map from the list.

If the map has not yet been set up, turn on **New** under **Browse From** and select the map type from the list. To set up the map, exit the Material/Map Browser and access the **Material Editor** . Click **Get Material** to access the Material/Map Browser again and turn on **Scene** under **Browse From**. Select the environment map from the list. The map appears in the current Material Editor slot, where you can edit it as usual with the Material Editor.

You can also assign a map by dragging the map from the Material Editor to the **Environment Color Map** button. In this case, you will be prompted to choose whether to use a copy or instance of the map for the environment map.

When you assign a map, the **Use Map** checkbox is automatically checked. When **Use Map** is checked, the map is used. If **Use Map** is unchecked, the **Fog** effect uses the **Color** as the fog color.

Environment Opacity Map Allows you to use a map to set the density of the fog. The grayscale values of the map are used to determine the density in different areas. Whiter areas make denser fog, while darker areas make lighter fog. Very dark areas eliminate fog altogether.

With **Standard** fog, the map works with the **Near %** and **Far %** to determine the density of the fog at any point between the camera's **Near** and **Far** range.

With **Layered** fog, the map works with the **Density** percentage to determine the density of the fog at any point.

An **Environment Opacity Map** is set up in the same way as an **Environment Color Map**. See **Environment Color Map** above for information on assigning the map.

Fog Background When checked, the background is fogged along with objects in the scene. When unchecked, fog does not affect the background.

Type Choose **Standard** or **Layered** fog. **Standard** fog works with the camera ranges to determine where the fog begins and ends in relationship to the camera position. **Standard** fog moves with the camera, and always appears to be in front of the camera.

Layered fog creates a slab of fog parallel to the Top viewport. The height and location of **Layered** fog is set under the **Layered** section of the dialog, and is not dependent on the camera location.

If you choose **Standard** fog as the **Type**, the **Standard** section of the rollout is accessible.

Exponential When checked, fog density increases exponentially with distance. Exponential fog usually represents real fog more accurately and correctly renders transparent objects inside the fog. On the down side, exponential fog sometimes renders as unrealistic color bands. Unless you have transparent objects inside the fog, leave **Exponential** unchecked.

Near % **Far %**	The percentage of total fog density at the camera's **Near** and **Far** ranges. If an **Environmental Opacity Map** has been specified, the density of the fog in the rendering is also affected by the map. For information on setting up a camera's **Near** and **Far** range, see the **Target Spot** entry in the *Panels* section of this book.

If you choose **Layered** fog as the type, the **Layered** section of the rollout can be accessed. Layered fog is a slab of fog extending infinitely on the World XY plane, also called the home grid. This means the fog slab is always parallel to the Top viewport. Layered fog can be seen only in a rendered camera view and is not actually visible in viewports.

The height and location of layered fog is set with the **Top** and **Bottom** values. The illustration below shows how layered fog is situated in relationship to the Left viewport. Note that layered fog is not actually visible in the Left viewport.

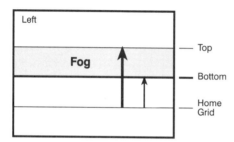

Top **Bottom**	Sets the top and bottom edge of the layered fog in units above the XY plane of the Home Grid (World XY plane).
Density	Sets the overall density of the layered fog. If an **Environmental Opacity Map** has been specified, the density of the fog in the rendering is affected by the map.
Falloff	Causes the fog to dissipate to zero density at the **Top** or **Bottom** of the fog. Choosing **None** causes no dissipation.
Horizon Noise	The top or bottom edge of the fog is visible at the camera's horizon line. When **Horizon Noise** is off, the edge of the fog renders as a straight edge. When **Horizon Noise** is on, the edge of the fog at the horizon line is made fuzzy with the use of a noise effect.
Size	Sets the size of the horizon noise effect.

Angle

Sets the amount of horizon noise using an angle. Imagine a large sphere passing through the horizon line with the camera at its center. Horizon noise appears along the virtual sphere over the number of degrees set by the **Angle** value.

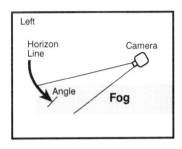

Angles from 5 to 10 create a realistic effect. With larger values such as 60, most or all of the fog in the scene is replaced with horizon noise. An **Angle** of 0 is the equivalent of turning off the **Horizon Noise** checkbox.

Horizon noise is always created in the direction away from the center of the camera view. When the camera and target are level or nearly level, the horizon line is near the center of the camera view. When this occurs while the camera is inside the layered fog and **Horizon Noise** is turned on, the fog will part horizontally and horizon noise will be created for both the upper and lower portions of the fog.

Phase

Sets the phase for the noise effect. This value can be animated to change the noise over time.

Volume Fog

Volume Fog makes thick, controllable fog throughout the scene or in an atmospheric gizmo.

To access the Volume Fog Parameters rollout, add a **Volume Fog** effect and highlight it on the list.

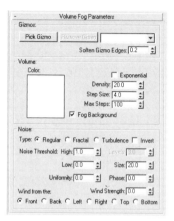

Pick Gizmo   You have the option of picking a gizmo to contain the volume fog. If you don't pick a gizmo, the fog will fill the entire scene. In order to pick a gizmo, you must first create it.

To create a gizmo, go to the **Create** panel 🔧, click

Helpers 🔩 and choose **Atmospheric Apparatus** from the pulldown list. Create a **BoxGizmo**, **CylGizmo** or **SphereGizmo**. The shape of the gizmo determines the overall shape of the volume fog. Gizmos can be scaled to further customize the object shape. For more information on creating an atmospheric apparatus, see the **BoxGizmo**, **CylGizmo** or **SphereGizmo** entries in the *Panels* section of this book.

Click **Pick Gizmo** and pick one or more atmospheric gizmos from the scene. You can also press the <H> key to pick objects from a list.

If you pick more than one gizmo, the same fog settings will be applied to each one. To create multiple **Volume Fog** effects with different settings, click the **Add** button under Atmosphere to add more **Volume Fog** effects to the list, and choose a different gizmo for each one. Set the values for each individual gizmo.

Menus

Remove Gizmo	To remove a gizmo from the selection list, use the pulldown list to select the object and click **Remove Gizmo**.
Soften Gizmo Edges	A value from 0.0 to 1.0 that determines how much the fog will be feathered at the gizmo edges. A value of 0.0 produces no softening, while 1.0 is the maximum softening level.
Color	Sets the color of the fog.
Exponential	When checked, fog density increases exponentially with distance. Exponential fog usually represents real fog more accurately and correctly renders transparent objects inside the fog. A disadvantage of exponential fog is that it sometimes renders as unrealistic color bands. Unless you have transparent objects inside the fog,, keep **Exponential** unchecked.
Density	Sets the overall density of the fog. High values create a nearly opaque fog.
Step Size	Sets the sampling step size for the fog, which affects the "graininess" of its appearance. Higher values produce coarse, lumpy fog, while lower values produce fine fog. This value is not used when a gizmo is selected to contain the fog effect.
Max Steps	The number of sampling steps are limited to the **Max Step** value. If you're not using a gizmo, fog goes infinitely in all directions. This value limits the number of calculations so they don't go to infinity too. **Max Steps** can be set lower if you are using a gizmo to contain the fog. Higher values take longer to render, but create a more accurate picture. **Max Steps** can be set low when **Density** is low.
Fog Background	When this checkbox is on, fog obscures the background map or color. If it is off, fog does not affect the background map or color.

The settings under the **Noise** section use noise to set the fog's density, giving it a more random, realistic look.

Type	Choose the type of noise. These types of noise are the same as those used for Noise maps. For information on noise types, see the **Noise** entry in the *Mat Editor* section of this book.

Invert   When checked, the current noise effect is inverted. Fog areas that noise caused to be dense will become lighter, and vice versa. Choose this option to see an alternate version of your noise effect.

High   A value from 0.0 to 1.0 that sets the limit for sparse fog. At 1.0, parts of the fog can be completely transparent. Set this value below 1.0 to keep the fog in a denser range.

Low   A value from 0.0 to 1.0 that sets the limit for dense parts of the fog. At 0.0, parts of the fog can be completely opaque. Set this value above 0.0 to force the densest parts of the fog to be partially transparent.

Uniformity   A value from -1.0 to 1.0 that "filters" the noise effect. The smaller the value, the smaller the blobs of fog. At -1.0, fog is a fine mist. Higher values remove random specks and yield bigger blobs. If **Uniformity** is set below 0.0, you might have to increase **Density** to see the fog.

Levels   A value from 1 to 6 that sets the number of times noise is calculated. This checkbox is only available with **Fractal** or **Turbulence** type fog.

Size   Sets the size of blobs within the fog.

Phase   Sets a number for calculating different versions of the noise function. Animate the **Phase** value to create churning fog. If **Wind Strength** is 0.0, an animated **Phase** value creates churning fog. If **Wind Strength** is greater than 0.0, the fog animates in the direction specified, making the fog appear to be blown by the wind.

Wind Strength   Sets the strength of the wind for the animated **Phase** value. A wind effect is created only for values above 0.0. The **Phase** value must be animated in order to see the fog move.

Wind from the   Sets the source direction of the wind relative to orthographic views.

Volume Light

A **Volume Light** adds fog or "thickness" to a light. Any kind of light can be set up as a volume light. Volume lights render only in Perspective or camera views.

Before accessing the **Volume Light** controls, create at least one light in your scene. Volume lights work with light attenuation. For the best results, set the **Start** and **End** points for **Far** attenuation on your light's **Create** or **Modify** panel, and turn on **Use** for **Far**. For information on attenuation, see the information on the **Start** and **End** parameters under the **Target Spot** entry in the *Panels* section of this book.

To access the Volume Light Parameters rollout, add a **Volume Light** effect and highlight it on the list.

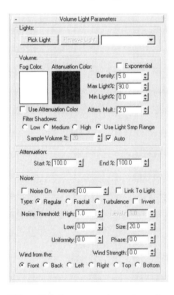

The **Lights** section is where you choose the light to receive the Volume Light effect. At least one light must be chosen in order for the effect to show up in the rendered scene.

Pick Light	Click **Pick Light** and select one or more lights from the scene. You can also press the **<H>** key to pick from a list.
Remove Light	To remove a light from the effect, select the light from the pulldown list and click **Remove Light**.

The **Volume** section is for setting the color and apparent thickness or volume of the light.

Fog Color	Sets the color of the fog in the light. The light's own color is colored with this light. If you want to use the light's color for the foggy effect in the light, leave this color as white.
Attenuation Color	The color of the light can be gradually changed into this color over the light's attenuation range. Turn on the **Use Attenuation Color** checkbox to use this color.
Use Attenuation Color	Enables the use of the **Attenuation Color** to gradually change the color of the light over the attenuation range.
Exponential	When checked, volume light density increases exponentially with distance. **Exponential** usually represents real volume lights more accurately and correctly renders transparent objects inside the light. A disadvantage of exponential volume light is that it sometimes renders as unrealistic color bands. Unless you have transparent objects within the light's range, keep **Exponential** unchecked.
Density	Sets the overall density of the volume light. High values create a nearly opaque light.
Max Light %	The maximum percentage of the light's total luminance (**V** value) that will appear in the volume light.
Min Light %	The minimum percentage of the light's luminance that will appear in the volume light. If attenuation has not been enabled for the light, this minimum light will permeate the entire scene.
Atten. Mult	Multiplies the **Attenuation Color.**
Filter Shadows	Sets the degree of sampling automatically. **Medium** and **High** take longer to render than **Low**. **Low** is fine for 8-bit images and AVIs, but might cause visible bands in high resolution images. **Use Light Smp Range** uses the **Smp Range** setting for the light. If the volume light casts shadows, use this option.
Sample Volume %	The rate of sampling for the volume. A percentage of 1 yield the lowest quality, while 100% yields the highest quality. Rendering time increases as this number goes higher.

Auto	When this checkbox is turned on, the **Sample Volume %** value is controlled by the **Filter Shadows** setting. **Low** sets the sample volume to 8, **Medium** to 25 and **High** to 50.

The **Attenuation** section sets the percentage of the Volume Light effect at the start and end of the light's attenuation range.

Start %	The percent of the light's **Far** attenuation **Start** point at which the volume light begins to attenuate. A value of 0 gives you a smooth glow with no hotspots.
End %	The percent of the light's **Far** attenuation **End** point at which the volume light glow ends. Set this value below 100% to keep the glow from going as far as the light.

The settings under the **Noise** section are for giving the Volume Light a random foggy effect, and for animating the light effect with wind.

Type	Choose the type of noise. These types of noise are the same as those used for Noise maps. For information on noise types, see the **Noise** entry in the *Mat Editor* section of this book.
Invert	When checked, the current noise effect is inverted. Areas that noise caused to be dense will become lighter, and vice versa. Choose this option to see an alternate version of your noise effect.
High	A value from 0.0 to 1.0 that sets the limit for sparse fog. At 1.0, parts of the fog can be completely transparent. Set this value below 1.0 to keep the fog in a denser range.
Low	A value from 0.0 to 1.0 that sets the limit for dense parts of the fog. At 0.0, parts of the fog can be completely opaque. Set this value above 0.0 to force the densest parts of the fog to be partially transparent.
Uniformity	A value from -1.0 to 1.0 that "filters" the noise effect. The smaller the value, the smaller the blobs of fog. At -1.0, fog is a fine mist. Higher values remove random specks and yield bigger blobs. If **Uniformity** is set below 0.0, you might have to increase **Density** to see the fog.
Levels	A value from 1 to 6 that sets the number of times noise is calculated. This checkbox is only available with **Fractal** or **Turbulence** fog.
Size	Sets the size of blobs within the fog.

Menus

Phase

Sets a number for calculating different versions of the noise function. Animate the **Phase** value to create churning fog. If **Wind Strength** is 0.0, an animated **Phase** value creates churning fog. If **Wind Strength** is greater than 0.0, the fog animates in the direction specified, making the fog appear to be blown by the wind.

Wind Strength

Sets the strength of the wind for the animated **Phase** value. A wind effect is created only for values above 0.0. The **Phase** value must be animated in order to see the fog move.

Wind from the

Sets the source direction of the wind relative to orthographic views.

Exit

Exits 3D Studio MAX.

General Usage

When you have finished working with 3D Studio MAX, choose *File/Exit* from the menu. If you haven't saved your most recent changes to the scene, the following dialog appears.

Dialog Options

Yes   Click **Yes** to save your work before exiting. If you haven't assigned a filename to the scene, you will be prompted to enter a filename. For information on entering a filename and choosing a subdirectory for saving the file, see the **Save** entry in this section of the book.

No   Click **No** to exit 3D Studio MAX without saving your work.

Cancel   Cancels the exit process and returns to 3D Studio MAX.

Expert Mode

Toggles the display to Expert Mode.

General Usage

Choose *Views/Expert Mode* from the menu to turn on **Expert Mode**. All menus, buttons and panels disappear and can no longer be accessed. Viewports are enlarged to fill the screen.

In **Expert Mode**, only keyboard shortcuts are available for use.

Click **Cancel** at the lower right of the screen to exit Expert Mode and return to the usual 3D Studio MAX display.

Explode

Ungroups objects in a group, including all nested groups.

General Usage

Select a group. Choose *Group/Explode* from the menu. All groups are ungrouped, including nested groups. All objects in the groups remain selected.

Usage Notes

A *nested* group is a group within a group. To see how this works, create four objects in the scene. Select two objects and choose *Group/Group* from the menu. Accept the default group name.

With the group still selected, add another object to the selection set. Choose *Group/Group* again to group the first group with the selected object.

Add another object to the selection set, and choose *Group/Group* again to make another group.

You now have three groups. The last group created contains two nested groups.

With the group still selected, choose *Group/Ungroup*. The last group created is ungrouped, but the two previously created groups remain grouped.

Choose *Group/Explode*. All nested groups are ungrouped.

Exports the scene to a file format other than *.max* format.

General Usage

Choose *File/Export* from the menu. A file selector appears. Choose a file format and enter or choose a filename. Click **Save** to export the file.

Usage Notes

The file export feature of 3D Studio MAX is useful when you want to use a model made with MAX with another 3D program. Most 3D programs will load files created in *.dxf* format. This format, however, does not save lights, cameras, or material and mapping assignments.

AutoCAD *.dwg* files can be read by AutoCAD and other CAD programs that support this format. Lights, cameras and materials are not retained with this format.

3D Studio DOS *.3ds* format can be read by 3D Studio DOS R4 and many other 3D programs, often with lights, cameras, and materials intact.

For information on the current export formats supported by 3D Studio MAX, consult the online help available from *Help/Online Reference* on the menu.

Fetch

Restores the scene to the state it was in when a Hold was last performed.

General Usage

In order to use **Fetch**, you must first perform a **Hold**. Choose *Edit/Hold* from the menu. The current scene is saved to a temporary file. You can then continue working, making changes to the scene.

If you later want to restore the held scene, choose *Edit/Fetch* from the menu. A dialog appears.

Click **Yes** to restore the held file, or **No** to return to the scene without restoring the file.

Note that if you restore a held file, it is the same as loading a new file Changes made to the scene are lost unless you saved your work prior to fetching the file.

Sample Usage

When you choose *Edit/Hold*, the current scene is saved to the file *maxhold.mx* in the Scenes directory. The file *maxhold.mx* is an ordinary MAX scene file. If you want to save a held file for later use, you can rename the file *maxhold.mx* to a file with the extension *.max*. This will cause the file to appear in the file window when you choose *File/Open*. Rename the file with the File Manager under Windows NT 3.51 or Windows Explorer under Windows NT 4.0.

Customizes the grid spacing and snap settings.

General Usage

Choose *Views/Grid and Snap Settings* from the menu. The Grid and Snap Settings dialog appears.

This dialog has three tabs: **Snaps**, **Options** and **Home Grid**.

Snaps

Under the **Snaps** tab you can choose the drawing elements to which you want to snap while drawing. Enable only the snap types that you think you'll be using frequently. If all snap types are chosen, you might find that too much snapping takes place, making it difficult to draw objects on the screen.

Snap is on when the **3D Snap Toggle** , **2.5D Snap Toggle** or **2D Snap Toggle** is turned on.

When **Standard** is selected from the pulldown, you can choose to snap to any of several standard drawing elements.

Grid Points   Snaps to intersecting points of grid lines. If other snap types are selected in addition to **Grid Points**, the other snap types take priority.

Pivot   Snaps to pivot points of objects.

Perpendicular Snaps to a point on a spline that is perpendicular to the previous drawing point.

Vertex   Snaps to vertices of mesh objects. To snap to spline vertices, use **Endpoint**.

Edge   Snaps along the length of face edges, whether visible or invisible. If you use **Edge**, turn off **Edges Only** on the Display Floater to see all face edges.

Face	Snaps to any point on the surface of a face. Backfaces are not considered during snapping.
Grid Lines	Snaps to grid lines. If other snap types are selected in addition to **Grid Lines**, the other snap types take priority.
Bounding Box	Snaps to one of the eight corners of an object's bounding box.
Tangent	Snaps to a point on a spline that is tangent to the previous drawing point.
Endpoint	Snaps to endpoints on meshes or splines. On a mesh, **Endpoint** causes a snap to a vertex while highlighting the face whose edge is being snapped to. On splines, **Endpoint** causes a snap to a vertex.
Midpoint	Snaps to the middle of edges on meshes and splines. On a mesh, **Midpoint** snaps to the middle of a face edge. On a spline, **Midpoint** snaps to the middle of a segment.
Center Face	Snaps to the center of faces. In 3D Studio MAX, all faces are triangular. A four-sided "face" is actually a polygon made up of two or more faces. To see triangular faces, turn off **Edges Only** on the Display Floater.
Clear All	Clears all snap settings.
Override OFF	Displays the current status of the override snap system. While you're drawing, the override snap systems allows you to temporarily override all snap types with one snap type. The override is good for one mouse click.

To use override, begin drawing on the screen as usual. Just before you want to utilize a different snap type, perform a <Shift> <Right-click>. A dialog appears on the screen where you can select Standard or NURBS snap types. The last selected snap type is also displayed, and you can also choose **None** to disable all snap types.

Select a snap type. If the Grid and Snap Settings dialog is currently displayed, you will see the **Override OFF** display change to the snap type you have just selected.

Move the cursor. The new snap type is now in effect, and all other snap types are disabled. Click the mouse. The temporary snap type no longer overrides other settings, and the snap types set on the **Snaps** tab of the Grid and Snap Settings dialog are in effect again.

Snap override can be used even if snap is not enabled.

You can also choose **NURBS** from the pulldown menu to set snaps for several NURBS drawing elements. These snaps will snap to NURBS elements when drawing both NURBS and non-NURBS splines and surfaces.

CV	Snaps to points on the control lattice of a CV surface.
Curve Center	Snaps to the midpoint on a curve.
Curve Tangent	Snaps to a curve point that is tangent to the previous drawing point.
Curve End	Snaps to the end of a curve.
Surf Normal	Snaps to a point on a surface that is perpendicular to the previous drawing point.
Point	Snaps to points on a point surface.
Curve Normal	Snaps to a point on a curve that is perpendicular to the previous drawing point.
Curve Edge	Snaps to any point on the curve itself.
Surf Center	Snaps to the center of a surface.
Surf Edge	Snaps to the edges of a surface.

Menus

Options

The settings on the **Options** tab affect all snap functions.

Display	When **Display** is on, snap guides appear when drawing. Snap guides are lines that appear around or on snap elements when a snap is possible. For example, when drawing with **Pivot** on, whenever your cursor is near another object, a box appears around the object to indicate that you can now snap to the object's pivot point.
Size	A small box appears around the drawing cursor when a snap occurs. **Size** sets the size of the box in pixels. The color swatch to the right of **Size** sets the color of the snap box and snap guides.
Snap Strength	Sets the snap strength in pixels, ranging from 1 to 20. When the cursor is near a snap point by a distance less than or equal to the snap strength, the cursor snaps to the snap point. You can think of snap strength as the radius of a circle with the snap point as the center. When the cursor is within the circle, the cursor snaps to the snap point.
Snap Values	The **Angle (deg)** value sets the increment for **Angle Snap Toggle**. The **Percent** value sets the increment for scaling and other operations that use percentages. In order to implement the **Angle (deg)** value, the Angle Snap Toggle must be on. In order to use the **Percent** value, the **Percent Snap** button must be on.
Use Axis Constraints	When this checkbox is on, the axis constraints on the Toolbar are used to limit snap functions. For example, if **Restrict to XY** is on, snap functions will operate only along the XY plane. When this checkbox is off, snap operates in any dimension.

Home Grid

The options under the **Home Grid** tab set up grid spacing for the world grid.

Grid Spacing	The amount of space between grid lines, in the current units. Grid lines are used in conjunction with snap tools. If you plan to use the snap tools, the grid spacing should be set to the minimum desired snap unit.
	As you zoom in and out of a viewport, grid spacing is multiplied and divided by 10 to maintain visible grid lines. The Grids value displayed on the prompt line displays the current grid setting. For example, if **Grid Spacing** is set to 10 and you zoom out from a viewport, you will eventually see the Grids value on the prompt line change to 100. This indicates that the grid lines are now 100 units apart.
	When the visible grid lines increase in distance, you can still snap to the original grid spacing, although it will be difficult to see exactly where you are. In this case, watch the prompt line to see where the cursor is located.
Major Lines Every Nth	Major lines are dark gray lines on the grid. These lines denote intervals along the grid, making it easier to keep track of the cursor's location along the grid lines. Values of 4, 5, 10 and 12 work well for most applications.
Inhibit Grid Subdivision Below Grid Spacing	As you zoom into a viewport, grid lines subdivide so that a workable system of grid lines is always visible. When this checkbox is checked, grid subdivision lines will not appear at levels below the **Grid Spacing** setting when you zoom into a viewport. This ensures that when snap is on, all visible grid lines can be used as snap references.

Dynamic Update	When **Active Viewport** is checked, any changes made to values for **Grid Spacing** and **Major Lines Every Nth** are updated immediately in just the active viewport. If **All Viewports** is selected, all viewports are updated.

Usage Notes

The snap settings on the **Snaps** tab are enabled by turning on **3D Snap Toggle** ⬚, **2.5D Snap Toggle** ⬚ or **2D Snap Toggle** ⬚ on the Toolbar.

While snap is enabled and the **Animate** button is off, you can click on an object's snap point and move, rotate or scale the object about the snap point.

The Grid and Snap Settings dialog can be accessed by right-clicking on any snap button such as **3D Snap Toggle** ⬚ or **Angle Snap Toggle** ⬚.

If you create an object without snap on and then later move the object with **Grid Lines** snap on, the object will snap to a distance equal to the distance between grid lines, not the grid lines themselves.

Groups two or more objects so they can be manipulated as one object.

General Usage

Select two or more objects. Choose *Group/Group* from the menu. The Group dialog appears.

Enter a name for the group, or click **OK** to accept the default group name.

After objects have been grouped, you can click on one object to select the entire group. You can also transform or modify the group as you would one object.

You can later ungroup the objects with the *Group/Ungroup* menu option. You can also work with individual objects in the group by opening the group with *Group/Open*, or remove an object from the group with *Group/Detach*.

Sample Usage

Groups are useful for temporarily attaching a set of objects for a particular function. This example illustrates how a group can be used to manipulate the size of a modifier gizmo.

Create two or more objects in the scene. Select all the objects and choose *Group/Group* from the menu. Accept the default group name.

With the group still selected, access the Modify panel and click **Bend**. Change the **Angle** spinner to a large value such as 90 so you can clearly see the effect of the bend. The gizmo surrounds the entire group and all objects are bent at once.

Hold

Saves the scene to a temporary file so **Fetch** can be used later on.

Menus

General Usage

Choose *Edit/Hold* from the menu. The current scene is saved to the temporary file *maxhold.mx* in your AutoBackup directory. You can then continue working, making changes to the scene.

If you later want to restore the held scene, choose *Edit/Fetch* from the menu.

Sample Usage

There are many operations in 3D Studio MAX which cannot be undone. For this reason, it is important that you use **Hold** and **Save** regularly.

The file *maxhold.mx* is an ordinary MAX scene file. If you want to save a held file for later use, you can rename the file *maxhold.mx* to a file with the extension *.max*. This will cause the file to appear in the file window when you choose *File/Open*. Rename the file with the File Manager under Windows NT 3.51 or Windows Explorer under Windows NT 4.0.

The file *maxhold.mx* is saved to the AutoBackup directory set with *File/Configure Paths*, under the **General** tab.

You can also set up MAX to automatically save files at regular intervals. For information on this procedure, look under the **Preferences** heading in the *Menus* section of this book, under the **Files** tab heading.

Import

Merges objects from a 3D scene in a format other than 3D Studio MAX's native format.

General Usage

Choose *File/Import* from the menu. A file selector appears.

Choose an import file type from the **Files of Type** list. Select a file to import. With all import types, a dialog appears asking you if you want to merge or replace the current scene.

If the import type requires further information, a dialog will appear with options for import. Each type of file import has a different dialog. See the headings below for details on dialog options.

If you are using an import plug-in to import a file, further dialogs and entries may display. Consult the import plug-in documentation for details.

If the merged scene and the current scene were created with different unit scales, incoming objects may need to be resized to match the current scene.

For information on importing a specific file type, consult the 3D Studio MAX online help, available by choosing *Help/Online Reference* from the menu.

Insert Tracks

Imports animation from an object in a saved file to an object in the current scene.

General Usage

Select the object in the current scene to which you want to import animation. Choose *File/Insert Tracks* from the menu. A file selector appears. Select the file containing the object with the desired animation.

The Insert Tracks dialog appears.

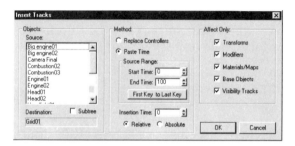

Dialog Options

Source   Displays all objects in the selected file, regardless of whether they have animation or controllers assigned to them. If an object name matches the selected object in the current scene, the object name is highlighted.

Destination   Displays the name of the selected object in the current scene. This object cannot be changed here. To change the destination object, you must start over and select the object before choosing *File/Insert Tracks* from the menu.

Subtree   When this option is checked, any animation on child objects linked to the source object is imported to the destination object's children. If the source and destination objects have more than one set of linked children, the order in which children were linked to the parent is used to determine which linked chains receive which animation, not the object names. If both the source and destination objects have many linked children, you must take care to link objects in exactly the same order in both files.

Replace Controllers	Applies all the source object's controllers to the destination object. The controller must be compatible with the destination object in order to be applied. For example, any position controller can be applied to the destination object as every object has a position track. However, if the source object has an animated modifier track such as **Taper**, and the destination object has no **Taper** modifier applied, the Taper track will have nowhere to go, and will not be applied to the destination object.
	When **Replace Controllers** is checked, the remaining options in the Method section are disabled, and all controllers and animation on the destination object are replaced with those from the source object.
Paste Time	Inserts animation from a specified time range. All controllers must be compatible with the destination object in order to be applied. If a controller is not compatible, it is skipped.
Start Time	The beginning of the range in the source file that you wish to import. The default is the first key in the source file.
End Time	The end of the range in the source file that you wish to import. The default is the last key in the source file.
First Key to Last Key	Restores the default values for **Start Time** and **End Time**.
Insertion	The frame in the current scene at which you want to begin inserting the imported animation to the destination object.
Relative	Causes changes to animation tracks relative to the destination object's current position and orientation. When this button is on, each source and destination track is examined to determine the value at the **Insertion** frame. The **Insertion** frame value in each source track is adjusted to match the **Insertion** frame key in the corresponding destination track. The same adjustment is then applied to all subsequent source keys in that track. This means that the position and orientation of the destination object will be the same on the **Insertion** frame as it was before tracks were imported, but will change on subsequent frames to reflect the source animation. Compare with **Absolute**.

Absolute   When this button is on, each source track completely replaces each destination track. Source track values are not adjusted, but are inserted exactly as is. Compare with **Relative**.

Transforms   Imports only transform animation (moving, rotating and scaling), which is applied to the Position, Rotation and Scale tracks of an object, respectively. Every object has these tracks by default.

Modifiers   Imports only modifier information. In order for modifier settings and animation to be imported, the modifier must already be applied to the destination object before the *Insert Tracks* operation was begun.

Materials/ Maps   Imports the source object's materials and any animation associated with the materials.

Base Objects Imports base object parameters such as radius, height and number of segments. Only parameters that match both objects are imported.

Visibility Tracks   Imports **Visibility** track information. **Visibility** tracks are used to make an object disappear and reappear over the course of an animation. A **Visibility** track exists for an object only if the track is created for the object in Track View. If both the source and destination objects have a **Visibility** Track, then the track is imported. For information on **Visibility** tracks, see the **Track View** entry in the *Toolbar* section of this book.

Accesses tutorials for 3D Studio MAX.

General Usage

Choose *Help/Learning 3D Studio MAX* from the menu. Click on a tutorial topic to see a tutorial on the selected subject.

Make Preview

Creates a preview of the scene's animation in AVI format.

General Usage

Activate the viewport you want to preview. Choose *Rendering/Make Preview* from the menu. The Make Preview dialog appears.

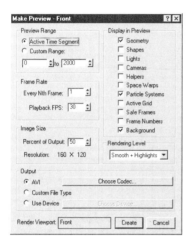

Select the types of objects to show in the preview under the Display in Preview section. Click **Create** to make the preview. When the preview is complete, you can play back the preview with the Media Player.

By default, the Media Player will automatically load when the preview is complete. Click the Play button on the Media Player to see the preview. If you don't want the Media Player to automatically appear when you make a preview, you can turn off this option by choosing *File/Preferences*. Under the **General** tab, turn off the **AutoPlay Preview File** checkbox.

A preview is saved to the file *_scene.avi* in the Previews directory. Every time a preview is rendered, it overwrites the previous preview. If you want to save a preview, you can rename *_scene.avi* before rendering another preview. The preview file can be copied to another file with the *Rendering/Rename Preview* menu option. You can also copy or rename the file with Windows Explorer or File Manager. The preview file *_scene.avi* remains saved regardless of whether a new scene is loaded or a *File/Reset* is performed.

When you save the scene to a *.max* file, the settings on the Make Preview dialog are saved along with it. Scenes saved from 3D Studio MAX R1.2 do not have preview settings saved. When you load a scene saved from R1.2, the Make Preview settings will remain the same as they were before loading.

Dialog Options

Active Time Segment When this option is checked, the currently active time segment renders in the preview. The active time segment is set by clicking **Time Configuration** 🔲 and entering a **Start Time** and **End Time** under the Animation section of the Time Configuration dialog.

Custom Range When this option is checked, the range of frames entered will be rendered.

Every Nth Frame Sets the increment for frames to be rendered. For example, if the number 5 is entered here, every 5th frame will be rendered.

Playback FPS Sets the playback speed for the preview in frames per second.

Display in Preview Sets the elements to appear in the preview. Rendering with only **Geometry** checked will most closely approximate the final rendering. Other elements such as lights, cameras and helpers can also be checked to assist in checking the animation for accuracy. Turning on **Frame Numbers** causes the frame number to be rendered at the upper left corner of each frame of the preview. **Background** causes the viewport background to be displayed in the preview. Note that the background used for previews is the viewport background set up with the *Views/Background Image* menu option, not the final rendering background set up with the *Rendering/Environment* menu option.

Rendering Level Sets the rendering level for the preview. Click the down arrow to choose from a list of shading levels. These levels are the same as those available for viewport display. For more information on rendering levels, see the information about the settings on the **Rendering Method** tab under the **Viewport Configuration** entry.

Image Size The **Percent of Output** value sets the size of the preview as a percentage of the final output size. To see or change the final output size, choose *Rendering/Render* from the menu or click **Render Scene** 🔲 to access the Render Scene dialog. The final rendering resolution is set under the **Output Size** section of the dialog.

The **Resolution** of the preview is automatically calculated from the final output size and the **Percent of Output** value, and cannot be changed directly.

Output   Choose **AVI** to render to the default AVI file type. You can also click **Choose Codec** to select a different type of AVI file.

To render to another file format, choose **Custom Output**. When you click **Create** to start creating the preview, you will be prompted for a file type and filename. If the file type is a single image file type such as TIF or BMP, each frame of the preview will render to a separate file. Files will be sequentially numbered in the order rendered.

Choose **Use Device** to record the preview to an external device such as a digital recorder. Click **Choose Device** to select and set up the device.

Render Viewport   Displays the name of the currently active viewport. This viewport will be rendered for the preview. You cannot change the viewport on this dialog. To change the viewport, exit the dialog, choose the correct viewport and select *Rendering/Make Preview* again.

Create   Begins the preview creation process. The preview renders on the screen, displaying each frame as it renders. A progress bar across the bottom of the screen indicates the progress of the preview rendering.

To stop preview creation, press **<Esc>**. A dialog appears.

Choose **Stop & Play** to cancel rendering and load the Media Player to view the partial preview. This option is available only if the **AutoPlay Preview File** checkbox under the **General** tab is turned on. This tab can be accessed by choosing *File/Preferences* from the menu.

Click **Stop & Don't Play** to cancel rendering without displaying the Media Player. Click **Don't Stop** to resume rendering the preview.

Matches the selected camera to a Perspective, Spotlight or Camera view.

General Usage

Select a camera and activate a non-orthographic viewport such as a Perspective, Spotlight or Camera view. Choose *Views/Match Camera to View* from the menu. The camera is repositioned to exactly match the selected viewport.

Usage Notes

This option cannot be used with orthographic viewports such as Top, Front, Left or User views.

This option is useful when you have set up a Perspective view and now want a Camera view to match it exactly.

Material Editor

Accesses the Material Editor.

General Usage

Choose *Tools/Material Editor* from the menu. The Material Editor dialog appears.

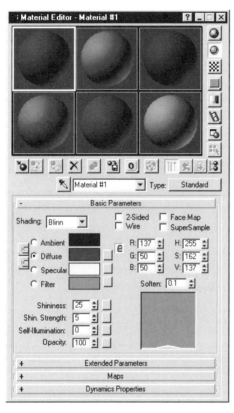

The many options on the Material Editor dialog are described in the *Mat Editor* section of this book.

Choosing *Tools/Material Editor* from the menu is the equivalent of clicking the **Material Editor** button from the Toolbar.

Material/Map Browser

Accesses the Material/Map Browser.

General Usage

Choose *Material/Map Browser* from the *Tools* menu. The Material/Map Browser appears.

The Material/Map Browser can be used with the Material Editor to load, save and change materials. It can also be used to access a map for use as a background or environment map.

For information on how to use the Material/Map Browser, see the **Material/Map Browser** entry in the *Mat Editor* section.

Merge

Merges objects from a MAX scene into the current scene.

General Usage

Choose *File/Merge* from the menu. A file selector appears. Select the file you wish to merge. An object selector appears. Select the objects you wish to merge into the current scene.

If an incoming object has the same name as an object already in the scene, a dialog appears.

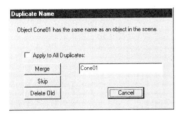

Here you can choose how to deal with the incoming object. If an incoming object holds a material that is also used in the current scene, a dialog appears.

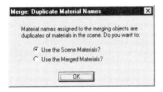

Here you can choose to use the materials from the current scene or the merged scene.

Duplicate Name

Apply to All Duplicates	Turn on this checkbox to apply a **Merge**, **Skip** or **Delete Old** action to all objects with duplicate names. If you choose to enter a new name for the incoming object, this checkbox is disabled.
Merge	Merges the object using the name currently displayed. You can change the incoming object name by typing in a new object name before clicking **Merge**. If you click **Merge** without changing the name, the object is merged with its original name, meaning there are now two objects with the same name in the scene.
Skip	Skips the object and does not merge it with the scene.

Delete Old   Deletes the object in the current scene and merges the incoming object.

Cancel   Cancels the entire merge operation, returning the scene to the state it was in before the merge operation was started.

Duplicate Material Names

When objects are merged into a scene and some incoming material names are the same, you must choose whether to use the materials from the current scene or from the incoming objects. You cannot make this choice on an object-by-object basis. If you wish to merge some materials but not all, you can save the materials in a material library before the merge and reassign them after the merge. For information on working with material libraries, see the **Material/Map Browser** entry in the *Mat Editor* section of this book.

Use the Scene Materials?   Click this button to use the materials from the current scene, not the incoming materials.

Use the Merged Materials?   Click this button to use materials from incoming objects.

Mirror

Mirrors selected objects along any axis.

General Usage

Menus

Select one or more objects, and activate the coordinate system you wish to use for mirroring. Choose *Tools/Mirror* from the menu. The Mirror dialog appears.

Choose an axis or axes over which to mirror the selected object. To copy and mirror an object at the same time, choose **Copy**, **Instance** or **Reference** under Clone Selection.

For information on how to use this dialog, see the **Mirror Selected Objects** entry in the *Toolbar* section of this book.

Clears objects, animation and/or hierarchy information from the scene.

General Usage

Choose *File/New* from the menu. If you haven't saved your current changes, a dialog appears.

Click **Yes** to save the file, **No** to load a new file without saving your work, or **Cancel** to cancel the operation. If you click **Yes**, and the scene has not yet been saved, a file selector appears where you can enter or choose a filename to save the file.

If you click **Yes** or **No**, another dialog appears. This dialog is described below.

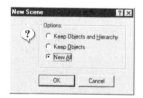

Dialog Options

Keep Objects and Hierarchy   Retains objects and hierarchical linking information, but deletes all animation from the scene.

Keep Objects   Deletes animation and hierarchical linking information from the scene, but retains objects.

New All   Deletes all objects, animation and hierarchical linking information from the scene.

Usage Notes

When you want to start a new scene, you can choose either *File/New* or *File/Reset*.

File/Reset loads the file *maxstart.max* and uses settings from this file. *File/New* retains certain settings such as the state of floating windows and the last entry for creation parameters. Performing a *File/New* is more like deleting all objects in the scene, while *File/Reset* replaces the entire scene setup.

New Track View

Opens a new Track View window.

General Usage

Choose *New Track View* from the *Track View* menu. A new Track View window is opened and displayed on the screen.

The default name of **Untitled-#** is automatically assigned to the Track View window, where **#** is the next sequential number for all the Track View windows you have created. The name is shown at the upper right of the Track View window. To change the name, access the name field and type in a new name.

When you are finished working with the Track View window, you can close it or minimize it. If you close it, you can later access it from the Track View menu. If you minimize it, you can choose it from its minimized position, usually at the bottom left of the screen. In order to delete a Track View window, you must first close it.

For information on Track View functions, see the **Track View** entry in the *Toolbar* section of this book.

Displays online help.

General Usage

Choose *Help/Online Reference* from the menu. The 3D Studio MAX Online Reference appears. Choose a topic, or use **Index** or **Search** to look for a specific option.

Menus

Open (File)

Loads a MAX scene file.

General Usage

Choose *File/Open* from the menu. If you are currently editing a scene with unsaved changes, a dialog appears.

Click **Yes** to save the file, **No** to load a new file without saving your work, or **Cancel** to cancel the operation. If you click **Yes**, a file selector appears where you can enter or choose a filename to save the file.

If you click **Yes** or **No**, a file selector appears. Only files saved in MAX format can be opened here. These files usually have the extension *.max*.

To open a file with a format other than the MAX format, use *File/Import*. To merge objects from a file with the current scene, use *File/Merge*.

Temporarily opens a group so the objects can be manipulated individually.

General Usage

Select a group. Choose *Group/Open* from the menu. The group is opened. Select an object and transform or modify it as desired. When finished, choose *Group/Close* to close the group.

Sample Usage

Select one or more objects in the scene. Create a group with *Group/Group*. Open the group with *Group/Open*. Select and move one or more of the objects. Choose *Group/Close* from the menu to close the group again.

The *Group/Open* option differs from *Group/Explode* and *Group/Ungroup*. *Group/Explode* and *Group/Ungroup* permanently disband the group, while *Group/Open* temporarily allows objects to be manipulated individually.

Open Track View

Opens a previously created Track View window.

General Usage

Choose *Open Track View* from the *Track View* menu. The last Track View window used appears on the screen. If no Track View window has been previously created, a new Track View window is opened.

You can also select a Track View window directly from the *Track View* menu. Names of closed Track View windows appear on the menu below *Delete Track View*. Minimized Track View windows do not appear.

For information on Track View functions, see the **Track View** entry in the *Toolbar* section of this book.

Positions a light to create a highlight on an object, or places an object to reflect in another object.

General Usage

Select a light. Choose *Tools/Place Highlight* from the menu. In the viewport in which you want the highlight to appear, click and drag on the object that will receive the highlight. A normal arrow appears on the object. Drag the cursor until the normal appears for the face which is to receive the highlight. The light moves as you move the cursor over the object, retaining its distance from the object. Release the mouse to set the light's position.

In the same way, you can set up an object to be reflected in another object. Select the object to be reflected. Click **Place Highlight**. In the viewport in which you want the reflection to appear, click and drag on the object to receive the reflection. The blue normal arrow appears on the object, and the object being reflected moves around as you drag the cursor. Drag until the arrow appears on the face to receive the reflection, and release the mouse. In order for the reflection to appear in renderings, the reflective object must be assigned a material with a **Reflect/Refract** map set up as the **Reflection** map. See the **Standard** and **Reflect/Refract** entries in the *Mat Editor* section for information on how to set up a material of this kind.

To place a highlight or reflection, you can also choose **Place Highlight** from the **Align** flyout on the Toolbar.

Usage Notes

When you want to control how a highlight or reflection appears in a final rendering, use the camera viewport to set up the highlight or reflection. Watch the face normal in the camera viewport to see where the highlight or reflection will fall from the camera's view.

Technically speaking, this feature aligns the local Z axis of one object with a face normal of another object. *Tools/Place Highlight* is not limited to lights or reflective objects, but can be used with any two objects.

Pop-up Menus

In many areas of 3D Studio MAX, you can right-click to access a pop-up menu.

Viewport Name

When a viewport name is right-clicked, a pop-up menu appears with options for controlling the viewport display. Many of these options are available on the Viewport Configuration dialog, which can be accessed by choosing Views/Viewport Configuration from the menu.

Smooth + Highlights Wireframe Other Edged Faces	Determines how objects will be displayed in the viewport. A checkmark appears next to the currently selected option. Move the cursor to **Other** to select from several display options. These options can also be selected on the **Rendering Method** tab of the Viewport Configuration dialog. See the **Viewport Configuration** entry for descriptions of these options.
Show Grid	When checked, the active grid appears in the viewport.
Show Background	When checked, the background set up with the *Views/ Background Image* menu option appears in the viewport. This option is available only when a background has been set up.
Show Safe Frames	When checked, safe frames appear in the viewport. Safe frame dimensions can be set under the **Safe Frames** tab of the Viewport Configuration dialog.
Texture Correction Disable View	See the **Viewport Configuration** entry for descriptions of these options.
Views	Displays a list where you can choose any view for the current viewport.

Swap Layouts	Swaps layouts A and B. Layouts are selected under the **Layout** tab of the Viewport Configuration dialog.
Undo	Undoes the last change to the viewport. This option performs the same function as the *Views/Undo* menu option.
Redo	Redoes the last change to the viewport. This option performs the same function as the *Views/Redo* menu option.
Configure	Accesses the Viewport Configuration dialog.

Object

When a selected object is right-clicked, a pop-up menu appears.

Move **Rotate** **Scale**	Select one of the transforms to make it active. Selecting a transform here is the equivalent of clicking the transform button on the Toolbar.
Transform	Move the cursor to **Transform** to select from a list of axis restrictions. Selecting an axis restriction here is the equivalent of selecting it on the Toolbar.
Select / Deselect Children	Selects or deselects any child objects of the currently selected object.
Properties	Accesses the Object Properties dialog. See the **Properties** entry in this section of the book for information on this dialog.

Preferences

Allows you to determine the settings for a number of aspects of MAX such as viewport layout and keyboard shortcuts.

General Usage

Choose *File/Preferences* from the menu. The Preference Settings dialog appears. This dialog has nine tabs for working with different areas of 3D Studio MAX.

General

The options under the **General** tab set preferences for previews, units, spinner use, log file, editing undo and the 3D Studio MAX screen display.

AutoPlay Preview File   Causes the preview to automatically play immediately after it is created with *Rendering/Make Preview*.

Optimize for 8-bit Display   Optimizes 3DS MAX for 256-color display. Check this option if your video card is limited to 256 colors, to make the display look better. Checking this option decreases system performance.

1 Unit =   Sets the real world size of one unit in 3D Studio MAX. There is rarely a reason to change the default scale of 1 unit = 1 inch. Do not confuse this scale with the working unit scale, which is set with the Units Setup dialog. To access this dialog, choose *Views/Units Setup* from the menu.

This scale is the one used to resize objects when files are imported or exported from MAX. MAX uses this scale, plus the working unit scale on the Units Setup dialog, to accurately resize objects. If you plan to share your files with other 3D Studio MAX users, leave the system unit scale as is so all files will use the same base scale.

If you plan to import or export files to other programs, be sure to note the system unit scale used by the other program. Although 3D Studio MAX will prompt you when an incoming file has a unit scale different from your settings, a file with a radically different scale can cause problems even if rescaled upon import. For example, camera clipping planes may clip the model if the scale has suddenly become much smaller.

Menus

Automatic Unit Conversion — When checked, merged objects from a file with a different unit scale are automatically scaled to the unit scale of the current scene. If you want to change the scale of objects merged from files with a different unit scale, turn this checkbox off. When you merge such a file, you will be presented with options on how to merge the incoming objects.

Spinner Precision — Determines the number of decimal places to which spinner values can be set. Setting this number to a lower value such as 0 or 1 will allow you to make large changes to a spinner value simply by clicking on one of the spinner arrows a few times. However, these lower values limit the accuracy of spinner values.

Spinner Snap — Sets the change increment for spinner values when a spinner arrow is clicked. The **Spinner Snap** increment is in effect only when the **Use Spinner Snap** checkbox is on, and only applies to clicks on spinner arrows, not click-and-drag or direct type-in.

Use Spinner Snap — Enables use of the **Spinner Snap** increment. When this checkbox is on, clicking a spinner arrow causes the value to jump to the current value plus or minus the **Spinner Snap** value.

The settings under **Log File Maintenance** are used to control how the log file is saved and maintained. The log file is *max.log* in the Network directory under the main 3D Studio MAX directory. In addition to network information, the log file contains useful information about files loaded and rendered.

Never delete log — Saves the log indefinitely. If you select this option, be sure to erase some or all of the log periodically.

Maintain only # days — Maintains the log for only the specified number of days.

Maintain only # Kbytes — When the log file reaches the specified number of bytes in size, older entries are deleted until the file size falls below the specified size.

Errors — When checked, any fatal error during rendering does not generate an alert dialog, but instead is written to the log. When unchecked, a fatal error generates an alert dialog and halts rendering on the machine experiencing the error.

Menus

Info	When checked, any non-fatal error during rendering does not generate an alert dialog, but instead is written to the log. When unchecked, any non-fatal error generates an alert dialog and halts rendering on the machine experiencing the error.
Debug	When unchecked, debug messages are written to the log instead of generating an alert dialog that halts rendering.

Under the **UI Display** section, you can control how the main screen is displayed.

Viewport Tooltips	When checked, the cursor displays each object name as the cursor passes over the object.
Display NU Scale Warning	When **Display NU Scale Warning** is checked, a warning is displayed when either **Select and Non-uniform Scale** or **Select and Squash** is selected from the Toolbar.
Short Toolbar	Displays a shorter Toolbar that fits a 800x600 screen resolution.
Righthand Command Panel	When checked, the command panel is displayed at the right side of the screen. When unchecked, the command panel is displayed at the left side of the screen.
Flyout Time	Sets the time, in milliseconds, between the mouse click-and-hold on a flyout and the flyout display.
Undo Levels	Sets the number of times that actions can be undone with the **Undo** button or the *Edit/Undo [last action]* menu option.

A higher number such as 20 gives you a lot of freedom when creating scenes, but uses up more memory than a lower value such as 5. Set this value to the lowest number you can be comfortable with.

The number of undo levels set here refers only to creation and editing of objects, not to view changes.

Rendering

The options under the **Rendering** tab set preferences for all aspects of rendering output. The minimum angle separation for light hotspot and falloff is also set here.

Video color checking is a process where colors that do not display well on video are detected and flagged during rendering. These colors are generally very bright colors, such as pure red or pure blue, which tend to smear or fuzz when the image is played from videotape. The **Video Color Check** options indicate the method for displaying non-video color pixels in a rendering. Color checking is performed only when the **Video Color Check** button on the Render Scene dialog is turned on during rendering.

For more information on video color checking, see **Video Color Checking** in the *Appendix*.

Flag with black	Turning on this checkbox causes non-video pixels in the rendering to turn black in the rendered image. This option does not correct video colors, but is intended to point out problem areas which will need to be addressed by altering materials or lighting in the scene.
Scale Luma	Corrects any unsafe video colors in the rendering by reducing the lightness (whiteness) of the color to within the safe video color range.
Scale Saturation	Corrects any unsafe video colors in the rendering by reducing the saturation (brightness) of the color to within the safe video color range.
NTSC	Uses the NTSC standard for color checking. NTSC is used in the United States and Canada.
PAL	Uses the PAL standard for color checking. PAL is used in Europe, Asia, Australia and South America.

Dithering is the process of mixing colors or scattering pixels to smooth a transition from one area of an image to another. Dithering is especially important when rendering to 8-bit images, where limited colors are used to represent the entire spectrum. The **Output Dithering** section of the dialog sets the methods for dithering colors during rendering.

True Color	Dithers true color (24-bit) images during rendering.
Paletted	Dithers 8-bit (256 color) images during rendering.

Menus

Field order is the order in which fields are rendered when the **Render to Fields** checkbox on the Render Scene dialog is checked. Field order is only a concern when you are rendering for broadcast quality video. For information on what fields are and how they are used, see the **Rendering Concepts** section of the *Appendix*.

All PC-based video playback systems such as the Perception and Targa 2000 use odd-numbered fields first. If you will be taking your rendered frames to a service bureau for output to video, check with the bureau first before choosing a field order.

Odd	Renders the odd-numbered lines of the video image first. This is the usual setting for rendering from low to high numbered frames, for playback in the same order.
Even	Renders the even-numbered lines of the video image first. Turn on **Even** if you are rendering from low to high numbered frames but plan to play them backward.

Super black is a sub-black color used in compositing. Super black is in effect only when the **Super Black** checkbox on the Render Scene dialog is on. For information on super black, see the **Rendering Concepts** section of the *Appendix*.

Threshold	Sets the minimum intensity (**V** value) for black colors on rendered objects. For example, if **Threshold** is 15, the **V** value of the color of black objects and dark areas of the screen will never drop below 15. The background area will render as pure black with a **V** value of 0.
Angle Separation	Determines the minimum separation angle between the hotspot and falloff of spot and direct lights. The default is 2.0. Change this value to 0.0 if you want to set up lights with the same hotspot and falloff angles, effectively creating a light with no falloff.
Don't Antialias Against Background	When this option is unchecked, geometry is not antialiased against the background. Leave this option unchecked unless you have a specific reason for checking it.
Default Ambient Light Color	Sets the ambient light color for each rendered scene. Ambient light is overall light that permeates the scene. In life, high ambient light is evident at dawn and dusk, when a dim light illuminates everything with the same intensity. High ambient light renders every object with nearly the same intensity, eliminating contrast in the

rendered image. The **Default Ambient Light Color** should be kept low to avoid washing out the scene.

Maximum
The **Pixel Size Limit** determines the maximum pixel size for rendered images. The pixel size can be set on the Render Scene dialog just before rendering. For information on pixel size and how it affects rendering, see the see the **Rendering Concepts** section of the *Appendix.*

Nth Serial Numbering
When this checkbox is off and the **Every Nth Frame** value on the Render Scene dialog is greater than 1, filenames are numbered with increments of the **Every Nth Frame** value. When this checkbox is on, filenames for a sequence of frames are numbered with increments of 1 regardless of the **Every Nth Frame** value.

The settings in the **Render Termination Alert** section cause a beep or other sound to alert you when rendering is complete.

Beep
Causes a beep when rendering is complete.

Frequency
Sets the frequency of the beep.

Duration
Sets the duration of the beep in milliseconds. One thousand milliseconds is one second.

Play Sound
Causes the sound file chosen with the **Choose Sound** button to be played when rendering is complete.

Choose Sound
Selects a file for the alert sound such as WAV or AVI. Only the sound portion of AVI or other multimedia files will be played when rendering is complete.

The settings in the **Current Renderer** section allow you to choose a renderer for production or draft rendering. 3D Studio MAX comes with the Default Scanline Renderer only. Third-party renderers such as radiosity renderers can be installed for further options.

Assign
To choose a renderer, click the **Assign** button across from **Production** or **Draft**. Any installed third-party renderers will appear on the list along with the Default Scanline Renderer. Choose a renderer and click **OK**.

Note that changing the renderer will also change the second rollout on the Render Scene dialog to offer rendering options particular to the chosen renderer.

Multi-threading is the dividing up of rendering tasks to different processors.

On	Turns on multi-threading, which uses more than one processor for rendering, if more than one processor is installed in the system.

Inverse Kinematics

The **Inverse Kinematics** options set standards and values for IK. For information on inverse kinematics, see the **Inverse Kinematics on/off toggle** entry in the *Toolbar* section of this book. You can also see the **IK** entry in the *Panels* section for a definition of *Applied IK* and information on how to bind objects to an IK chain.

Position	When an end effector (last child in an IK chain) is bound to another object, this value determines how close the end effector has to get to the follow object to be considered a valid solution. If the end effector is not bound, this value determines how close the end effector has to be to the cursor position. The value is expressed in the current units. Smaller values make more accurate animation, but take longer to solve.
Rotation	When an end effector (last child in an IK chain) is bound to another object, this value sets how closely the end effector's orientation has to match the follow object's orientation to be considered a valid solution. The value is expressed in degrees. Smaller values are more accurate, but take longer to solve.
Iterations	Sets the maximum number of times IK is calculated to come up with a solution. Higher values make a more accurate solution, but take longer to calculate.
Use secondary threshold	When checked, an extra check is performed during IK calculations. If the IK chain can never reach the specified thresholds, such as when the IK chain is stretched to its limits but the end effector still cannot touch the follow object, the IK solution is considered complete, and calculation ceases.
Always transform children of the world	When this checkbox is checked, when you select and transform a root object (child of the world) directly, its IK constraints are ignored. Uncheck this checkbox if you want to avoid transforming objects without linked children when working with IK.

Animation

The options under the **Animation** tab control the display of animated objects. You can also load a sound plug-in and set up animation controllers.

The **Key Bracket Display** section sets how and when key brackets will be displayed. A key bracket is a white frame around an object, displayed when a current frame is a transform key frame for the object.

All Objects   Displays key brackets around all objects with a transform key at the current frame.

Selected Objects   Displays key brackets only on selected objects with a transform key at the current frame.

None   Displays no key brackets at any time.

Use Current Transform   When checked, display of key brackets is limited to keys for the currently selected transform. For example, if the **Select and Move** ⊕ transform is currently selected, only objects with Position keys will have key brackets. If the **Selected Objects** option is on, only objects with keys on the currently selected transform will have key brackets.

Position   When checked, only objects with Position keys will appear with key brackets when on a key frame. A Position key is set with **Select and Move**. This checkbox is only available when **Use Current Transform** is unchecked.

Rotation   When checked, only objects with Rotation keys will appear with key brackets when on a key frame. A Rotation key is set with **Select and Rotate** ⟳. This checkbox is only available when **Use Current Transform** checkbox is unchecked.

Scale   When checked, only objects with **Scale** keys will appear with key brackets when on a key frame. A **Scale** key is set with options under the **Select and Uniform Scale** ▣ flyout. This checkbox is only available when **Use Current Transform** is unchecked.

Local Center During Animate   When checked, objects are animated around their local centers during animation.

The **Midi Time Slider Control** options allow you to use a midi device to control the time slider.

On   Turns on the use of the midi device.

Setup   Click **Setup** to select and set up a midi device for controlling the time slider.

The **Sound Plug-In** section allows you to select a sound plug-in for use with 3D Studio MAX.

Assign   Selects a sound plug-in. Click **Assign** to select a sound plug-in from a list.

The **Controller Defaults** section sets default values for all animation controllers that support defaults.

Set Defaults   Displays a dialog listing all controllers that can have default values set. Select a controller from the list and click **Set**. A dialog appears where you can set the default values.

Restore to   Resets all controllers to their default settings.
Factory
Settings

Keyboard

The **Keyboard** options allow you to set custom keyboard shortcuts. Many users find keyboard shortcuts to be an effective time-saving device when working in 3D Studio MAX.

Main U.I.   Displays and allows changes to menu, button and panel keyboard shortcuts for the main user interface.

Track View   Displays and allows changes to keyboard shortcuts for the Track View window's buttons and functions.

Material   Displays and allows changes to keyboard shortcuts for
Editor   the Material Editor window's buttons and functions.

Video Post   Displays and allows changes to keyboard shortcuts for the Video Post window's buttons and functions.

Use Only   When this checkbox is on, all shortcuts for Track View,
Main U.I.   Material Editor and Video Post are disabled, and only
Shortcuts   the main user interface shortcuts are enabled.

The **Command** section lists all commands of the selected type for which keyboard shortcuts can be created. To set a keyboard shortcut for a function, scroll through the list and highlight the function.

Current Shortcut Displays the current keyboard shortcut for the selected function, if any.

Remove Removes the currently selected keyboard shortcut. This button is available only when a function with a keyboard shortcut has been selected.

Save List to File Saves the current list of keyboard shortcuts in all four categories to a file. When this button is pressed, a file selector appears, allowing you to specify the filename and directory for the file. The file is saved as an ASCII text listing, with a tab separating each function and its keyboard shortcut, if any.

Reset Category Removes all user-defined shortcuts in the current category.

Press Key Click this button to specify the keyboard shortcut with the keyboard. When you click **Press Key**, a message appears requesting that you press a key.

Press a key or key combination. A key combination can consist of the **<Ctrl>**, **<Alt>** or **<Shift>** key plus any other key. You can also click **Cancel** to cancel the operation.

Once you have pressed a key or key combination, the key appears in the entry box below **Press Key**. If you used the **<Ctrl>**, **<Alt >** or **<Shift >** key to specify the keyboard shortcut, the appropriate checkbox is checked below the entry box.

You can also specify a key by typing the key letter or name in the entry box. For example, the **<F4>** key is specified by typing in the letters **F4**. Non-alphanumeric keys, such as **<spacebar>** or **<left arrow>**, can be specified by clicking the down arrow next to the entry box and selecting from the list.

Ctrl Check this checkbox to specify a key combination using the **<Ctrl>** key. This checkbox is automatically turned on when you click **Press Key** and specify a keyboard shortcut which includes the **<Ctrl>** key.

Alt	Check this checkbox to specify a key combination using the **<Alt>** key. This checkbox is automatically turned on when you click **Press Key** and specify a keyboard shortcut which includes the **<Alt>** key.
Shift	Check this checkbox to specify a key combination using the **<Shift>** key. This checkbox is automatically turned on when you click **Press Key** and specify a keyboard shortcut which includes the **<Shift>** key.
Assign	Click this button to assign the specified keyboard shortcut to the selected function. This button is only available when both a function has been selected and a keyboard shortcut has been specified.
Set	Displays the name of the current key assignment set. To create a new set, type in a new name and click **Save**.
Save	Saves the current key assignments to the name specified under Set.
Delete	Deletes the current key assignment set.

Files

The options under the **Files** tab pertain to file saving and archiving. The **File Handling** options control how files are saved with the *File/Save* menu option.

Backup on Save	When this option is checked, any time you save a file with the *File/Save* menu option and there is a previously saved file of the same name, the previously saved file is saved as a backup file with the name *maxback.bak*. The directory to which the file is saved is specified with the *File/Configure Paths* menu option as the **AutoBackup** directory. You can retrieve this backup file by renaming it to a file with the extension *.max* and loading it as you would any other scene file. The file can be renamed using File Manager under Windows NT or Windows Explorer under Windows 95.
Increment on Save	When this option is checked, any time you save a file with the *File/Save* menu option, the filename is incremented with a two-digit number. For example, if you load or create a file called *house.max* then save the

Menus

file, the new file will be saved as *house01.max*. The next time you save the file it will be called *house02.max*. This is the equivalent of using the *File/Save As* option and clicking the + button to save the file.

Recent Files in File Menu   Sets the number of recently used files that appear at the bottom of the *File* menu.

Save Viewport Thumbnail Image   Saves a 64-pixel thumbnail of the active viewport when a *.max* file is saved. Saving a thumbnail adds about 9K to a *.max* file. Thumbnails can be viewed with the **Asset Manager** under the **Utilities** panel .

The **AutoBackup** options set up automatic saves at specified intervals. Your scene will be automatically saved at intervals specified by the **Backup Interval (minutes)** option. The directory to which the backup files are saved is specified with the *File/Configure Paths* menu option as the **AutoBackup** directory. The backup files are called *autobak#.mx*, where # is the number of the backup. Up to 9 backup files can be stored at once. A backup file can be retrieved by renaming it to a filename with a *.max* extension and loading the file with *File/Open*.

Enable   Enables automatic backup.

Number of Autobak files   Sets the number of backup files. This number can range from 1 to 9.

Backup Interval (minutes)   Sets the number of minutes between automatic backups . This number can range from 0.01 to 480.

Under **Archive System**, you choose the archiving program that is used with the *File/Archive* menu option.

Program   Enter the command name for your preferred archiving program. You can precede the command with any necessary command line options.

Gamma

Gamma settings affect the brightness and contrast of images going into and out of 3D Studio MAX. The options on the **Gamma** tab adjust gamma values for input images, output images and monitor display.

Gamma settings affect the apparent brightness and contrast of images going into and out of 3D Studio MAX. *Gamma*, the Greek letter G, is an engineering term for a way of defining how dark and light areas of an image are interpreted. Understanding the technical definition of gamma is not as important as understanding how to set gamma values for the best images possible.

For many applications, you will never have to adjust or use gamma values. If your images look fine to you, leave the gamma values as they are. You might try adjusting the gamma values to see if it makes an improvement in your final output to a printer or videotape. The truest test if for you to look at the images at the output end and decide if they look good to you.

If you are printing images with a four-color process or are recording animation to videotape, you might find that the images are consistently too bright or too dark. Increasing or decreasing the **Output Gamma** value sometimes corrects this problem.

You might also find that bitmaps coming into your computer are consistently too washed-out or too dark. In this case, changing the **Input Gamma** value can make a difference.

Unfortunately, there is no magic solution for finding the right gamma values; trial and error is the only way to be sure. If you're planning to send your images out to a service bureau for output for the first time, try a number of different gamma values and send the images to the bureau beforehand to have them test the files. Images to be shown on a computer monitor often look best with an **Output Gamma** of 1.5 to 2.0. Images for four-color printing processes sometimes require an **Output Gamma** of up to 2.0, while video images have been known to look best with an **Output Gamma** of 2.0 or 2.2.

Raising the **Output Gamma** value does make rendered images lighter, but it is not a solution for underlit images. A too-high **Output Gamma** value results in washed-out, low contrast images.

Enable Gamma Correction	Applies gamma adjustments made on this dialog.
Load Enable State with MAX Files	Saves and loads the enable state with each *.max* file. If you load a file with a state different than the current state, you will be prompted to choose whether you want gamma enabled.

Under the **Display** section, a gray square is displayed inside another gray square. When the two squares are as close in color as possible, the gamma values are set correctly.

Gamma	The current gamma value. This value can only be changed if **Enable Gamma Correction** is on. Increase or decrease this value until the two squares are as close in color as possible. Values between 1.0 and 2.0 usually work best.
Input Gamma	Some bitmap types, such as *.tga*, have their own gamma value built in. For file types with no inherent gamma value, the gamma value is set with the **Input Gamma** value. Leave this value at 1.0 unless experimentation has shown you that another value works better.
Output Gamma	Sets the gamma value for rendered images. This value may be used by the device or program that receives the file, such as video output hardware.

Viewports

The **Viewports** tab has viewport display options for all viewports.

Use Dual Planes	Enables the use of a front/back plane drawing system when redrawing a viewport. With this system, the selected object is redrawn as it is manipulated, while other objects are not redrawn. This system provides the fastest redraw times under most circumstances.
Zoom About Mouse Point	When this checkbox is on, the **Zoom** button $\boxed{Q}$ zooms about the mouse point. When it is off, zooms are performed about the viewport center.
Draw Links as Lines	When unchecked, the **Display Links** checkbox on the **Display** panel displays links between objects as bones. When checked, links are displayed as lines.
Backface Cull on Object Creation	When checked, backfaces on each object are culled, or removed from view, as the object is created. **Backface Cull** can be turned on or off for individual objects on the **Display Floater** or on the Object Properties dialog, which can be accessed with the *Edit/Properties* menu option.
Attenuate Lights	When checked, attenuated lights are approximated in shaded viewports.
Mask	When checked, the area outside the Live Area safe

Menus

Viewport to Safe Region	frame is displayed as blank in the viewport. This option has an effect only if **Show Safe Frames in Active View** under the **Safe Frames** tab of the Viewport Configuration dialog is checked. To access this dialog, select *Views/ Viewport Configuration* from the menu.
Update background images while playing	When checked, the viewport background is updated during playback of animation from the viewport. If a viewport background is animated, uncheck **Real Time** under the Time Configuration dialog for smooth playback. The Time Configuration dialog can be accessed with the **Time Configuration** button ⌷⊕ on the Toolbar.
Filter Environment Backgrounds	When an environment background is used as a viewport background, checking this checkbox filters the image. Filtering is a method of antialiasing and smoothing an image as it is resized. Filtering takes time and slows down viewport redraw time, so check this checkbox on only when absolutely necessary. This checkbox has an effect on the background display only when **Use Environment Background** on the Viewport Background dialog is checked. To access this dialog, choose the *Views/Background Image* menu option.
Low-Res Environment Background	When this checkbox is checked and an environment background is used, the background is resized to half its original size, then magnified to the size of the viewport. Although it results in a rougher image, this process greatly improves viewport redraw speed. This checkbox has an effect on the background display only when **Use Environment Background** is checked on the Viewport Background dialog is on. To access this dialog, choose the *Views/Background Image* menu option.
Grid Nudge	Sets the distance an object is nudged when an arrow key is pressed.
Non-scaling object size	Sets the display size of cameras, lights, and other objects that don't change in size when viewports are zoomed. The default size is 1. A value of 2 doubles the default size.

The settings under **Move/Rotate Transforms** control how the mouse will respond when moving and rotating objects.

Menus

Intersection   When this option is chosen, movement in non-ortho-graphic viewports works with the mouse's position on the screen. This makes it easy to move the object, but as the object moves toward the horizon it moves faster and faster.

Projection   When this option is chosen, the mouse position is projected onto the active grid. This makes its motion at the horizon more stable, but it can be difficult to move the object when the viewport is nearly perpendicular to the plane.

Persp Sensitivity   Sets mouse sensitivity when objects are moved in the Perspective View.

Rotation Increment   Sets the rotation increment when the mouse is moved one pixel. Set this value lower if you need finer rotation.

The options under **Ghosting** control the display of ghosts. Ghosts are displays of animated object positions and deformations before and/or after the current frame.

Ghosting Frames   Sets the number of ghosts before and/or after the current frame.

Display Nth Frame   Sets the ghost frame increment. Ghosts will be displayed at the object's animated position every Nth frame.

Ghost Before Current Frame   Turn this option on to display object positions only before the current frame.

Ghost After Current Frame   Turn this option on to display object positions only after the current frame.

Ghost Before and After   Turn this option on to display object positions both before and after the current frame.

Ghost in Wireframe   When this option is checked, ghosts are displayed as wireframes in shaded viewports. By default, the ghost wireframe color is black. When unchecked, ghosts are shaded with the ghost wireframe color in shaded viewports.

Show Frame Numbers   Displays each ghost's frame number next to the ghost object.

The options in the **Middle Mouse Button** section control how the middle mouse button is utilized in 3D Studio MAX.

Pan/Zoom   When this option is active, the middle mouse button pans the viewport. If you have a Microsoft Intellimouse, the middle wheel zooms the viewport.

Stroke   Enables the use of strokes as keyboard shortcuts. For information on strokes and how they are used, see the **Strokes** entry in the *Panels* section.

The **Display Drivers** section is used to choose and configure a display driver for 3D Studio MAX.

Currently Installed Driver   Displays the current display driver.

Choose Driver   Click this button to select a different display driver.

Configure Driver   Displays options for configuring the chosen driver. See the documentation for 3D Studio MAX or your driver for details on options.

Colors

On the **Colors** tab, you can set colors for nearly every aspect of 3D Studio MAX.

Main UI   Sets colors for various aspects of the user interface.

Objects   Sets colors for different types of objects found in 3D Studio MAX.

Gizmos & Apparati   Sets colors for gizmos and apparatus objects.

Grids   Sets the color and intensity of the drawing grid. Selecting **Grid Intensity** uses gray for the grid, and sets the darkness of the gray with the **Grid Intensity** value. **Invert** inverts dark colors to light and vice versa. **Reset Grid Intensity** resets the grid intensity to the default value of 158.

You can also select **Grid Color** to use a custom grid color. **Reset Item** resets the grid color to the default color.

Properties

Sets properties for the selected object.

General Usage

Select one or more objects. Choose *Properties* from the *Edit* menu. The Object Properties dialog appears.

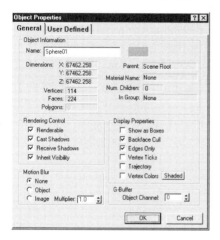

You can also access this dialog by right-clicking on the selected object and choosing *Properties* from the pop-up menu.

General

Name   The name of the selected object. If more than one object is selected, then **Name** is **Multiple Selected**.

Dimensions   The object's dimensions along each of the object's local axes. **Dimensions** are not changed by scale transforms, but might be changed by modifiers such as **Bend** or **Taper**.

Vertices   The number of vertices on the object at the current frame.

Faces   The number of faces on the object at the current frame.

Polygons   The number of polygons at the current frame. This value appears only for shapes.

Parent   If the object is the child of another object, the name of the parent is displayed here. If it is not linked to another object, the words **Scene Root** appear here.

Material Name The name of the material assigned to the object.

Num. Children	The number of hierarchically linked children this object has.
In Group	If the object is part of a group, the group name is displayed here. If the object is not part of a group, the word **None** appears here.
Renderable	When checked, the object renders in the scene as usual. When unchecked, the object is treated like a dummy object, where it doesn't render but still can be used to control the motion of other objects. When this checkbox is off, the object also will not cast shadows in the scene.

In order for shapes to render, both this checkbox and the **Renderable** checkbox on the shape's **Modify** panel must be checked. |
Cast Shadows	When checked, the object casts shadows when both a shadow-casting light shines upon it, and an object in the scene is positioned to receive the shadow.
Receive Shadows	When checked, the object can receive shadows from a shadow-casting spotlight when another object in the scene is positioned to cast a shadow upon it.
Inherit Visibility	When checked, the object inherits its parent's **Visibility** track, if the parent object has one. See the **Track View** entry in the *Toolbar* section of this book for information on **Visibility** tracks.
Motion Blur	Choose **Object** to apply object motion blur to the object, **Image** to apply image motion blur, or **None** to apply no motion blur. The **Multiplier** sets the strength of the smear for image motion blur. For an explanation of motion blur types and parameters, see the **Render** entry under **Object Motion Blur** and **Image Motion Blur**.

Under **Display Properties**, you can determine how the object will be displayed in viewports. These settings are similar to those found on the

Display Floater and the **Display** panel .

Show as Boxes Displays the selected objects as bounding boxes.

Backface Cull Eliminates the display of faces behind other faces in wireframe display.

Menus

Edges Only   Displays edges only, rather than the crossline separating triangular faces.

Vertex Ticks   Displays tick marks at each vertex.

Trajectory   Displays the animation trajectories of selected objects.

Vertex Colors   In shaded viewports, shades the objects with vertex colors. Vertex colors can be set under the Vertex sub-object level of an editable mesh.

To assign vertex colors, first create an object, then click

Edit Stack ⬛ and choose *Editable Mesh* from the pop-up menu. The object is converted. Click **Sub-object** and select Vertex as the sub-object level. Select vertices on the object. Under the Vertex Color rollout, click the color swatch to change the color of selected vertices. Vertex colors appear only in shaded viewports.

Shaded   In shaded viewports, turning on **Shaded** shades the object with vertex colors where colors have been changed from the default color, and shades the rest of the object with the object color.

In the **G-Buffer** section, you can set the object channel for the selected object.

Object Channel   Sets the object channel number for the object. Object channels are used in Video Post for assigning effects to specific objects. For more information on object channels, see the **Material Effects Channel** entry in the *Mat Editor* section.

User Defined

On the **User Defined** tab, information about this object can be entered.

Usage Notes

The Object Properties dialog can also be accessed by selecting and right-clicking on an object to access a pop-up menu, and choosing **Properties** from the pop-up menu.

Redo [*last action*]

Reverses the last *Edit/Undo [last action]* command.

General Usage

Perform an operation such as selecting or moving an object. Choose *Edit/Undo [last action]* to undo the operation. Choose *Edit/Redo [last action]* to redo the command.

You can also click the **Redo** button on the Toolbar to perform the same operation. If you right-click on the **Redo** button, a list of recently undone actions appears, allowing you to choose several actions to undo at once.

Usage Notes

While *Edit/Redo [last action]* redoes the last Undo for selections, transforms and modifiers, *Views/Redo [last view change]* redoes the last Undo for viewport changes such as zooms and pans.

Redo [*last view change*]

Reverses the last *Views/Undo [last view change]* command.

General Usage

Perform a viewport operation such as zooming or panning. Choose *Views/Undo [last view change]* to undo the operation. Choose *Views/Redo [last view change]* to redo the change.

Usage Notes

While *Views/Redo [last view change]* redoes the last Undo for viewport operations, *Edit/Redo [last action]* redoes the last Undo for object modifications.

Redraw All Views

Redraws geometry in all viewports.

Menus

General Usage

Choose *Views/Redraw All Views* from the menu. Each viewport is redrawn.

Sample Usage

In the course of using 3D Studio MAX, you may find that parts of the scene occasionally disappear. This is the result of MAX's attempt to redraw quickly with your video display card. The objects are still in the scene, but are not properly displayed.

To rectify this situation, choose *Views/Redraw All Views*. The missing objects and parts of objects will reappear.

If objects or parts of objects do not reappear, some of your objects may have normals pointing away from the viewport. This happens when faces are removed from objects, or when an open shape is used for lofting. To make both sides of the object appear in the viewport, choose *Views/Viewport Configuration* from the menu. Turn on the **Force 2-Sided** and **All Viewports** checkboxes. The objects should now appear in all viewports.

If you experience this problem frequently, your video display card may not be set up correctly. Consult your video display card installation manual and the 3D Studio MAX Installation Guide.

Rename Preview

Saves the most recent preview to a specified filename.

General Usage

Choose *Rendering/Rename Preview* from the menu. A file selector appears. Enter a new name for the preview. A copy of the preview will be made with the new filename.

The default directory for previews is the Previews directory. You can change this directory on the file selector.

To play the renamed preview, choose *View File* from the *File* menu and select the preview. The Media Player appears and the file is loaded. Click the play button to view the preview.

Sample Usage

When you create a preview with *Rendering/Make Preview*, 3D Studio MAX creates a preview of the current scene in the file *_scene.max*. This file is saved in the Previews directory. This preview can be viewed at any time, even if you load a different MAX scene.

The next time you make a preview, the previous preview is deleted and the new preview is saved as *_scene.max*. Use *Rendering/Rename Preview* to save a preview for later viewing and prevent it from being wiped out by the next preview creation.

Render

Accesses the Render Scene dialog to begin rendering a scene.

General Usage

Choose *Rendering/Render* from the menu. The Render Scene dialog appears.

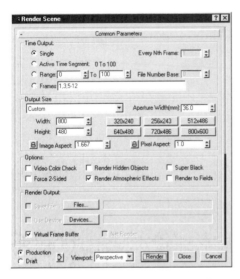

The Render Scene dialog can also be accessed by clicking the Render Scene button.

Dialog Options

Single	Renders the current frame only.
Active Time Segment	Renders the entire animation.
Range	Renders a range of frames.
Frames	Renders individual frames and/or a range of frames. You can enter several individual frames or ranges of frames, separated by a comma.
Every Nth Frame	Renders every Nth frame. For example, if this value is 5, every 5th frame is rendered. Files are numbered according to the **Nth Serial Numbering** setting under *File/Preferences* on the **Rendering** tab.

File Number Base   When a **Range** is entered, the **File Number Base** deter mines the first number to be used for sequentially numbered files. The **File Number Base** is added to the frame number to determine the sequential file number. The **File Number Base** can be a negative number.

If the Range is 0 to 50 and the **File Base Number** is 20, the first filename will be *file0020*. If the range is 50 to 100 and the **File Number Base** is -50, the first filename will be *file0000*.

In the **Output Size** section, you can select from a pulldown list of several standard film and video image formats. As you choose an image format, the six buttons change to show six resolution choices for that format. You can also type in a different **Width** or **Height**. For all image formats except Custom, **Aperture Width**, **Image Aspect** and **Pixel Aspect** are preset and cannot be changed, and the **Width** and **Height** change to reflect the **Image Aspect** when one or the other is changed. With a Custom image format, you can change these values.

Aperture Width   This value is used to calculate the relationship between the camera's lens length and FOV. For standard formats chosen from the pulldown list, this value is preset and cannot be changed.

If you already have a camera view set up and the camera view is active when you enter the Render Scene dialog, this value will affect the camera's lens length, but not its FOV. The view in the camera viewport is effectively unchanged, except that now the lens length is changed so it works with the chosen format.

When you choose a Custom format, you can change the **Aperture Width** directly to affect the lens length on the current camera if a camera view is active. The relationship between the lens length and FOV for any cameras created in the future will also be affected.

When working with Custom format, leave the **Aperture Width** at its default value of 36, or select a standard format similar to your custom format to get an idea of what the Aperture Width should be. In general, the **Aperture Width** should range between 30 and 60.

Image Aspect   Sets the ratio of image **Width** to image **Height**. With formats other than Custom, this value cannot be changed. With Custom format, changing the **Image Aspect** adjusts the **Height** of the image to the new ratio. Click the lock button next to **Image Aspect** to lock this value while changing the **Width** and **Height**.

Pixel Aspect   Sets the ratio of pixel width and height. A standard pixel is square, yielding a **Pixel Aspect** of 1. With formats other than Custom, this value cannot be changed. With Custom format, changing the **Pixel Aspect** adjusts the **Image Aspect** as well. Click the lock button next to **Pixel Aspect** to lock this value while changing other values. When using a Custom format, leave **Pixel Aspect** at 1 unless you have a specific reason for changing it.

The settings in the **Options** section set various options for rendering.

Video Color Check   Enables video color checking in the rendered image. Video color checking checks the image for colors that are "safe" to be displayed on video. Non-safe colors are flagged or changed according to the settings under *File/ Preferences* on the **Rendering** tab. See the **Video Color Checking** entry in the *Appendix.*

Force 2-Sided   Forces rendering of both sides of all objects regardless of the direction of surface normals. This option is useful when both the inside and outside of objects show in the scene, or when a model has been imported from another program and face normals are pointing in all directions. If you render an image of an imported model and pieces of the model seem to be missing, use **Force 2-Sided** when rendering to get both sides of all faces to render.

Render Hidden Objects   Renders hidden objects.

Render Atmospheric Effects   Renders atmospheric effects such as fog, volume lights and combustion.

Super Black   Renders super black whenever background black appears in the scene. Super black is a sub-black color used in compositing. For further information on super black, see the **Rendering Concepts** entry in the *Appendix.*

Render to Fields   Renders fields. *Fields* are odd and even lines of a video image. If you are rendering video animation, turn on **Render to Fields** to make your animation look smoother. If you are rendering to an AVI, do not turn on **Render to**

Menus

Fields. For information on what fields are and how they are used, see **Rendering Concepts** in the *Appendix*.

Render Output
Click **Files** to select or enter the filename to render to. Choosing a single-image format such as TGA or TIF when rendering a series of frames will create a series of sequentially numbered files. The **Save File** checkbox is automatically turned on when you choose a filename. You can uncheck **Save File** to render to the screen only, and the filename will be retained.

You can also click **Device** to render to a specific device. Select the device. Depending on the device you choose, there may be setup options that can be accessed with the **Setup** button. The **Use Device** option is automatically checked on when you select a device.

Virtual Frame Buffer
The Virtual Frame Buffer (VFB) is the window that displays the rendered image on your screen during rendering. When checked, the VFB displays during rendering. Turning off the VFB saves rendering time.

Net Render
Turn on this checkbox to render to your network.

Production / Draft
3D Studio MAX allows two complete rendering setups, **Production** and **Draft**. One set of parameters may be kept on the Render Scene dialog for **Production**, while a completely different set is kept for **Draft**.

The **Production** settings are meant to be used for a final rendering, while the **Draft** settings are for trial renderings and tests. When you go from test renderings to final rendering, resetting several parameters manually can lead to mistakes. Use **Production** and **Draft** settings to keep your two sets of parameters separate, and switch between the two as needed.

The **Copy Render Params** button ⧉ copies all parameters from the unselected setup to the selected setup.

Viewport
Sets the viewport to render. By default, the active viewport is displayed. You can also select a different viewport with the pulldown menu.

Render
Renders the scene.

Close
Closes the Render Scene dialog without rendering.

Cancel
Cancels all changes to the Render Scene dialog.

When the **Render Scene** dialog is scrolled down, the Default Scanline Renderer rollout appears.

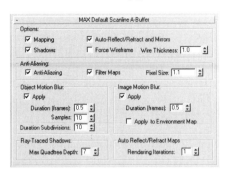

The settings under **Options** can be used to save rendering time during draft renderings.

Mapping — Renders maps.

Shadows — Renders shadow map or ray-trace shadows.

Auto-Reflect/ Refract and Mirrors — Renders materials with **Reflect**/**Refract** maps and/or **Flat Mirror** maps.

Force Wireframes — Renders all objects as wireframes. The wireframe color is the **Diffuse** color of the object's material, or, if no material has been assigned, the color of the object.

Wire Thickness — Sets the thickness of the wireframe in pixels when **Force Wireframe** is on.

Anti-Aliasing — Antialiases the rendered image.

Filter Maps — Most bitmaps used in materials must be scaled up or down in size by the renderer during rendering. *Filtering* determines how to treat the bitmap's detail during the scaling process. When **Filter Maps** is checked, the filtering method chosen for each bitmap in the Material Editor is used for filtering. The filtering method is set on the Bitmap Parameters rollout with the **Filtering** option. If **Filter Maps** is unchecked, no filtering takes place, and the pixels on the bitmap may appear rough or chunky. Unchecking **Filter Maps** improves rendering time.

Pixel Size — 3D Studio MAX performs antialiasing on rendered images with the help of the **Pixel Size**. This is the size of the area used for averaging colors during rendering. If **Pixel Size** is 1.3, for example, an area 1.3 times the size of a pixel area is used to average colors during rendering. A **Pixel Size** of 1.0 performs no antialiasing during rendering.

Higher **Pixel Size** values smooth out a rendering more than lower values, but may make the rendering appear fuzzy. In general, use higher values such as 1.3, 1.5 or 2.0 when rendering high-resolution scenes, or scenes with obvious straight edges that move a great deal. Use lower values for still images. Experimentation may be necessary to determine the optimum **Pixel Size** for your rendering.

The settings under **Object Motion Blur** apply mtoin blur to individual objects. In order for an object to blur with these settings, the object properties must be set for object motion blur with the *Edit/Properties* menu option. On the **General** tab under the Motion Blur section, select the **Object** option.

Apply   Enables object motion blur.

Duration   To calculate motion blur, the renderer looks at the object's position for a specified interval before and after the current frame and creates copies of the object based on the object's movement during this time. **Duration** sets the interval that will be used. For example, a **Duration** of 0.5 looks at the object's position for 0.25 of a frame before and 0.25 of a frame after the current frame. Although **Duration** can range from 0.0 to 99.0, fractional values such as 0.5 work best.

Samples   Sets the opacity of the motion blur copies that appear in the final rendering. **Samples** divided by **Duration Subdivisions** gives you the fraction of the original object's intensity that is used to paste pictures of the object copies. For example, if **Samples** is 9 and **Duration Subdivisions** is 12, each object copy appears at 0.75 of the original object's opacity. **Samples** cannot be larger than **Duration Subdivisions**.

Duration   To calculate motion blur, the renderer divides **Duration**
Subdivisions   by **Duration Subdivisions** to get a finer interval, and makes a copy of the object at each interval. For example, if **Duration** is 0.5 and **Duration Subdivisions** is 10, copies of the object are created at the object's position at 10 intervals. Each interval is 0.05 of a frame long (0.5 divided by 10). Half of these copies are before the object's current position, and half are after. A picture of each

copy is then pasted into the rendering with the opacity set by the **Samples** value.

Image Motion Blur uses each object's position before the current frame to calculate motion blur. Compare with Object Motion Blur, which uses each object's position both before and after the current frame. Image Motion Blur takes less time to render than Object Motion Blur, but does not produce realistic results when the object is moving along a curved path. In order for an object to use Image Motion Blur, the object properties must be set for image motion blur under the *Edit/Properties* menu option.

Apply	Enables image motion blur.
Duration	To calculate motion blur, the renderer looks at the object's position for a specified interval before the current frame and creates a blur based on the object's movement during this time. **Duration** sets the interval used. For example, a value of 0.5 looks at the object's position for 0.5 of a frame before the current frame.
Apply to Environment Map	Applies motion blur to background maps when the camera moves. This option works only with environment type background maps. In other words, at the map level on the Material Editor under the Coordinates rollout, the **Environ** button must be on, and either **Spherical Environment, Cylindrical Environment** or **Shrink-wrap Environment** must be selected from the **Mapping** pulldown.

The remaining options refer to ray-traced shadows and reflection maps.

Max Quadtree Depth	When ray-traced shadows are generated, a "tree" of bouncing light is used to calculate the shadows. The higher this value is, the deeper the tree. Higher values use more RAM, and thus take less time to render ray-traced shadows. However, this leaves less RAM to render the rest of the scene. Change this value only when you want to fine-tune the trade-off between RAM and shadow render time. In most cases, this value should be left at its default value of 7. Very low values such as 1 or 2 will cause very long render times for ray-traced shadows.
Rendering Iterations	Sets the number of times the reflection will "bounce" between **Reflect/Refract** reflections. Set this number to a high value such as 5 or 8 to "bounce" the reflection back and forth and create a more realistic image. Higher values increase rendering time.

Replaces objects in the current scene with objects with the same name from another MAX file.

General Usage

Load or create a scene. Choose *File/Replace* from the menu. A file selector appears. Select a file containing objects with the same name as the current scene. An object selection dialog appears. Only objects in the incoming scene with the same name as an object in the current scene are listed on the selector. Choose the objects you wish to replace.

Click **OK** to begin the replace operation. Another dialog appears.

Click **Yes** to replace materials from the current scene with incoming materials, or click **No** to leave the materials in the current scene.

Replacing objects is a technique used by animators who must create large, complex models. While a portion of the model is being worked on, other parts of the model can be replaced with simpler versions of complex parts. When the model is finished, the simple objects can then be replaced with the complex objects for the final rendering.

Usage Notes

If an object being replaced has instances, then all instances are replaced with the incoming object.

Replacing works with object names. If more than one object in the current scene has the same name as a replacement object, all objects with that name will be replaced.

The **Replace** and **Merge** menu options are similar in function, but have a few distinct differences. **Replace** replaces an object and its modifiers with the incoming object, but the replacement object's transforms, animation, bound space warps and relationship in a linked hierarchy do not come into the scene. These attributes are retained on the original object. Only geometry and modifiers, and possibly materials, are brought into the current scene to replace an object.

On the other hand, when a **Merge** deletes an old object, the object is replaced with the object from the incoming scene along with any animation, transforms and other attributes associated with the object.

Reset

Performs a reset, which clears the scene and loads default settings.

General Usage

Choose *File/Reset* from the menu.

If you haven't saved your current changes, a dialog appears asking if you want to save the file. Click **Yes** to save the file, **No** to load a new file without saving your work, or **Cancel** to cancel the operation. If you click **Yes**, a file selector appears where you can enter or choose a filename to save the file.

If you click **Yes** or **No**, or if you have not made any changes since your file was saved, a dialog appears.

When you reset, the file *maxstart.max* is automatically loaded. The reset settings are taken from this file.

Usage Notes

File/Reset loads the file *maxstart.max* and uses settings from this file. You can change the state of 3D Studio MAX after a reset by setting up your preferred settings and saving them to *maxstart.max* in the Maxstart path. By default, this path is the Scenes directory off your main 3D Studio MAX directory.

When you want to start a new scene, you can choose either *File/New* or *File/Reset*. *File/New* retains certain settings such as the state of floating windows and the last entry for creation parameters. Performing a *File/New* is more like deleting all objects in the scene, while *File/Reset* replaces the entire scene setup.

Reset Background Transform

Restores a zoomed or panned background image to its original position.

General Usage

Choose *Views/Background Image* to set up a background image. On the Viewport Background dialog, turn on the **Match Bitmap** button or **Match Rendering Output** button. Make sure the **Display Background** checkbox is on, and turn on the **Lock Zoom/Pan** checkbox.

The background image appears in the viewport at the selected aspect ratio. Because **Lock Zoom/Pan** is on, you can zoom or pan the background along with the objects in the viewport.

After zooming or panning, choose *Views/Reset Background Transform*. The background is restored to its original position in the viewport.

Usage Notes

The *Reset Background Transform* menu option is available only when the background image has been set up as described above.

Restore Active [*current*] View

Restores a previously saved view.

General Usage

Use viewport navigation tools such as **Region Zoom** 🔍 and **Pan** ✋ to get the desired view of the scene. Choose *Views/Save Active View* from the menu. The active viewport is saved. Use viewport navigation tools to change the view. Choose *Views/Restore Active View* from the menu to restore the saved view.

Usage Notes

In order to use *Views/Restore Active View*, you must first save a viewport display with *Views/Save Active View*. In order to restore the view, you must activate the same viewport that was active when you saved the view. If a different viewport is active, the *Views/Restore Active View* option will not be available from the menu.

When the view is restored, objects themselves are not altered. Any objects that were moved, transformed or modified since the view was saved remain in their current states. Only the view is restored to the saved position.

Saves the current scene to a *.max* file, including all materials and animation.

General Usage

Choose *File/Save* from the menu. If you have already saved the file, the file is saved again with the current filename. If you're saving a file for the first time, a file selector appears.

Enter the name of the file in the **File name** entry box. Click on **Save** or press **<Enter>** to save the file.

You can also choose an existing filename from the displayed list. If you do so, you will be prompted to confirm that you want to replace the existing file.

Save Active View

Saves the currently active view.

General Usage

Use viewport navigation tools such as **Region Zoom** 🔍 and **Pan** 🖐 to get the desired view of the scene. Choose *Views/Save Active View* from the menu. The active viewport is saved. Use viewport navigation tools to change the view. Choose *Views/Restore Active View* from the menu to restore the saved view.

Saves the current scene with a new filename.

General Usage

Choose *File/Save As* from the menu. A file selector appears. Enter or choose a filename for the file.

File/Save As works differently from *File/Save*. *File/Save* saves the file with its current filename, while *File/Save As* allows you to enter or choose another filename.

You can also make use of the + button on the file selector. The + button adds a number to the current filename before saving it. If there is already a number at the end of the filename, the + button increments the number by 1 before saving. This feature is useful when you want to save a series of files as you work.

Save Selected

Saves selected objects to a separate file.

General Usage

Choose *File/Save Selected* from the menu. A file selector appears. Enter or choose a filename. All selected objects are saved in the new file. The current scene and its name are unaffected.

This option is available only when one or more objects are selected.

Select All

Selects all objects in the scene.

General Usage

Choose *Edit/Select All* from the menu. All unfrozen, unhidden objects specified by the **Selection Filter** are selected. For example, when the **Selection Filter** is limited to cameras, all unhidden cameras are selected.

The **Selection Filter** is located at the top left of the Toolbar, just to the left of the **Select by Name** button . For information on the **Selection Filter,** see the **Selection Filter** entry in the *Toolbar* section of this book.

Select By Color / Name

Selects objects by object color or by name.

General Usage

To choose objects by name, choose *Edit/Select by/Name* from the menu. The Select Objects dialog appears. For information on using this dialog, see the **Select by Name** entry in the *Toolbar* section of this book.

To choose objects by color, Click *Edit/Select by/Color*. A Color Pick cursor appears on the screen. Click on an object with the color by which you want to select. All objects with that color are selected. This option selects by object color, not material color.

Usage Notes

When creating a complex scene, it can be helpful to group objects by color. For example, in an architectural interior, all chairs could have the same object color. Then you can select all chairs instantly by choosing *Edit/Select by Color* and clicking on one chair.

If you plan to group objects by color, turn off the **Assign Random Colors** checkbox on the **Object Color** dialog. You can access this dialog by clicking on the object's color swatch. This color swatch can be found under the Name and Color rollout on the Create panel, or at the top of any other panel. When the **Assign Random Colors** checkbox is turned off, the current color is assigned to all objects you create.

Inverts the current selection.

General Usage

Select one or more objects. Choose *Edit/Select Invert* from the menu. Selected objects are unselected, while unselected objects become selected.

Usage Notes

Edit/Select Invert works with the **Selection Filter**. Only object types specified by the **Selection Filter** are selected or unselected. For example, when the **Selection Filter** is limited to **Geometry**, all selections of geometry objects are inverted, but lights, cameras and space warps are not selected or unselected.

The **Selection Filter** is located at the top left of the Toolbar, just to the

left of the **Select by Name** button . For information on the **Selection Filter,** see the **Selection Filter** entry in the *Toolbar* section of this book.

Selection Floater

Displays the Selection Floater.

General Usage

Choose *Selection Floater* from the *Tools* menu. The Selection Floater appears.

This dialog is a floater, which means it can remain on the screen while you continue working on the scene. You can then use the Selection Floater at any time to change the current selection.

The Selection Floater is exactly the same as the Select Objects dialog that appears when you choose *Edit/Select by Name* from the menu or

click the **Select by Name** button ![button] on the Toolbar. For information on using this dialog, see the **Select by Name** entry in the *Toolbar* section of this book.

Unselects all objects.

General Usage

Choose *Edit/Select None* from the menu. All objects specified by the **Selection Filter** become unselected. For example, when the **Selection Filter** is limited to **Geometry**, all geometry is unselected.

The **Selection Filter** is located at the top left of the Toolbar, just to the left of the **Select by Name** button ▣. For information on the **Selection Filter,** see the **Selection Filter** entry in the *Toolbar* section of this book.

Shade Selected

Shades selected objects in all viewports.

General Usage

Choose *Shade Selected* from the *Views* menu. Selected objects are shaded in all viewports. As you change your object selection, the selected objects become shaded, while unselected objects change back to the type of display set for the viewport.

Show Axis Icon

Toggles the display of object axis icons.

General Usage

Choose *Views/Show Axis Icon* from the menu. If this option was checked, it will become unchecked, and axis icons will not appear for objects as they are selected. If this option was unchecked, it will become checked, and axis icons will appear for objects as they are selected.

Usage Notes

An *axis icon* is a set of XYZ axes for each object. These axes provide a reference for working with objects. Axis constraints such as the **Restrict to X** $\boxed{X}$ and **Restrict to YZ** $\boxed{YZ}$ buttons use the object's axes to determine how the object will respond to transforms and modifiers. When *Views/Show Axis Icon* is checked, an object's axis icon appears when the object is selected.

Each object has a set of axes that refer to its individual orientation in space. These axes are referred to as the object's *local axes*. At all times, a reference coordinate system is also active in 3D Studio MAX, which may or may not match the object's local axes. The current coordinate system determines which set of axes will be displayed when *Views/Show Axis Icon* is checked.

To change the current coordinate system, locate the **Reference Coordinate System** pulldown at the top center of the Toolbar. Click on the down arrow to choose a coordinate system. If **Local** is selected and *Views/Show Axis Icon* is checked, an object's local axes are displayed on the screen when the object is selected. For more information on coordinate systems, see the **Reference Coordinate System** entry in the *Toolbar* section of this book.

When a new coordinate system is chosen, the axis icon for each object might change. In addition, the **View** and **Screen** coordinate systems cause the axis icons to change from one viewport to another.

Show Dependencies

Causes dependent objects to turn green when another dependent object is selected.

Menus

General Usage

Create an instance or reference of an object. Choose *Views/Show Dependencies* from the menu to turn this option on. When either the original object or instance is selected, the other object turns green to show the two objects are dependent on one another.

If you select more than one object and apply a modifier to the selection, all objects in the selection are now dependent on one another. Selecting any of these objects will cause all the other objects to turn green.

Show Ghosting

Toggles the display of ghost objects.

General Usage

Choose *Show Ghosting* from the *Views* menu. Ghosts for the selected object only are displayed.

Usage Notes

Ghosting is the display of wireframe "ghosts" to represent the object at different stages of the animation.

To set up ghost display, choose *File/Preferences* and access the **Viewports** tab. Set the number and frequency of ghosts under the Ghosting section. You will see ghosts only if you choose to display ghosts before, after or both before and after the current frame. You can also set up the color of ghosts under the **Color** tab.

After choosing *Views/Show Ghosting*, you might have to reselect the object or move the time slider to see the ghosts for the first time.

Show Home Grid

Toggles the display of the home grid in the current viewport.

General Usage

By default, a checkmark appears next to the *Views/Grids/Show Home Grid* menu option, meaning this option is on and the home grid is currently displayed in the active viewport.

Choose *Views/Grids/Show Home Grid* from the menu to turn off this option. The home grid disappears from the current viewport. Choose *Views/Grids/Show Home Grid* from the menu again to turn it back on.

This option can be on or off for each viewport separately.

Show Key Times

Toggles the display of key times on a trajectory.

General Usage

Animate an object with two or more keys. Select the object and go to the **Display** panel [■]. On the Display Properties rollout, check **Trajectories**. The trajectory for the object appears in all viewports.

Choose *Show Key Times* from the *Views* menu. The time for each key on the trajectory path appears next to the key.

When you access the *Views* menu again, a checkmark appears next to *Show Key Times* to indicate that this option is on. Choose this option again to turn it off.

Usage Notes

In order to see the key times when the *Show Key Times* option is checked, the object's trajectory must be displayed. You can display an object's trajectory in any of four ways.

- Select the object. Go to the **Display** panel. On the Display Properties rollout, check **Trajectory**.

- Select the object. Choose *Tools/Display Floater* from the menu. Access the **Object Level** tab and check **Trajectory**.

- Select the object. Choose *Edit/Properties* from the menu, or right-click on the object and choose *Properties* from the pop-up menu. Under the **Display Properties** section, check **Trajectory**.

- Select the object. Go to the **Motion** panel [◎] and click **Trajectories**. The selected object's trajectory appears on the screen.

With the first three methods, the trajectory remains on the screen regardless of whether the object remains selected. With the last method, the trajectory disappears when the object is no longer selected.

Show Last Rendering

Displays the last rendered image.

General Usage

Choose *Show Last Rendering* from the *Rendering* menu. The last rendered image is displayed.

This option is available only when an image has already been rendered.

Creates clones of an animated object's state at specified intervals.

General Usage

Transform an object over time. This can be done by moving, rotating or scaling the object at various frames with the **Animate** button turned on.

Choose *Tools/Snapshot.* The Snapshot dialog appears.

You can take one snapshot of the object at the current frame, or you can make a series of snapshots over a range of frames. You can also choose whether the snapshot objects will be copies, instances, references or collapsed mesh objects.

You can also create snapshots with the **Snapshot** button ⚔ from the

Array flyout ⚙ on the Toolbar.

Dialog Options

Single	Creates one snapshot object based on the object's transformed state at the current frame.
Range	Creates a series of snapshots over a range of frames. Enter the range of frames in the **From** and **To** fields, and enter the number of copies next to **Copies**. A copy will be made at the **From** frame, at the **To** frame, and at regularly spaced frames in between. For example, if **From** and **To** are 0 to 100 and **Copies** is 6, snapshots will be created at frames 0, 20, 40, 60, 80 and 100.
Copy **Instance** **Reference**	Makes each snapshot a copy, instance or reference of the source object. For information on these clone types, see the **Clone** entry in this section of the book.
Mesh	Makes each snapshot clone a collapsed mesh version of the source object.

Usage Notes

Snapshot makes clones of the source object at selected intervals. The picture below shows five snapshots of a teapot that has been moved and rotated over time.

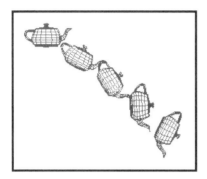

If the object doesn't transform during the animation, Snapshot simply creates clones of the object at the same location as the original object.

Each snapshot contains all animated modifier information. If animated modifiers have been applied to the source object, the same animated modifiers will be present on each clone, and will continue to animate over time.

Snapshot can also be used with two or more selected objects at a time.

Displays information about the current scene.

General Usage

Choose *File/Summary Info* from the menu. The Summary Info dialog appears.

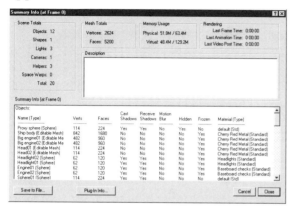

Dialog Options

Scene Totals   Lists the total number of each object type in the scene.

Mesh Totals   Lists the total number of vertices and faces in the scene.

Memory Usage   **Physical** refers to the total RAM used by 3D Studio MAX, followed by the total RAM available. **Virtual** refers to the total page size used followed by the total page size available.

Rendering   Provides the time elapsed for the last frame, animation and video post rendering times.

Description   Enter a file description. This entry is useful when more than one person will access a file, and file details must be passed on from one person to the next.

Summary Info   Lists all objects in the scene, including hidden and frozen objects, and some of their attributes. Objects are arranged by type. At the bottom of the list are materials and environment elements.

Object attributes cannot be changed on this dialog. To change attributes, close this dialog, select an object and choose *Edit/Properties*.

Save to File   Saves the information under **Summary Info** to a text file. When you click **Save to File**, a file selector appears. Enter a filename. The file is saved as a text file with the extension *.txt.*

Plug-In Info   When you click **Plug-In Info**, a dialog appears listing all the plug-ins installed in 3D Studio MAX.

Show Details displays further information about each plug-in. **Show Used Only** displays only the plug-ins used in the current scene. **Close** closes this dialog and returns you to the Summary Info dialog.

If you click **Show Used Only,** you may be surprised to find that your scene contains many more plug-ins than you remember using. From a programming perspective, many of 3D Studio MAX's functions are actually plug-ins.

Cancel   Cancels and exits the dialog without saving changes to the **Description** field.

Close   Saves changes to the **Description** field and exits the dialog.

Displays a floating entry box for transforms (move, rotate, scale).

General Usage

Choose *Transform Type-In* from the *Tools* menu. The Transform Type-In dialog appears. The title of the dialog changes depending on the current transform.

A transform must be chosen in order to use this dialog. The Transform Type-In dialog can also be accessed by right-clicking on a transform button. The Transform Type-In dialog can be left on the screen while you perform other tasks.

Move Transform Type-In

If **Select and Move** ⊕ is selected, the dialog values refer to the position of the object. This dialog can also be accessed by right-clicking on **Select and Move**.

Absolute: World — The placement of the object in world space, expressed in units. More specifically, these values represent the offset of the object's pivot point from the world coordinate system's origin along the X, Y and Z axes.

To change an object's position in world space, enter a new value for **X**, **Y** or **Z**. Note that the world coordinate system axes are used here regardless of the current coordinate system.

Offset — Offsets the current position of the object by the specified number of units in the current coordinate system. When you enter an offset value and press **<Enter>**, the object's position is immediately updated and all values return to zero. The values under **Absolute:World** are updated to reflect the object's new position.

Next to the word **Offset**, the current coordinate system name is displayed. The offset value works on the current coordinate system, which may not be the World coordinate system. If it is not, values typed in for **Offset** may update a different value under **Absolute:World**. For example, typing in a **Y** value of 50 under **Offset** might increase the **X** or **Z** value under **Absolute:World** rather than the **Y** value.

If you are using the **View** coordinate system, the name **World** or **Screen** will appear next to **Offset** rather than **View**. This is because the **View** coordinate system actually uses the Screen coordinate system in orthographic viewports and World for non-orthographic views.

Rotate Transform Type-In

If **Select and Rotate** [icon] is selected, the dialog values refer to angles of rotation. This dialog can also be accessed by right-clicking on **Select and Rotate**.

Absolute: World   The rotation of the object in relationship to the world coordinate system's X, Y and Z axes.

To change an object's orientation in world space, enter a new value for **X**, **Y** or **Z**. Note that the world coordinate system axes are used here regardless of the current coordinate system.

Offset   Offsets the object's current rotation by the specified number of degrees in the current coordinate system. When you enter an offset value and press **<Enter>**, the object's rotation is immediately updated and all values return to zero. The values under **Absolute:World** are updated to reflect the object's new rotation.

Next to the word **Offset**, the current coordinate system name is displayed. The offset value works on the current coordinate system, which may not be the World coordinate system. If it is not, values typed in for **Offset** may update a different value under **Absolute:World**. For example, typing in a **Y** value of 45 under **Offset** might increase the **X** or **Z** value under **Absolute:World** by 45 rather than the **Y** value.

If you are using the **View** coordinate system, the name **World** or **Screen** will appear next to **Offset** rather than **View**. This is because the **View** coordinate system actually uses the Screen coordinate system in orthographic viewports and World for non-orthographic views.

Scale Transform Type-In

If **Select and Uniform Scale** 🔲, **Select and Non-uniform Scale** 🔲 or **Select and Squash** 🔲 are selected, the dialog values refer to a scale percentage of the original object size. This dialog can also be accessed by right-clicking on the scale button.

Absolute:
Local
The object's scale in relationship to its size when created. The Scale Transform Type-In always uses the object's local scale as the absolute reference point. A value of 100 is the equivalent of the object's original size along the specified axis.

To change an object's overall scale directly, enter a new value for **X**, **Y** or **Z**. Note that the object's local axes are used here regardless of the current coordinate system.

Offset
Offsets the object's current scale by the specified percentage. When you enter an offset value and press **<Enter>**, the object's scale is immediately updated and all values return to 100. The values under **Absolute:Local** are updated to reflect the object's new scale.

The value 100 refers to the object's current size. To increase the object to 1-1/2 times its current size, enter 150. To cut the object size in half, enter 50.

If **Select and Uniform Scale** is selected, only one value appears on the dialog under **Offset**. If **Select and Non-uniform Scale** or **Select and Squash** is selected, **X**, **Y** and **Z** percentages can be entered.

Next to the word **Offset**, the current coordinate system name is displayed. The offset value works on the current coordinate system, which may not be the **Local** coordinate system. If it is not, values typed in for **Offset** may update a different value under **Absolute:Local**. For example, typing in a **Y** value of 200 under **Offset** might increase any of the **X, Y** or **Z** values under **Absolute:Local**.

If you are using the **View** coordinate system, the name **World** or **Screen** will appear next to **Offset** rather than **View**. This is because the **View** coordinate system actually uses the **Screen** coordinate system in orthographic viewports and **World** for non-orthographic views.

Undo [*last action*]

Reverses the last operation.

General Usage

Perform an operation such as selecting or moving an object. Choose *Edit/Undo [last action]* from the menu to undo the operation. You can choose *Edit/Undo* repeatedly to undo several consecutive operations.

You can also click the **Undo** button on the Toolbar to perform the same operation. Right-clicking this button displays a list of recent actions, and allows you to choose several actions to undo at once.

Usage Notes

Most object creation and modification operations can be undone. The following are examples of operations that can be undone with *Edit/Undo*:

- Transforms
- Object selection
- Object deletion
- Changes to parameters on modifier rollouts
- Applying a modifier
- Removing a modifier from the stack

Examples of actions that cannot be undone:

- Saving a file
- Applying a material
- Changing a parameter on a dialog

If you are about to perform an operation that cannot be undone, you might want to perform a *Hold* on your scene. A *Hold* puts the entire scene in a temporary file. Perform a hold by choosing *Edit/Hold* from the menu. If you later want to revert to the held file, choose *Edit/Fetch*.

By default, you can undo up to 20 operations. Setting the number of undo levels to a lower number takes up less memory, and can improve the performance of 3D Studio MAX. You can change this number by choosing *File/Preferences* from the menu. Click the **General** tab and change the **Undo Levels** value under the **Scene Undo** section.

While *Edit/Undo* undoes the last operation, *Views/Undo* undoes the last viewport change such as a zoom or pan.

Undo [*last view change*]

Reverses the last view change.

General Usage

Perform a viewport operation such as zooming or panning. Choose *Views/Undo [last view change]* to undo the operation.

Usage Notes

The *Views/Undo* command stores a separate list of viewport changes for each viewport. The last viewport change shown on the menu refers to the last change to the current viewport.

You can undo the last 20 view changes by repeatedly choosing *Views/ Undo*. To redo a view change, choose *Views/Redo*.

While *Views/Undo* undoes the last view change, *Edit/Undo* undoes the last object modification.

Ungroup

Ungroups the selected group.

General Usage

Select a previously grouped object set. Choose *Ungroup* from the *Group* menu. The objects no longer belong to a group and can be manipulated separately.

Usage Notes

If you want to work with grouped objects individually without ungrouping the group, you can open the group temporarily with *Group/Open*. This option allows you to work with objects separately. Choose *Group/Close* to close the group when you're done.

Sets up the units display for drawing in 3D Studio MAX.

General Usage

Choose *Units Setup* from the *Views* menu. The Units Setup dialog appears.

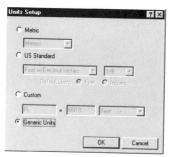

Dialog Options

You can choose one of four options for the units used in 3D Studio MAX.

Metric   Choose a metric unit from the pulldown menu. When drawing objects, one unit will be equal to the metric unit chosen. When you specify units while drawing, each value will be followed by the abbreviation for the metric unit. For example, if you have chosen centimeters as the base unit and you enter the value 40 while creating an object, the value will automatically be changed to 40CM.

US Standard   Choose a type of unit from the pulldown menu. If you choose a unit type with the word **Fractional** in it, the fractions entry box to the right of the pulldown is enabled. Choose a fractional unit that will equal one unit when drawing.

The **Feet** and **Inches** options are enabled when you choose a unit type containing both feet and inches. When you enter a number and press **<Enter>** without specifying feet or inches, the system will assume either **Feet** or **Inches** depending on which option is turned on. For example, if you enter 5 as a radius value while **Feet** is selected on in this dialog, 3D Studio MAX will redisplay this value as 5'. If the **Inches** option is turned on, this value will be redisplayed as 5".

When drawing, US standard measurements are always expressed with the ' symbol for feet and " for inches.

Custom   Allows entry of custom units. At the left, enter a two or three character abbreviation for your custom units. In the second entry box, enter a numeric value. In the third, choose a standard unit from the pulldown. Example: FA = 6 Feet. The character abbreviation appears after all entered values when drawing.

Generic Units   Uses the system unit scale of 1 unit = 1 inch. When **Generic Units** is selected, no units abbreviation appears after entered values. If you are not required to use technical precision in your drawings, use **Generic Units**.

Usage Notes

When you switch from one unit setup to another, all values in the scene automatically convert to the new setup.

This dialog sets up how the units will display in 3D Studio MAX, but does not affect the actual scale of the geometry. The actual scale of the geometry, which defaults to 1 unit = 1 inch, is set with the **System Unit Scale** value on the **General** tab under *File/Preferences*. Unless you have a very good reason to do so, don't change this value as it will prevent you from sharing files with other users or merging your own files together.

Because a 3D Studio MAX drawing is not a physical object but rather a bunch of numbers and data, the system unit scale doesn't need to have anything to do with reality. The main reason for the system unit scale is to have a standard base unit for files. This way, when you share your files with other 3D Studio MAX users, you will all be working from the same base unit, even if one drawing uses centimeters and the other uses feet.

Update Background Image

Updates the background image after a change to map or rendering parameters.

General Usage

Choose *Views/Background Image* to set up a background image. Make sure **Display Background** is checked. The background image appears in the viewport.

Change the image in a paint program or other external program.

Choose *Views/Update Background Image.* The background is updated to reflect the changes to the image.

Usage Notes

It is only necessary to update the background when you have made changes to the background image in a paint program or other external program.

Update During Spinner Drag

Toggles the screen update method used when spinners are dragged.

General Usage

Choose *Update During Spinner Drag* from the *Views* menu to turn it on or off.

Update During Spinner Drag affects how the screen updates when you drag on a spinner to change a value. When *Update During Spinner Drag* is on, the screen is updated as you drag the spinner. When *Update During Spinner Drag* is off, the screen is updated only when you release the mouse button.

It can speed up your drawing time to turn off *Update During Spinner Drag* when working with complex models or with spinner parameters that require a lot of processing when moving from one value to the next.

Video Post

Accesses the Video Post window.

General Usage

Choose *Rendering/Video Post* from the menu. The Video Post window appears.

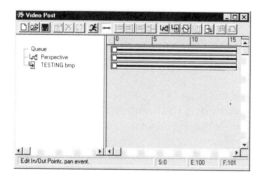

Video Post is used to composite images, scene elements and effects, much like a video post production suite.

Video Post uses a sequence called a *queue* to composite images and effects. Each action on the queue is called an *event*. As each event is added to the queue, it appears at the left of the Video Post window. When the queue is rendered, Video Post starts at the top of the queue and moves downward.

The numbers across the top of the window refer to frame numbers. Across from each event name is a range bar. Be default, each event occurs over the time range of the current animation. The range can be edited so an event only occurs across specified frames. Ranges appear as blue bars. When an event is selected, its range bar appears as a red bar.

Some events are dependent on other events. A dependent event is called a *child* event. A child event appears on the queue indented below its parent. A child event is created by highlighting an event and choosing another event. The highlighted event becomes the child, and the new event is the parent. When a parent event is executed on the queue, it uses the child events to create the effect before moving on to the next parent event in the queue.

To add the 3D scene to the queue, click **Add Scene Event** . To add an image to the queue, click **Add Image Input Event** . Filters are added with **Add Image Filter Event** , and the output of the queue is set with **Add Image Output Event** .

As each event is added, a dialog appears with options for the event. The queue can consist of any combination of events.

To render the Video Post queue, click **Execute Sequence** .

Video Post Toolbar

New Sequence — Clears the queue for a new sequence.

Open Sequence — Opens a previously saved Video Post queue file.

Save Sequence — Saves the current sequence in a Video Post file. Video Post files are saved with the extension *.vpx*.

Edit Current Event — To edit an event, highlight the event on the queue and click **Edit Current Event**. The same dialog that appeared when the event was first added is redisplayed, and options can be changed.

Delete Event — Removes the highlighted event from the queue.

Swap Events — To swap the position of two events in the queue, high light one event, then hold down the <Ctrl> key and highlight another event. Click **Swap Events** to swap the events. This button is available only when two events are highlighted.

Execute Sequence — Executes the Video Post sequence.

Edit Range Bar — When this mode is on, the length of a range bar can be edited by clicking and dragging on either end of the bar.

Align Selected Left — Shifts the range bars of all selected events so they all start at the first frame of the range bar selected last. The length of shifted range bars is not changed. This option is available only when two or more events are selected.

Align Selected Right — Shifts the range bars of all selected events so they all end at the last frame of the range bar selected last. The length of shifted range bars is not changed. This option is available only when two or more events are selected.

🔲	**Make Selected Same Size**	Makes all selected range bars the same length. The length used is the length of range bar selected last. Each selected range bar retains the same starting frame, and its length is extended to the right. This option is available only when two or more events are selected.

🔲	**Abut Selected**	Shifts selected range bars so each one starts when the previous one finishes. The length of range bars is not changed. This option is available only when two or more events are selected.
🔲	**Add Scene Event**	Adds an event to the queue for rendering the current scene. The viewport name appears in the queue.
🔲	**Add Image Input Event**	Adds an image file to the queue. The image file can be a still image, an animation file or an *.ifl* listing of images.
🔲	**Add Image Filter Event**	Adds a filter event to the queue. A filter can be used to affect an image in any number of ways, from altering the contrast to adding lens flares. See your 3D Studio MAX documentation for a list of available filters and how they are used

To add a filter to a specific image, highlight the image and click **Add Image Filter Event**. Select the filter and set up any parameters associated with it. When you click **OK** to exit the dialog, the queue displays the image as an indented entry under the filter event. This indicates that the image is a child of the filter, and thus is affected by the filter.

You can also add a filter event to the queue without highlighting an image. In this case, the filter has no child events, and affects the image at that point in the queue.

🔲	**Add Image Layer Event**	Adds a compositor event to two events. This option is used to composite two images with an alpha channel, or with another type of compositor. To use this option, highlight two events in the queue, and click **Add Image Layer Event**. Choose the layer event type from the pulldown list. To composite two images, choose **Alpha Compositor**.

When you click **OK** to exit the dialog, the two highlighted events become indented child events of the layer event in the queue.

Menus

⊞	**Add Image Output Event**	Sets the output file for the rendered queue. You must place an output event at the end of the queue in order for the rendered Video Post output to be saved.
⊠	**Add External Event**	Adds an external event to the queue. This option allows you to call an external program for image processing. An external event is always a child event.
⊡	**Add Loop Event**	Loops an event. A loop event has one child event, which it loops.

Sample Usages

Video Post can be used to composite two images.

1. Access the Video Post window. Click **Add Image Input Event** 🔳. Click **Files** and choose any image from the file selector.

2. Click **Add Image Input Event** again. Click **Files** and choose the file *daisy.tif*. This image, which shows a daisy, has an alpha channel.

3. Highlight both events on the queue. Click **Add Image Layer Event** 🔳. From the pulldown, choose **Alpha Compositor**.

You have just instructed Video Post to composite the two images, using the alpha channel of the second image, *daisy.tif*, to layer the second image over the first.

4. Click on an area of the queue to unselect all queue events. Click **Add Image Output Event** ⊞. Click **Files** and enter a filename.

5. Click **Execute Sequence** 🔳. Turn on the **Single** option. Click **Render**.

Each image appears as it is processed. After a few moments, the composited image appears, with the daisy placed over the first image.

Video Post can also be used to put a glow on one or more objects in a scene.

1. Click **New Sequence** 🔳 to start a new sequence.

2. Create two identical materials in the Material Editor, each with different names. Change the **Material Effects Channel** of one material to 1. To do this, click and hold on the **Material Effects Channel** button. Drag the cursor to the number 1 and release the mouse. The **Material Effects Channel** button changes to display the number 1.

3. Create two spheres and assign one of the new materials to each one. Note which sphere has the material with **Material Effects Channel** 1.

4. Activate the Perspective viewport. Click **Quick Render**  to render the scene. The two spheres render as expected.

5. Access the Video Post window. Click **Add Scene Event** 🖼. A dialog appears. Click **OK** to accept the default settings.

The scene view to render has been added to the Video Post queue, indicated by the **Perspective** entry in the queue.

6. Highlight the Perspective listing in the Video Post queue. Click **Add Image Filter Event** 🖼. The Add Image Filter Event dialog appears. Choose **Lens Effects Glow** from the pulldown.

7. Click the **Setup** button. The Lens Effects Glow dialog appears.

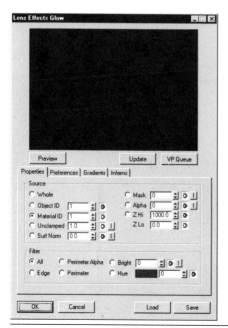

Under the Source section, choose **Material ID** and leave this number at 1. Click **OK** to accept the **Lens Effects Glow** settings.

Even though the parameter on the Lens Effects Glow dialog is labeled **Material ID**, it refers to the **Material Effects Channel** for the material on the object, not the object faces' material ID.

8. On the Add Image Filter Event dialog, change the **Label** to Glowing Ball. Click **OK** to accept the settings.

9. On the Video Post window, click **Execute Sequence** ⚹. The Execute Video Post dialog appears. Under Time Output, click **Single** to render only the current frame. Click **Render**.

The renderer makes two passes, one to render the objects and another to create the glow. The glow does not appear until the rendering is finished. When rendering is complete, the sphere with the material with **Material Effects Channel** 1 will have a glow.

Displays an image or animation file, and information about the file.

General Usage

Choose *View File* from the *File* menu. A file selector appears, with additional options.

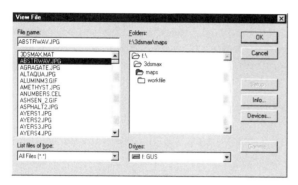

Here you can select a file to view. Highlight the filename and double-click or click **OK** to see the file.

Dialog Options

OK
Displays the selected file.

Cancel
Exits without displaying the file.

Setup
Sets up the device from which to get the file. This option is available only if a device has been set up with the **Device** option.

Info
Displays information about the file such as resolution, file type and size, and date saved.

Device
Indicates a device from which to get the image. For example, you might want to view a file from a digital recorder.

Gamma
Sets the gamma level for the image being viewed. This option is available only when **Enable Gamma Correction** is checked on the **Gamma** tab of the Preferences Settings dialog. To access this dialog, choose *File/Preferences* from the menu.

When **Gamma** is clicked, a dialog appears.

The **Use image's own gamma** option will display the image with its own native gamma value. **Use system default gamma** uses the **Output Gamma** setting on the Preference Settings dialog under the **Gamma** tab. **Override** uses an entered value as the image's output gamma.

Usage Notes

The selected image appears when you click **OK** or when you double-click on the filename.

If you choose to view an AVI or other animation file, the Media Player appears. Click **Play** ▶ to view the file.

If you choose a still image or file list such as an *.ifl* file, the Virtual Frame Buffer (VFB) appears. The VFB is a viewing window for files. Although you can change the display to view the file in many different ways, you cannot save the changes on this window.

The red, green and blue buttons at the upper left display the image's red, green and blue channels. You can also view the image's alpha channel, or see it as a monochrome image.

For information on the file and its colors, right-click and drag across the image. A window appears with information about the file, and also about each color over which the cursor moves. When you release the mouse, the last color accessed appears in a color swatch at the upper right of the VFB window. You can copy this color to another color swatch on another window or panel, or you can click the color to see the Color Selector and get the color's values. Any changes to the color on the Color Selector are not recorded on the image.

For more information on the VFB and how it works, see the **Rendering Concepts** section of the *Appendix*.

Viewport Configuration

Allows you to set up a number of viewport parameters such as layout and display levels.

General Usage

Choose *Views/Viewport Configuration* from the menu. The Viewport Configuration dialog appears.

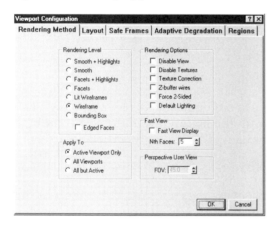

There are five tabs for working with five different aspects of viewport configuration.

Rendering Method	Controls how the objects will be displayed in viewports while you're creating and editing the scene.
Layout	Sets up the viewport layout.
Safe Frames	Sets the size of the video safe area, allowing you to see where cropping could occur during videotape recording or playback.
Adaptive Degradation	Controls how MAX will respond when onscreen animation playback cannot keep up with the chosen viewport display method and frame rate.
Regions	Sets the region size for region and blowup rendering.

Rendering Method

The **Rendering Method** tab controls the display and shading of objects on the screen, not in rendered images.

The options under the **Rendering Level** section set the level of display for objects in the viewport. Only one option may be active at a time.

Smooth + Highlights	Displays objects with smooth shading, which smooths rounded areas. Highlights are displayed where they naturally fall according to the current lighting setup and the view.
Smooth	Displays objects with smooth shading.
Facets + Highlights	Displays objects with faceted shading. On rounded objects, each facet is shaded separately. Highlights are displayed where they naturally fall according to the current lighting setup and the view.
Facets	Displays objects with faceted shading.
Lit Wireframes	Displays objects as wireframes with color variations. The wireframe is lit with the lighting from the scene, and the wireframe color changes accordingly across the object.
Wireframe	Displays the wireframe segments making up the objects.
Bounding Box	Displays each object as a box showing the extents of the object.
Edged Faces	When checked, object wireframes are displayed along with shading. This option is available for shaded viewports only.

The options under the **Apply** section allow you to choose the viewports that will be affected by the Rendering Level. Only one option may be active at a time.

Active Viewport Only	Choose this option to apply settings to only the active viewport.
All Viewports	Choose this option to apply the settings to all viewports.
All but Active	Choose this option to apply the settings to all viewports except the active viewport.

The options under the **Rendering Options** section further customize the viewport display.

Disable View	Disables the viewport. Changes from other viewports will not appear in a disabled viewport, but you can still access the viewport and make changes to the scene in the viewport.

Disable Textures	This option prevents bitmaps from appearing on objects in the viewport. A bitmap appears on an object only when a material containing a bitmap has been applied to the object and the **Show Map in Viewport** button in the Material Editor has been clicked for that bitmap. The viewport must also have a smooth or facet rendering level in order to display a bitmap.
Texture Correction	Some maps appear distorted when displayed in viewports with the Material Editor's **Show Map in Viewport** button. Turning on **Texture Correction** for the viewport makes maps appear more accurately, but slows down viewport redraw time.
Z-buffer wires	Uses Z-buffer (depth) information to redraw the screen. The use of this option eliminates some redraw problems.
Force 2-Sided	Displays all faces on an object regardless of the direction in which the normals point. Turn on this option if some objects partially appear, or don't appear at all, due to normals pointing away from the view.
Default Lighting	Forces the use of default lighting even when lights have been placed in the scene.

The options under **Fast View** cause objects to redraw with fewer faces, making redraw time faster.

Fast View Display	Enables Fast View for viewports indicated under the Apply section.
Nth Faces	Drops every Nth face when **Fast View Display** is on. For example, a value of 5 drops every 5th face.
Perspective User View	Sets the initial field of view (FOV) for a Perspective view.

Layout

The **Layout** tab offers several ways to arrange the viewport layout.

Along the top of the **Layout** tab are 14 layouts from which you can choose. When you select a layout, it replaces the current layout at the lower left of the screen.

You can also change the viewport types in the new layout. Click on a viewport label to select from a list.

Most viewport types are tied to a specific view of the world coordinate system, and cannot be changed. For example, the Top view always looks down the Z axis of the world coordinate system.

You can set up two layouts and switch between them quickly by selecting **Layout A** or **Layout B** under the Current Layout section.

Safe Frames

Safe frames are used when rendering images that will eventually be viewed on videotape or television, where the edges of the image often "bleed" over the edge of the viewing area and cannot be seen. This is particularly important for titles, which cannot be read if they are too close to the edge of the image. A safe frame encloses the area that will definitely be viewable on a television or video monitor. The **Safe Frames** tab is used to set up these safe frame areas.

Live Area Turn on this checkbox to outline the area of the viewport that will render. The aspect ratio is set on the Render Scene dialog which can be accessed with the

Render Scene button ▦. The **Live Area** bounding area may not fit exactly in the viewport depending on the aspect ratio. The longest dimension of the resolution is fit just into the viewport, while the other dimension may be smaller than the viewport size depending on the aspect ratio.

Action Safe Check this option to outline the area of the viewport that is safe for action and movement in the scene. This area will most likely be visible on television and video monitors, with possible minor distortion near the edges.

Title Safe Check this option to outline the area of the viewport that is safe for titles. Because titles can distort near the edges of a television or video monitor, and because titles are difficult to read near the edges of a screen, this is the smallest area. If you are placing titles in your scene, make sure they fall within the **Title Safe** area.

User Safe Check this option to enable a user-definable safe area.

12-Field Grid Check this option to set up a *field grid* in the viewport. A field grid is a reference grid used by some video directors to give directions to artists. Choose between a 4x3 grid or a 12x9 grid.

Percent Reductions Sets the percent of reduction from the **Live Area** for each type of safe area. Turn on **Lock** to keep the horizontal and vertical reduction percentages the same, or turn off **Lock** to enter different **Horizontal** and **Vertical** percentages.

Show Safe Frames in Active View Displays the selected safe frame areas in the current viewport.

Default Settings Restores the default settings to the **Safe Frames** tab.

Adaptive Degradation

Adaptive degradation controls how objects appear when MAX cannot keep up the shading level while changes occur onscreen. The controls on the **Adaptive Degradation** tab control how much the display of objects degrades when redraw at the chosen rendering level is not possible. These controls are in effect only when the **Degradation Override** button is on ⬚. For more information, see the **Degradation Override** entry in the *Toolbar* section.

General Degradation Controls how far the rendering level will degrade during animation playback in non-current viewports. Playback appears in non-current viewports only when the **Active Viewport Only** checkbox on the Time Configuration dialog is off. Select more than one level to cause the system to drop back to successively lower levels as redraw takes more and more time. Set this rendering level as low (toward the bottom of the list) as you can while still being able to see what's happening on the screen. See the **Rendering Method** tab for descriptions of rendering levels.

Active Degradation Controls how far the rendering level will degrade during animation playback in the active viewport.

Maintain FPS FPS stands for frames per second. When MAX cannot maintain this display speed, the rendering level drops to the next lower level.

Reset on Mouse Up	Resets the rendering level to the highest level when the mouse is released. If unchecked, the system stays at a lower rendering level once it has been reached.
Show rebuild cursor	Displays a cursor when the viewport is recalculating the viewport display.
Update Time	Sets the interval between updates during viewport shading.
Interrupt Time	Sets the interval between checks for a mouse click during viewport shading.

Regions

The controls under **Regions** deal with regions for various functions.

Blowup Region	Sets the coordinates of the **Top**, **Left**, **Right** and **Bottom** corners of the bounding box for Blowup renderings. The bounding box appears in the viewport when **Render Type** is set to Blowup and a rendering is started.
Sub Region	Sets the coordinates of the **Top**, **Left**, **Right** and **Bottom** corners of the bounding box for Region renderings. The bounding box appears in the rendering viewport when the **Render Type** pulldown is set to Region and a rendering is started.
Virtual Viewport	The Virtual Viewport is a zoomed in view of a viewport primarily for zooming into a camera view to check an object's position against a background. The Virtual Viewport is enabled only if you are using an OpenGL driver as your display driver. To see which driver you are using, choose *Help/About 3D Studio MAX* from the menu and check the **Driver** listed at the top of the window. To use the Virtual Viewport, activate the viewport into which you want to zoom. Choose *Views/Viewport Configuration* and access the **Regions** tab. A reduced version of the image appears in the viewbox under the Virtual Viewport section. Use the **Zoom**, **X Offset** and **Y Offset** values to adjust the size and position of the viewbox. You can also drag the white window anywhere in the image.

Click **OK** when the viewbox shows the area into which you want to zoom. The active viewport is now zoomed in to the chosen area. Note that the viewport label may not show in the viewport when Virtual Viewport is active.

Plays the most recently made preview.

General Usage

Choose *Rendering/View Preview* from the menu. The Media Player appears with the most recently made preview loaded. Click the **Play** button ▶ to watch the preview.

A renamed preview cannot be viewed with this option. To view a renamed preview, choose *File/View File* and select the renamed preview from the file selector.

Window

Changes the current selection mode to window selection mode.

General Usage

Choose *Edit/Region/Window* from the menu. Window selection is now enabled.

When window selection mode is active, drawing a bounding area on the screen for selection will select only objects contained completely in the bounding box. Compare with crossing selection mode where objects touching the box are also selected.

The **Window/Crossing Selection** button at the bottom center of the screen reflects the current selection mode. When the button is depressed [image], window selection mode is active. When the button is not depressed [image], crossing selection mode is active. This button can also be used to set the current selection mode.

Sample Usage

Draw three or more objects on the screen. Choose *Edit/Region/Window* from the menu. Click **Select object** [image]. Make sure **Rectangular Selection Region** [image] is selected.

In any viewport, move the cursor to an area containing no objects. Click and drag in the viewport to create a bounding box. Objects contained inside the bounding box are selected.

Toolbar

This section describes the buttons and controls on the 3D Studio MAX Toolbar.

If you encounter a button in MAX and want to know more about its use, pass the cursor over the button to determine its label. Look up the label in this section of the book.

Buttons found on panels or dialogs are not covered here. Panel buttons are described in the *Panels* section of this book. Buttons on dialogs are covered under the menu or panel entry that accesses the dialog, in the *Menus* or *Panels* sections of this book.

The buttons described in this section of the book can be found in any of the following areas of MAX.

Top Toolbar The Toolbar across the top of the screen contains buttons for general usage of MAX. The Toolbar is always displayed on the screen.

Prompt Line The buttons along the center bottom of the screen are used for controlling selection and transformation of objects. These buttons, which include snap controls such as

Angle Snap Toggle , are always displayed on the screen.

Time Controls These buttons, found at the center bottom of the screen, control animation functions such as playback and current frame. Time control buttons are always displayed on the screen.

Viewport Navigation	Viewport navigation buttons are displayed at the lower right of the screen. Different buttons are available depending on the type of viewport currently active. For example, buttons for the Perspective viewport differ from buttons available when the Top viewport is active.
Track View	Track View buttons are listed under the **Track View** entry in this section.

Buttons in the following areas are found elsewhere in this book.

Loft Deformation	Loft deformation buttons are available only on deformation windows, which can be accessed under the **Modify** panel when a loft object is selected. Loft deformation buttons are listed under the **Loft** entry in the *Panels* section of this book.
Modify Panel	Buttons on the **Modify** panel, such as **Edit Stack** and **Make Unique** , are listed under the **Modify** entry in the *Panels* section.
Video Post	Video Post buttons are listed under the **Video Post** entry in the *Menus* section.
Material Editor	Material Editor buttons are listed in the *Mat Editor* section of this book.

Flyouts

A *flyout* is a series of buttons that can be accessed from one button. A button that can access a flyout has a small triangle at the lower right corner of the button. To access a flyout, click and hold on a button with a triangle. One or more other buttons appear, along with another copy of the flyout button. Drag the cursor to another button and release the mouse. The newly selected button appears in place of the originally selected flyout button. The newly selected button remains in place until the system is reset.

Unfortunately, you cannot see the names of flyout buttons while selecting from a flyout. You must choose the button from the flyout first, then pass the cursor over the button after it's selected to see its name.

In this book, for each button that accesses a flyout, an airplane ✈ appears across from the button name. When a button is accessed from a flyout, an airplane appears along with a picture of the button used to access the flyout.

 **Align**

Aligns one or more objects with another object.

General Usage

Select the object to be aligned. Click **Align** . The cursor turns into the Align cursor, which looks similar to the picture on the button.

When the cursor is moved over a selectable object, crosshairs appear near the Align cursor. Click on the target object, the object to which you wish to align the selected object. The Align Selection dialog appears.

Alignment takes place in relationship to the current **Reference Coordinate System**. Alignment can also be performed on two or more selected objects.

Dialog Options

Align Position	The **X Position**, **Y Position** and **Z Position** checkboxes move the current object's X, Y or Z alignment point to the X, Y or Z position of the target object's alignment point. Alignment points are set with the buttons under the **Current Object** and **Target Object** sections. Note that these options change the selected object's position, while the settings under **Align Orientation (Local)** change the selected object's local axis orientation.
Minimum Maximum	Uses the object's bounding box to set alignment points. The bounding box is created in the reference coordinate system. Along any axis, the reference coordinate axis will intersect the object's bounding box at two points, at the negative and positive ends of the axis. The **Minimum** option uses the negative intersection point along an axis for the alignment point, while **Maximum** uses the positive intersection point.

Minimum and **Maximum** are most useful when the target object is aligned with the reference coordinate system. In this way, you can use **Minimum** or **Maximum** to align edges of objects.

Center	Uses the center of the object as the alignment point.
Pivot Point	Uses the object's pivot point as the alignment point.
Align Orientation (Local)	Orients the current object's local axis with the target object's local axis. Choose **X Axis**, **Y, Axis** and/or **Z Axis** to orient one or more of the selected object's local axes. Compare with **Align Position** which affects the object's position rather than orientation.
Match Scale	If the target object has been scaled, choosing **X Axis**, **Y Axis** or **Z Axis** causes the selected object to be scaled to the same percentage along its local axes. Note that the target object's scale percentage is used, not its actual size, so using this option does not guarantee that both objects will be the same size. If the target object has not been scaled, choosing one of these options has no effect.
OK	Applies the current settings and exits the dialog.
Cancel	Cancels all settings and removes the dialog.
Apply	Applies the current settings to the selected object(s) and clears the dialog without exiting the dialog.

Usage Notes

Align can be used to align any object, including helpers, space warps and atmospheric gizmos.

You can also use this tool to align sub-objects and gizmos as well as objects. To align a gizmo, go to the sub-object level, select the gizmo and begin the alignment process. Only the gizmo will be aligned.

To align a sub-object, apply a **Mesh Select** modifier to the object and select faces to align. Apply an **XForm** modifier to the selection and begin the alignment process.

When aligning sub-objects and gizmos, the **Current Object** and **Match Scale** options are disabled. If you plan to align orientation, switch to the **Local** coordinate system first so the sub-object axes are properly oriented.

This option can also be accessed by choosing *Tools/Align* from the menu.

Moves a camera to point at a selected face.

General Usage

Select the camera to be aligned. From the **Align** flyout , choose

Align Camera . Click and drag over object faces in the scene until the desired face displays its normal as a blue arrow. Release the mouse. The camera moves to point at the selected face.

The camera maintains its distance from the target object. If the camera has a target, the target moves to the selected face.

Usage Notes

Align Camera can be used to align one object's faces with another. To do this, select any type of object, then click **Align Camera**. Move the mouse over another object to select the face and normal to which to align the selected object, and release the mouse. The object is aligned with the chosen face, but maintains its original distance from the face.

For information on face normals and what they are, see the **Normal** entry in the *Panels* section.

Toolbar

Toolbar

Aligns an object to the current view.

General Usage

Select an object to align. Activate the viewport to which you want to align the object. On the **Align** flyout 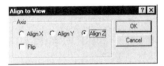, click **Align to View** ⊞. The Align to View dialog appears.

Dialog Options

Align X **Align Y** **Align Z**	Aligns the selected object's local X, Y or Z axis so it points out from the currently active viewport.
Flip	Flips the object so the selected axis points into the viewport.

Usage Notes

It is easier to choose the correct axis for alignment when the Local coordinate system is selected before clicking **Align to View**. To do this, access the **Reference Coordinate System** pulldown and choose **Local**. For more information, see the **Reference Coordinate System** entry.

Turns angle snap on and off.

General Usage

Click **Angle Snap Toggle** to turn it on. Use a tool that utilizes angle rotation such as **Select and Rotate** . Select and rotate an object. The object rotates in the degree increments set with the Grid and Snap Settings dialog.

Usage Notes

Angle Snap Toggle also causes other rotation tools such as **Orbit Camera** to snap to angle increments.

The angle snap increment is set with the **Angle (deg)** value on the Grid and Snap Settings dialog, under the **Options** tab. To access the Grid and Snap Settings dialog, right-click on **Angle Snap Toggle** or choose *Views/Grid and Snap Settings* from the menu.

Toolbar

Animate

Sets keys when objects are transformed or modified at the current frame.

General Usage

Move the time slider to a frame other than zero. Click the **Animate** button to turn it red and enable animation. Move, rotate or scale an object, or change a parameter on the panel. A key is created for the object at the current frame.

Click **Play Animation** to view the animation.

Usage Notes

When the **Animate** button is on, a red border appears around the current viewport to remind you that animation is enabled.

When an object is transformed at a frame other than zero and the **Animate** button is turned off, the transform is applied to the object at frame zero.

Keys set with the **Animate** button can be viewed and changed in the Track View window. See the **Track View** entry for information on changing keys.

Arc Rotate / Selected

Rotates the view around the center of the viewport or selected objects.

General Usage

Activate a viewport. Click **Arc Rotate** ⬛. A green trackball appears, with nodes at the top, bottom and sides. Move the cursor to one of the nodes, inside the trackball or outside to change the cursor. Click and drag to rotate the view. The view rotates around the view center.

You can also click and hold on **Arc Rotate** to access the flyout and select **Arc Rotate Selected** ⬛. **Arc Rotate Selected** works the same as **Arc Rotate**, except the view is rotated around selected objects.

Dragging inside the trackball rotates the view up and down and from left to right, but will not tilt the plane. Dragging outside the trackball rotates around an axis perpendicular to the screen, tilting the view plane.

If you move the cursor to the top or bottom trackball handles, the cursor changes to a vertical circular arrow. If you click and drag when this cursor is visible, the view rotates up and down. If you move the cursor to a side handle, the cursor becomes a horizontal vertical arrow. Click and drag on this cursor to rotate the view from side to side.

Arc Rotate and **Arc Rotate Selected** are not available in camera views. When you use one of these options on an orthographic viewport, the viewport automatically becomes a User view.

Sample Usage

Arc Rotate is particularly useful for getting an angled look at objects in a Perspective view. Click **Zoom Extents** ⬛ to get the entire scene into the viewport. Click **Arc Rotate** and move the cursor to the bottom trackball handle until the vertical circular arrow appears. Click and drag to rotate the view so you can see the top of the model.

 **Array**

Creates an array (a series of objects) from one or more objects.

General Usage

Select one or more objects. Click **Array** . The Array dialog appears.

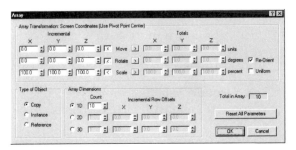

This dialog allows you to array objects in any direction with any transform.

Dialog Options

The settings in the **Array Transformation** section determine the transforms and values used to create the array.

Incremental	These values specify the increments between each array object. If you click the right arrow next to the transform name (**Move, Rotate, Scale**), then the **Totals** settings are used to specify the increments instead.
Totals	These values specify the distance, angle or scale percentage between objects as a total amount. The total distance, angle or scale is divided by the **Total in Array** to determine the increments between each object in the array.
	Note that if you leave all the **Incremental** and **Total** values at 0, you will end up with several objects piled on top of each other with exactly the same position, orientation and size.
X, Y, Z	Specifies the distance, angle or scale percentage along each of the specified axes. The axes of the current coordinate system are used.
Move	Under **Incremental**, this value specifies the distance between two sequential objects in the array along the specified axis. Under **Total**, this value specifies the total distance between the first and last objects in the array.

Rotate	Under **Incremental**, this value specifies the angle between two sequential objects in the array around the specified axis. Under **Total**, this value specifies the total angle between the first and last objects in the array.
	When you enter values for **Rotate**, the arrayed objects are rotated around the currently selected transform center. For example, if **Use Pivot Point Center** ⊞ is selected, the rotated array will be created around the pivot point of the original object, which may not be what you want.
Scale	Under **Incremental**, this value specifies the scale percentage increment between two sequential objects in the array along the specified axis. Under **Total**, this value specifies the total scale percentage between the first and last objects in the array.
Re-Orient	Rotates objects as they are arrayed. When unchecked, objects maintain the original object's orientation as they are rotated. This option has an effect only when objects are rotated as they are arrayed.
Uniform	When this checkbox is on, the **Y Scale** and **Z Scale** values are disabled. The **X** value is used for all three axes, creating a uniform scale on each arrayed object.
Type of Object	Specifies whether the arrayed objects will be copies, instances or references of the original object.

Under the **Array Dimensions** section, you can choose to create a 1D, 2D or 3D array. A 1D array creates objects in a straight line. A 2D array creates a series of objects with rows and columns, while a 3D array creates rows, columns and levels.

1D	Creates a one-dimensional array. **Count** is the total number of objects along the first dimension of the array. If **1D** is selected, the **Count** is also the total number in the array.
2D	Creates a two-dimensional array. **Count** is the number of columns in the second dimension. For example, for a 5x3 array, the **Count** for **1D** would be 5 while the **Count** for **2D** is 3. Note that if **2D** is chosen, you should also enter a value for the **1D Count** to get a 2D array. Leaving the **2D Count** at 1 creates a 1D array.

The **X, Y** and **Z Incremental Row Offsets** determine how far away from the **1D** array the second dimension of objects will be placed.

3D Creates a three-dimensional array. **Count** is the number of levels in the third dimension. For example, for a 5x3x4 array, the **Count** for **1D** would be 5, the **Count** for **2D** is 3 and the **Count** for **3D** would be 4. Note that if **3D** is chosen, you should also enter values for **1D** and **2D Count** values to get a 3D array. Leaving the **3D Count** at 1 creates a 2D array.

The **X, Y** and **Z Incremental Row Offsets** determine how far away from the 2D array the third dimension of objects will be placed.

Total in Array The total number of objects in the array. This number is automatically calculated as **1D Count** times the **2D Count** times the **3D Count**, and cannot be changed directly.

Reset All Parameters Resets all parameters to their default values. This operation cannot be undone.

OK Accepts the parameters and creates the array.

Cancel Cancels the array operation.

Binds an object to a space warp, causing the space warp to affect the object.

General Usage

Under the **Create** panel [icon], choose **Space Warps** [icon]. Create a space warp, then create an ordinary 3D object. Click **Bind to Space Warp** [icon]. Click and drag on the object. The cursor changes to look similar to the **Bind to Space Warp** button.

Drag the cursor to the space warp. When the cursor is over a space warp, the cubes on the cursor turn white. Release the mouse to bind the object to the space warp.

Usage Notes

Space warps act as a "force field" against an object, changing the object's form. Many space warps can be applied to one object, and many objects can be bound to the same space warp.

For more information on space warps, look under individual space warp names in the *Panels* section.

Toolbar

Sets the current selection region shape to a circle.

General Usage

Create two or more objects in the scene. Choose **Select object** . From the **Rectangular Selection Region** flyout , choose **Circular Selection Region** .Click and drag in any viewport to draw a circular selection region that includes at least one object.

If **Window Selection** is on, only objects inside the circle will be selected. If **Crossing Selection** is on, all objects touched or bounded by the region will be selected.

Sample Usage

The circular selection region is useful for selecting any objects or vertices in a circular arrangement. To see how this works, create a cylinder in the Top viewport with **Cap Segments** of 3. Under the **Modify** panel , choose the **Mesh Select** modifier to work with the cylinder at the vertex level.

Look at the cylinder in the Top viewport and consider how to select just the outer ring of vertices. To easily do this, change to **Circular Selection Region** and choose **Select object**. Click and drag in the Top viewport starting near the center of the cylinder. Draw the circle to include all vertices. Next, hold down the **<Alt>** key and click and drag at the center of the cylinder. Draw the circle only large enough to include the two inner rings of vertices. When you release the mouse button, the two inner rings of vertices will be deselected while the outer ring remains selected.

Changes the current selection mode to window selection mode.

General Usage

By default, crossing selection is on. Click the **Crossing Selection** button ![icon] to change to **Window Selection** ![icon]. Window selection is now enabled.

When crossing selection mode is active, drawing a bounding area on the screen for selection will selects all objects touched by the bounding box. Compare with window selection mode, where only objects inside the box are selected.

Toolbar

 **Degradation Override**

Turns adaptive degradation on and off.

General Usage

Click **Degradation Override** to turn it on or off .

Usage Notes

When you click the **Play** button , 3D Studio MAX attempts to play the animation with the specified shading method at the speed set with

Time Configuration . When **Real Time** is checked on the Time Configuration dialog, MAX attempts to play back animation at the speed set with the **Speed** option. Sometimes MAX cannot refresh the screen fast enough, particularly in shaded viewports, and so must find some way to cut down on processing time and make playback smoother. MAX can also have trouble keeping up the display when objects are being transformed or when views are zoomed, panned and rotated.

Under these circumstances, an *adaptive degradation* process is used to cause objects to be displayed with a lower display method so processing can go faster. For example, objects can be displayed as bounding boxes instead of shaded wireframes. Adaptive degradation reduces the display method until movement on the screen is smooth.

To set the display methods used, choose *Views/Viewport Configuration* from the menu to access the Viewport Configuration dialog. Click the **Adaptive Degradation** tab. Under **General Degradation** and **Active Degradation**, choose a series of display methods. You can choose as many display methods as you like. When 3D Studio MAX attempts to play an animation onscreen, it will drop down through the selected methods until it finds a display method it can use while still keeping up with changes on the screen.

The settings under **Active Degradation** apply to the active viewport, while the settings under **General Degradation** apply to the remaining viewports.

When **Real Time** is checked and **Degradation Override** is on , the settings under the **Adaptive Degradation** tab on the Viewport Configuration dialog are in effect. When **Real Time** is checked and **Degradation Override** is off , these settings have no effect. If **Real Time** is unchecked, adaptive degradation is not used, and MAX refreshes the screen with the selected display method no matter how long it takes.

When a light or camera viewport is active, the dolly buttons are available at the lower right of the screen. Dolly buttons move a camera or light along its line of sight.

In order to use the dolly buttons, click the button, then click and drag in the camera or light viewport. You cannot use these buttons to adjust cameras or lights in other viewports.

The camera dolly buttons work only in camera viewports.

	Dolly Camera	Moves the camera along its line of sight.

	Dolly Camera + Target	Moves both camera and target along the camera's line of sight. This option is available for target camera views only.

	Dolly Target	Moves the target along the camera's line of sight. This option is available for target camera views only.

The light dolly buttons work only in light viewports.

	Dolly Light	Moves the light along its shine direction.

	Dolly Light + Target	Moves both light and target along the light's shine direction. This option is available for target light views only.

	Dolly Target	Moves the target along the light's shine direction. This option is available for target light views only.

Usage Notes

To set up a camera viewport, create a camera in the scene. Press <C> on the keyboard to change the current viewport to the camera view.

To change a viewport to a light viewport, create a direct light or spotlight in the scene. Right-click on a viewport label to activate the pop-up menu. Move the cursor to *Views* and pick the name of the light from the top of the list that appears. The current viewport changes to show the light view.

⌗ Fence Selection Region

Sets the current selection region shape to a custom bounding area (fence).

General Usage

To use **Fence Selection Region** ⌗, choose it from the **Rectangular Selection Region** flyout ⌗.

Create two or more objects in the scene. Choose **Select object** ▶. Click and drag in any viewport to start the fence selection, then click to draw the points of the fence. Double-click to close the fence, or move the cursor to the start of the fence to display crosshairs and click.

If **Window Selection** ⌗ is on, only objects inside the fence will be selected. If **Crossing Selection** ⌗ is on, all objects touched or bounded by the region will be selected.

Changes the field of view in a camera or Perspective view.

General Usage

Activate a camera or Perspective view. Click **Field-of-View** . Click and drag in the viewport to change the field of view.

When you change the field of view in a camera view, the camera lens length changes accordingly.

Usage Notes

Changing the field of view in a camera view can be deceptive. It may appear that the camera is moving when in fact it is not.

Camera settings in 3D Studio MAX parallel real life. Each camera has a field of view determined by its lens length. A camera usually comes with a lens for "normal" use with a lens length of 35mm or 50mm. A 50mm lens length gives you a field of view of around 40 degrees.

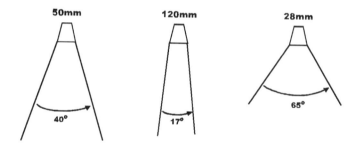

The longer the lens length, the smaller the field of view. A telephoto lens might have a lens length of 120mm, giving you a field of view of around 17 degrees. A wide-angle lens with a length of 28mm has a field of view of around 65 degrees.

With a lens length of 35mm or 50mm, faraway objects look smaller than objects close to the camera, and there is minimal distortion of objects at the edge of the scene. With a longer lens length and small field of view, it's hard to tell which objects are far away as they all look to be the same size. With a short lens length and large field of view, you get a greater "depth" in the scene as faraway objects are smaller, but objects at the edge of the scene are distorted. A fisheye lens is an example of a short lens length with a large field of view.

The **Perspective** button ⬦, available only with camera views, changes the camera's field of view while dollying the camera toward or away from the target. This retains roughly the same composition in the scene while changing the field of view. **Perspective** can be used only in a camera view.

You can also change the field of view in a Perspective view. To see the current field of view setting or to change the number directly, activate the Perspective view and right-click on the **Field-of-View** button. The Viewport Configuration dialog appears with the **Rendering Method** tab active. The **FOV** value at the lower right of the dialog displays the current field of view in degrees for the Perspective view. You can type in a new value for the field of view if you like.

Before using **Field-of-View,** be sure that you actually do want to change the field of view. If what you really want is to move a camera closer to or further away from its target, move the camera icon itself in an orthographic view, or use **Dolly Camera** ⬍ in the camera view. To zoom in a Perspective view, switch to a User view, zoom in or out, then change back to a Perspective view.

Toolbar

Go to End / Go to Start

Moves the time slider to the first or last animation frame.

General Usage

Click **Go to End** . The time slider moves to the last animation frame.

Click **Go to Start** . The time slider moves to the first animation frame.

Usage Notes

The length of the animation, and the start and end animation frames, can be set by clicking **Time Configuration** . Under the Animation section of the Time Configuration dialog, the **Start Time** and **End Time** values set the first and last frame numbers. The **Length** is automatically calculated when you change the **Start Time** or **End Time**. You can also change the **Length** directly, which changes the **End Time** accordingly.

Help

Accesses online help.

General Usage

Click **Help** ![help icon], then click anywhere on the screen. The online help window is displayed.

Help can also be accessed by choosing *Help/Online Reference* from the menu.

Turns IK on and off for hierarchically linked objects.

General Usage

Link one or more objects together. Turn on the **Inverse Kinematics on/off toggle** . The linked objects will now behave in accordance with the rules of inverse kinematics. Turn off the **Inverse Kinematics on/off toggle** to return to forward kinematics operation of linked objects.

Usage Notes

3D Studio MAX supports both forward kinematics and inverse kinematics. With forward kinematics, when a parent object is moved or rotated, each child object rotates or moves in exactly the same way.

With inverse kinematics, children still mimic parents' moves. However, children can also be moved and rotated, and parents will move and rotate just enough to keep the chain together.

In the example shown below, the root parent object is at the top. In the linked chain at left, the root was rotated, and all children rotated along with it. In the linked chain at right, the last child at the end of the chain was moved, and the rest of the chain followed, much like a real chain would. Note that the root moved the least of all the linked objects.

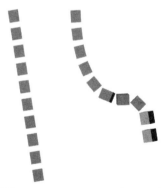

The movement and rotation of linked objects with inverse kinematics can be limited for more realistic animation. To set these limits, go to the **Hierarchy** panel and click **IK**. The settings on the Rotational Joints and Sliding Joints rollouts set limits for inverse kinematics.

See the **IK** entry in the *Panels* section for more information on inverse kinematics.

Toolbar

 **Key Mode Toggle**

Causes the **Next Frame** and **Previous Frame** buttons to become the **Next Key** and **Previous Key** buttons.

General Usage

Turn on the **Animate** button and animate one or more objects in the scene. Select an animated object. Select a transform such as **Select and Move** . Turn on **Key Mode Toggle** . Click **Next Key** . The time slider moves to the next frame on which there is a key for the selected object and selected transform.

Usage Notes

To specify the criteria 3D Studio MAX uses to find the next key, right-click on **Key Mode Toggle** . The Time Configuration dialog appears. The settings under the Key Steps section specify how MAX looks for the next key.

Light Falloff / Hotspot

Adjusts the light falloff or hotspot in a light viewport.

General Usage

Create a light with hotspot and falloff angles, such as a **Target Spot**. Make one of the viewports a light viewport. While the light viewport is active, click **Light Hotspot** ⊙ or **Light Falloff** ⊙. Click and drag in the light viewport to adjust the hotspot or falloff.

When **Light Falloff** is used, the falloff in the viewport will appear to remain the same as the hotspot is adjusted, but in fact the falloff changes and the hotspot remains the same. Because a light viewport is always about the same size as the falloff, the viewport size changes with the falloff, making the hotspot appear to change size.

Usage Notes

Light Hotspot and **Light Falloff** are available only in light viewports. To change a viewport to a light viewport, create a direct light or spot-light in the scene. Right-click on a viewport label to activate the pop-up menu. Move the cursor to *Views* and pick the name of the light from the top of the list that appears. The current viewport changes to show the light view.

As the light's hotspot and falloff are adjusted, the **Hotspot** and **Falloff** values on the light's panel change accordingly.

 Lock Selection

Locks the current selection of objects, faces, edges or vertices.

General Usage

Select one or more objects. Turn on **Lock Selection** 🔒.

To change the selection, turn off **Lock Selection** and select objects, or choose **Select by Name** 📇 from the Toolbar.

Sample Usage

Lock Selection is handy when you have several objects, faces and vertices on the screen. When **Lock Selection** is on, you are assured that your selection will remain intact while you transform or modify the selection. If you click in a viewport to transform your selection, **Lock Selection** prevents other elements from being accidentally selected.

 **Material Editor**

Accesses the Material Editor.

General Usage

Click **Material Editor** 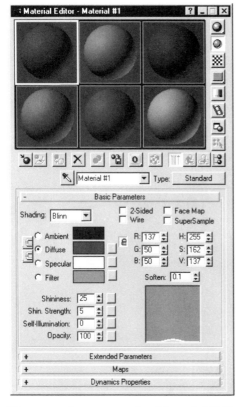. The Material Editor dialog appears.

Toolbar

You can also access the Material Editor by choosing *Tools/Material Editor* from the menu.

The many options on the Material Editor dialog are described under the **About the Material Editor** entry at the beginning of the *Mat Editor* section of this book.

⊞ Min/Max Toggle

Toggles the display in the current viewport between layout size and full screen.

General Usage

Click **Min/Max Toggle** ⊞. The current viewport expands to fill the entire display area. Click **Min/Max Toggle** again. The viewport returns to its layout display size.

Mirrors selected objects along any axis.

General Usage

Select one or more objects, and activate the coordinate system you wish to use for mirroring. Click **Mirror Selected Objects** .

Choose an axis or axes over which to mirror the selected object.

This dialog can also be accessed with the *Tools/Mirror* menu option.

Mirroring an object with the **Animate** button on creates a **Scale** key for the object. The animated mirror object will scale to nothing in the direction of the mirror, then scale up to the mirrored position.

Dialog Options

Mirror Axis
Choose an axis or axes over which to mirror the selected object. The axis or axes used are based on the current coordinate system.

Offset
Shifts the position of the mirrored objects by the **Offset** value. Objects are moved along the axis or axes selected.

Clone Selection
To mirror and clone objects at the same time, choose **Copy, Instance** or **Reference**. To mirror without copying, choose **None**. For information on these clone types, see the **Clone** entry in the *Menus* section.

Mirror IK Limits
If you have set up IK limits for the objects being mirrored, this option mirrors the IK limits along with the objects. This feature is for an animatable character with IK limits for each body part. You can mirror the body parts to the other side of the body along with IK limits when this checkbox is on. If no IK limits have been set for the objects being mirrored, this checkbox has no effect. See the **IK** entry in the *Panels* section.

Named Selection Sets

Names and accesses selection sets of objects or sub-objects.

General Usage

Select one or more objects. In the **Named Selection Sets** entry box, enter the name of the selection set and press **<Enter>**. The selection set is saved.

You can also save selection sets for sub-objects. To do this, apply a modifier which accesses sub-objects, such as **Mesh Select**. At the sub-object level, select faces, vertices or edges. Enter the selection set name in the **Named Selection Sets** entry box and press **<Enter>**.

To choose a named selection set, click the down arrow next to the entry box to access the pulldown list. Select the named selection set from the list. At the object level, only object selection sets are listed. At the sub-object level, only selection sets for the sub-object type are listed. For example, selection sets created at the face sub-object level are not listed when at the vertex sub-object level.

Usage Notes

The **Named Selection Sets** entry box is located to the left of the **Track View** button on the Toolbar.

You must press **<Enter>** immediately after entering a selection set name to ensure that it will be saved.

Face, vertex and edge selection sets can be passed from one sub-object level to the other with the **Mesh Select** modifier. See the **Mesh Select** entry for more information.

To edit or remove named selection sets, choose *Edit/Edit Named Selections* from the menu.

 Next Frame / Next Key

Moves the time slider to the next frame or next key.

General Usage

When **Key Mode Toggle** ⊶ is off, this button is the **Next Frame** button ▐▶. Click **Next Frame** to move the time slider to the next frame.

When **Key Mode Toggle** is on, this button becomes the **Next Key** button ▶▌. Select an object that has been animated, and click **Next Key** to move the time slider to the next frame containing a key for the selected object.

Usage Notes

When this button is the **Next Key** button, you can specify how 3D Studio MAX moves from one key frame to the next with the Time Configuration dialog. Click the **Time Configuration** button 🖭 to access the dialog. Under the **Key Steps** section at the bottom of the dialog, you can choose whether to move to key frames for selected objects only, or for only specific transforms. See the **Time Configuration** entry in this section of the book for more information on working with these settings.

Normal Align

Aligns an object with the normals of another object.

General Usage

Select the object to be aligned. Click **Normal Align** . Click and drag over the selected object until the desired normal appears as a blue arrow. Release the mouse. Click and drag over the object to which you want to align the selected object until the desired normal appears as a green arrow. Release the mouse.

The selected object jumps to align with the target object, with the two selected faces aligned. The **Normal Align** dialog appears.

This dialog can also be accessed by choosing *Tools/Align Normals* from the menu.

Dialog Options

Position Offset	The number of units to move the selected object along its local **X**, **Y** and **Z** axes.
Angle	The selected object's angle of rotation around the axis of the two normals.
Flip Normal	Aligns the selected object as if the selected normal were flipped. Turning on **Flip Normal** does not actually flip normals, just the alignment.
OK	Accepts the alignment settings.
Cancel Align	Cancels the entire alignment operation.

Usage Notes

You can use normal alignment on sub-objects as well as objects. You can also use **Normal Align** to align the local Z axis of helpers and other non-geometry objects with no faces.

For information on normals, see **Normal** in the *Panels* section.

Toolbar

Orbit Camera/Light

Orbits a camera or spotlight around its target.

General Usage

Activate a camera or light viewport. Click **Orbit Camera** or **Orbit Light** . Click and drag in the viewport to move the camera or light around its target while maintaining the distance from its target.

Usage Notes

Orbit Camera is only available in an active camera viewport. It cannot be used to position the camera in other viewports. Likewise, **Orbit Light** is only available when a light viewport is active, and cannot be used to adjust the light's position in other viewports.

Orbit Camera works with free cameras as well as target cameras. For a free camera, the **Target Distance** is used to set an imaginary target around which the camera orbits.

Orbit Light can be used with free spots and free direct lights. For a free spot or direct light, the **Target Distance** is used to set an imaginary target around which the light orbits.

Pan

Pans the viewport.

General Usage

Click **Pan** . Click and drag in a viewport. The view in the viewport moves in the direction in which it is dragged.

Orbits a camera target or light target around the camera or light.

General Usage

Activate a camera or light viewport. On the **Orbit Camera** or **Orbit Light** flyout, choose **Pan Camera** or **Pan Light**. Click and drag in the viewport to move the camera or light target while maintaining the target's distance from the camera or light.

Usage Notes

Pan Camera is only available in an active camera viewport. It cannot be used to position the camera target in other viewports. Likewise, **Pan Light** is only available when a light viewport is active, and cannot be used to adjust the light target's position in other viewports.

Pan Camera works with free cameras as well as target cameras. For a free camera, the **Target Distance** is used to set an imaginary target which orbits around the camera.

Pan Light can be used with free spots and free direct lights. For a free spot or direct light, the **Target Distance** is used to set an imaginary target which orbits around the light.

To pan a camera or light and its target together, use **Truck Camera** or **Truck Light**.

 **Percent Snap**

Turns percent snap on and off.

General Usage

Click **Percent Snap**  to turn it on. Use a tool that utilizes percentages such as **Select and Uniform Scale** 🔲. Select and scale an object. The object's scale is increased or decreased by increments set with the Grid and Snap Settings dialog.

Usage Notes

The percent snap increment is set with the **Percent** value on the Grid and Snap Settings dialog, under the **Options** tab. To access the Grid and Snap Settings dialog, right-click on **Percent Snap** or choose *Views/ Grid and Snap Settings* from the menu.

Changes a camera's position and field of view simultaneously.

General Usage

Activate a camera view. Click **Perspective**  and click and drag on the camera view. The camera moves toward or away from its target. The camera's field of view changes at the same time to keep the same composition in view.

Usage Notes

Perspective is useful for camera effects similar to director Alfred Hitchcock's famous field of view manipulations in the film *Vertigo*. There, the camera was moved away from the subject while the field of view was narrowed. A narrow field of view makes objects behind the subject appear to be closer than they really are. Moving the camera back while narrowing the field of view makes objects behind the subject appear to be "closing in" on it. **Perspective** can be used to animate the camera and create a similar effect.

Changing the field of view with **Perspective** also changes the camera's lens length. For more information on field of view, see the **Field-of-view** entry.

🄾 Place Highlight

Positions a light to create a highlight on an object, or places an object to reflect in another object.

General Usage

Select a light. From the **Align** flyout 🄾, choose **Place Highlight** 🄾. In the viewport in which you want the highlight to appear, click and drag on the object that will receive the highlight. A normal arrow appears on the object. Drag the cursor until the normal appears for the face which is to receive the highlight. The light moves as you move the cursor over the object, retaining its distance from the object. Release the mouse to set the light's position.

In the same way, you can set up an object to be reflected in another object. Select the object to be reflected. Click **Place Highlight**. In the viewport in which you want the reflection to appear, click and drag on the object to receive the reflection. The blue normal arrow appears on the object, and the object being reflected moves around as you drag the cursor. Drag until the arrow appears on the face to receive the reflection, and release the mouse. In order for the reflection to appear in renderings, the reflective object must be assigned a material with a **Reflect/Refract** map set up as the **Reflection** map. See the **Standard** and **Reflect/Refract** entries in the *Mat Editor* section for information on how to set up a material of this kind.

To place a highlight or reflection, you can also choose *Tools/Place Highlight* from the menu.

Usage Notes

When you want to control how a highlight or reflection appears in a final rendering, use the camera viewport to set up the highlight or reflection. Watch the face normal in the camera viewport to see where the highlight or reflection will fall in the camera view.

Technically speaking, this feature aligns the local Z axis of one object with a face normal of another object. For this reason, this feature is not limited to lights or reflective objects, but can be used with any two objects.

▶ Play Animation / ▷ Play Selected

Plays the current animation in viewports.

General Usage

Set up an animated sequence. Click **Play Animation** ▶. The animation plays in the current viewport or in all viewports. During playback,

the **Play Animation** button changes to the **Stop Animation** button ■. Click **Stop Animation** to stop playback.

You can also choose **Play Selected** ▷ from the **Play Animation** flyout to play animation for selected objects only. Any animated objects not currently selected disappear from the screen while the animation is played.

Usage Notes

Animation can be played in the active viewport or in all viewports. Playback viewports are set with the Time Configuration dialog. To access this dialog, right-click on **Play Animation** or **Play Selected**, or click the

Time Configuration button ⊡. For further information, see the **Time Configuration** entry.

◀‖ Previous Frame / ▎◀ Previous Key

Moves the time slider to the previous frame or previous key.

General Usage

When **Key Mode Toggle** ▣ is off, this button is the **Previous Frame**

button ◀‖. Click **Previous Frame** to move the time slider to the previous frame.

When **Key Mode Toggle** is on, this button becomes the **Previous Key**

button ▎◀. Select an object that has been animated, and click **Previous Key** to move the time slider to the previous frame containing a key for the selected object.

Usage Notes

When this button is the **Previous Key** button, you can specify how 3D Studio MAX moves from one key frame to another with the Time Configuration dialog. Click the **Time Configuration** button ▣ to access the dialog. Under the **Key Steps** section at the bottom of the dialog, you can choose whether to move to key frames for selected objects only, or for only specific transforms. See the **Time Configuration** entry in this section of the book for more information on working with these settings.

Renders the current viewport with the latest rendering settings.

General Usage

Activate the viewport you wish to render. Click **Quick Render (Production)**  to render the active viewport with the latest Production settings. From the **Quick Render (Production)** flyout, you can also choose **Quick Render (Draft)**  to render with the latest Draft settings.

For information on **Production** and **Draft** settings, see the **Render** entry in the *Menus* section, under the **Production/Draft** parameter description.

Toolbar

Rectangular Selection Region

Sets the current selection region shape to a rectangle.

General Usage

Create two or more objects in the scene. Choose **Select object** ![cursor icon]. Click and drag in any viewport to create the rectangular bounding box.

If **Window Selection** ![icon] is on, only objects inside the bounding region will be selected. If **Crossing Selection** ![icon] is on, all objects touched or bounded by the region will be selected.

Reverses the last **Undo** ↰ command.

General Usage

Perform an operation such as selecting or moving an object. Click **Undo** to undo the operation. Click **Redo** ↱ to redo the command.

Usage Notes

To redo several actions at once, right-click on the **Redo** button. A list of recently undone actions appears. Click on an action. The high-lighted action, and all those above it, will be redone when you click on the **Redo** button under the list.

Redo also redoes commands undone with the *Edit/Undo* menu option.

The **Redo** button performs the same function as the *Edit/Redo* menu option.

Reference Coordinate System

Sets the coordinate system for transforms.

General Usage

Select a transform. Click on the **Reference Coordinate System** pulldown, located to the right of the transform buttons. Choose a coordinate system. Select and transform objects as desired.

Pulldown Options

View　　Uses the **Screen** coordinate system in orthographic viewports, and the **World** coordinate system in non-orthographic views such as Perspective views.

Screen　　The coordinate system axes change as different viewports are activated so the coordinate system's XY plane is always parallel to the viewport, and the Z axis points into the screen.

World　　The XYZ axes are fixed, with the XY plane always parallel to the Top viewport and the Z axis pointing up out of the Top viewport, regardless of which viewport is active.

Local　　Uses the selected object's local axes. If more than one object is selected, each object transforms around its own axes. To change an object's local axes, change the pivot point axes. See **Pivot** in the *Panels* section for details.

Parent　　Uses the selected object's parent's local axes. If the object is not a linked child, the **Parent** coordinate system works the same as **World**.

Grid　　Uses the coordinate system of the active grid. When the default home grid is active, the **Grid** coordinate system works the same as **View**.

Pick　　Picks an object in the scene for use as the coordinate system, using the object's local axes. Choose **Pick**, then click on an object in the scene. The object's name appears on the **Reference Coordinate System** display box, and is displayed on the pulldown list below **Pick**.

Usage Notes

A coordinate system consists of three axes, X, Y and Z, along which objects are transformed. These axes work in conjunction with **Restrict to X/Y/Z** $\boxed{X}$ $\boxed{Y}$ $\boxed{Z}$ and **Restrict to XY/YX/ZX Plane** $\boxed{XY}$ $\boxed{YZ}$ $\boxed{ZX}$ to constrain transforms to one or two axes.

The **Use Pivot Point** ▢▢, **Use Selection Center** ▢ and **Use Transform Coordinate Center** ▢▢ buttons determine the center around which objects are scaled and rotated. If **Use Transform Center** is selected, the object is transformed around the **Reference Coordinate System** center.

A different coordinate system can be set for each transform. When a transform is selected, the coordinate system changes to the last coordinate system selected for that transform.

Toolbar

Sample Usage

Many users find the default coordinate system, **View**, to be most useful. Click **Select and Move** ▢ and turn on **Restrict to XY Plane**. Click **Select and Rotate** ▢ and turn on **Restrict to Z** ▢. Select the **View** coordinate system. When you choose Select and Move and click in an orthographic viewport, movement will always take place on the view's plane, making it easy for you to move objects around. When you choose **Select and Rotate**, rotation will always take place around the axis going into the viewport, making it easy to rotate objects exactly as you like.

Under various circumstances, changing the **Reference Coordinate System** can sometimes help you transform objects more easily. In this example, a rotated cylinder is more easily moved in the direction of its length with the **Local** coordinate system.

1. Create a cylinder of any size in the Top viewport.

2. Click **Select and Rotate** and turn on **Restrict to Z** if necessary. Make sure the **Reference Coordinate System** is set to **View**. In the Left viewport, rotate the cylinder by 30 degrees.

Suppose you now want to move the cylinder along its height. This will be much easier with the **Local** coordinate system.

3. Change the **Reference Coordinate system** to **Local**.

4. Choose **Select and Move** and turn on **Restrict to Z**.

5. Move the cylinder in any viewport. The cylinder moves in the direction of its height.

 # Region Zoom

Zooms into a specified region in an orthographic viewport.

General Usage

Activate an orthographic viewport such as Top, Front or Left. Click **Region Zoom** . Click and drag in the viewport to draw a rectangular region. The view zooms to the specified region.

Usage Notes

Region zoom is not available when a Perspective, camera or light view is active.

To zoom in on a Perspective view, temporarily change the view to a User view. Use **Region Zoom** to zoom in to the desired area, then switch back to a Perspective view.

Renders the last viewport rendered, with the most recent rendering settings.

General Usage

Render a scene with **Render Scene** ▣. After making changes to the scene, click **Render Last** ▣. Regardless of which viewport is currently active, the last rendered viewport will render with the most recent settings on the Render Scene dialog.

Usage Notes

Render Last renders one frame only, and does not save the rendered image to a file.

To render and save to a file, use **Render Scene** ▣. If an output filename has already been entered, you can also use **Quick Render (Production)** ▣ or **Quick Render (Draft)** ▣.

 **Render Scene**

Renders the scene.

General Usage

Set up the scene as desired. Click **Render Scene** 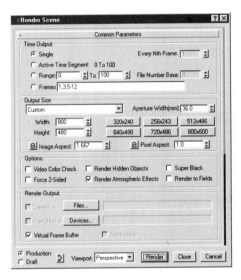. The Render Scene dialog appears.

Click **Render** to begin rendering.

For detailed information on the options on the Render Scene dialog, see the **Render** entry in the *Menus* section of this book.

Render Type

Sets the area of the viewport or the objects that will be rendered.

General Usage

Select a render type. Enact a rendering by choosing *Rendering/Render* from the menu, or by clicking **Render Scene** 🖼, **Quick Render (Production)** 🖼, **Quick Render (Draft)** 🖼 or **Render Last** 🖼.

If **Region** or **Blowup** were selected as the render type, adjust the selection region and click the **OK** button at the lower right of the viewport.

Pulldown Options

View   Renders the entire viewport. This is the default setting.

Selected   Renders selected objects only. The previously rendered image remains in the image display, and selected objects are rendered over it. If two or more objects are selected, the area between selected objects renders as the background.

Region   Renders the specified region. When rendering is enacted, a rectangular selection region appears in the current viewport. Move the handles at the corners of the selection region to surround the region you wish to render. Hold down the **<Ctrl>** key when moving handles to preserve the current aspect ratio. Click the **OK** button at the lower right of the viewport to start the rendering. The previously rendered image remains in the image display, and the selected region is rendered over it.

Blowup   Renders the specified region, blown up to the render resolution size. When rendering is enacted, a rectangular selection region appears in the current viewport. Move the handles at the corners of the selection region to surround the region you wish to render. The selection region is constrained to the current rendering aspect ratio. Click the **OK** button at the lower right of the viewport to start the rendering.

Usage Notes

The **Selected** and **Region** render types can save rendering time by rendering only objects or regions of the screen that have changed since the last render.

 **Restrict to X/Y/Z**

Restricts the current transform to the selected axis.

General Usage

Click on a transform to select it, then click **Restrict to X** $\boxed{X}$, **Restrict to Y** $\boxed{Y}$ or **Restrict to Z** $\boxed{Z}$. Select an object to transform, and transform the object. The transformation will be limited to the selected axis.

Usage Notes

The current coordinate system is used in conjunction with the axis restriction. For example, if the **Local** coordinate system is currently selected, the transformation will be restricted to the object's local X, Y or Z axis. See the **Reference Coordinate System** entry for further information on coordinate systems and their relationship to axis constraints.

Restrict to XY/YZ/ZX Plane

Restricts the current transform to the selected plane.

General Usage

Click on a transform to select it, then click **Restrict to XY Plane** XY,
Restrict to YZ Plane YZ or **Restrict to ZX Plane** ZX. Select an object
to transform, and transform the object. The transformation will be limited to the selected plane.

Usage Notes

The current coordinate system is used in conjunction with the plane restriction. For example, if the **Local** coordinate system is currently selected, the transformation will be restricted to the object's local XY, YZ or ZX plane. See the **Reference Coordinate System** entry for further information on coordinate systems and their relationship to axis constraints.

 **Roll Camera/Light**

Rotates a camera or light.

General Usage

Activate a camera viewport or light viewport. Click **Roll Camera** or **Roll Light** . Click and drag in the viewport to rotate the camera or light.

Usage Notes

Roll Camera is only available in an active camera viewport. It cannot be used to rotate the camera in other viewports. Likewise, **Roll Light** is only available when a light viewport is active, and cannot be used to adjust the light's rotation in other viewports.

Links one or more objects to another object.

General Usage

Select one or more objects. Click **Select and Link** . Move the cursor over one of the selected objects until the cursor changes to look similar to the **Select and Link** button. Click and drag the cursor over the parent object until one of the cubes on the cursor turns white.

To choose a parent object, you can also click **Select by Name** after clicking **Select and Link**, or you can press the <H> key on the keyboard.

Toolbar

Usage Notes

When you have finished using **Select and Link**, you should immediately choose another selection tool such as **Select object** or **Select and Move** to avoid accidentally linking objects. For example, if you leave **Select and Link** on and click **Select by Name**, 3D Studio MAX assumes you are choosing a parent object for the currently selected object.

To view the hierarchy of linked objects, click another selection button such as **Select object**, then click **Select by Name**. Turn on the **Display Subtree** checkbox at the lower left of the Select Objects window. Linked child objects are indented below their parents on the listing.

Select and Move

A transform that moves objects.

General Usage

Click **Select and Move** ⊕. Click and drag on an object to move it.

To move more than one object at a time, select the objects with any method. Click **Select and Move**. Move the cursor over the objects until the cursor changes to look like the **Select and Move** button. Click and drag to move all objects at once.

Usage Notes

You can access the Move Transform Type-In dialog by right-clicking on **Select and Move**. Here you can directly enter a position for an object. See the **Transform Type-In** entry in the *Menus* section for information on how to use the dialog.

Movement of objects is restricted by **Restrict to X/Y/Z** X Y Z and **Restrict to XY/YZ/ZX Plane** XY YZ ZX. These buttons work in conjunction with the **Reference Coordinate System** to constrain movement of objects. See the **Reference Coordinate System** entry for more information.

Select and Non-uniform Scale

A transform that scales objects on one or two axes.

General Usage

Click **Select and Non-uniform Scale** ⬛. A message appears, warning you to use an **XForm** modifier instead of **Select and Non-uniform Scale**. Click **Yes** to continue.

Choose **Restrict to X/Y/Z** X Y Z to constrain scaling to one axis or **Restrict to XY/YZ/ZX Plane** XY YZ ZX to constrain scaling to a plane. Click and drag on an object to scale it on the selected axis or plane.

To scale more than one object at a time, select the objects with any method. Click **Select and Non-uniform Scale**. Move the cursor over the objects until the cursor changes to look like the **Select and Non-uniform Scale** button. Click and drag to scale all objects at once.

Usage Notes

Select and Non-uniform Scale is a transform. When 3D Studio MAX builds an object, it applies all modifiers first, then transforms are applied. This sequence occurs even if you transform an object before applying modifiers to it. Sometimes this sequence causes difficulty with objects that are scaled with **Select and Non-uniform Scale**, then later linked or bound to a space warp. To avoid these problems, use an **XForm** modifier to scale the object instead. For information on this procedure, see the **XForm** entry in the *Panels* section.

You can disable the warning message that appears when you click **Select and Non-uniform Scale** and **Select and Squash** by unchecking the **Display NU Scale Warning** checkbox on the **General** tab of the Preference Settings dialog. This dialog can be accessed with the *File/Preferences* menu option.

To scale objects on all three axes at once, use the **Select and Uniform Scale** button ⬛.

 **Select and Rotate**

A transform that rotates objects.

General Usage

Click **Select and Rotate** ⟳. Click and drag on an object to rotate it.

To rotate more than one object at a time, select the objects with any method. Click **Select and Rotate**. Move the cursor over the objects until the cursor changes to look like the **Select and Rotate** button. Click and drag to rotate all objects at once.

Usage Notes

You can access the Rotate Transform Type-In dialog by right-clicking on **Select and Rotate**. Here you can directly enter the orientation for an object. See the **Transform Type-In** entry in the *Menus* section for information on how to use the dialog.

Rotation of objects is restricted by **Restrict to X/Y/Z** ⟨X⟩ ⟨Y⟩ ⟨Z⟩ and **Restrict to XY/YZ/ZX Plane** ⟨XY⟩ ⟨YZ⟩ ⟨ZX⟩. These buttons work in conjunction with the **Reference Coordinate System** to constrain rotation of objects. See the **Reference Coordinate System** entry for more information.

When rotating without the **Animate** button on, objects rotate about the axis point set by the **Use Pivot Point** ⟨⟩ button, the **Use Selection Center** ⟨⟩ button or the **Use Transform Coordinate Center** ⟨⟩ button. You can also cause rotation to be performed about a snap point on the object with **2D Snap** ⟨²⟩, **2.5D Snap** ⟨²·⁵⟩ or **3D Snap** ⟨³⟩. See these entries for information on how to use them for rotation.

When the **Animate** button is on, all rotation is performed about the object's pivot point. An object's pivot point can be changed under the **Hierarchy** panel ⟨⟩ with the **Pivot** option. See the **Pivot** entry in the *Panels* section for information on moving or rotating an object's pivot point.

A transform that scales an object up or down on one or two axes, and scales it in the other direction on the remaining axis or axes.

General Usage

Click **Select and Squash** 🔲. A message appears, warning you to use an **XForm** modifier instead of **Select and Squash**. Click **Yes** to continue. Choose **Restrict to X/Y/Z** **X** **Y** **Z** to control scaling on one axis or **Restrict to XY/YZ/ZX Plane** **XY** **YZ** **ZX** to control scaling on two axes. Click and drag on an object to scale it on the selected axis or axes. If you scale the object down, it scales up on the remaining axis or axes, effectively squashing the object. If you scale it up, it scales down on the remaining axis or axes.

To squash more than one object at a time, select the objects with any method. Click **Select and Squash**. Move the cursor over the objects until the cursor changes to look like the **Select and Squash** button. Click and drag to squash all objects at once.

Usage Notes

Select and Squash is a transform. When 3D Studio MAX builds an object, it applies all modifiers first, then transforms are applied. This sequence occurs even if you transform an object before applying modifiers to it. Sometimes this sequence causes difficulty with objects that are scaled with **Select and Squash**, then later linked or bound to a space warp. To avoid these problems, use an **XForm** modifier to scale the object instead. For information on this procedure, see the **XForm** entry in the *Panels* section.

You can disable the warning message that appears when you click **Select and Non-uniform Scale** and **Select and Squash** by unchecking the **Display NU Scale Warning** checkbox on the **General** tab of the Preference Settings dialog. This dialog can be accessed with the *File/Preferences* menu option.

To scale objects on all three axes at once, use the **Select and Uniform Scale** button 🔲. To scale objects on just one axis, use **Select and Non-uniform Scale** 🔲.

⬚ Select and Uniform Scale ✈

A transform that scales objects on all three axes.

General Usage

Click **Select and Uniform Scale** ⬚. Click and drag on an object to scale it on all three axes.

To scale more than one object at a time, select the objects with any method. Click **Select and Uniform Scale**. Move the cursor over the objects until the cursor changes to look like the **Select and Uniform Scale** button. Click and drag to scale all objects at once.

Usage Notes

When scaling without the **Animate** button on, objects are scaled about the axis point set by the **Use Pivot Point** ⬚ button, the **Use Selection Center** ⬚ button or the **Use Transform Coordinate Center** ⬚ button. You can also scale about a snap point on the object with **2D Snap** ⬚, **2.5D Snap** ⬚ or **3D Snap** ⬚.

When the **Animate** button is on, all scaling is performed about the object's pivot point. An object's pivot point can be changed under the **Hierarchy** panel ⬚ with the **Pivot** option. See the **Pivot** entry in the *Panels* section for information on moving or rotating an object's pivot point.

To scale objects on one or two axes, use the **Select and Non-uniform Scale** button ⬚. This button can be accessed from the **Select and Uniform Scale** flyout.

Displays a list of object names so objects can be selected by name.

General Usage

Click **Select by Name** . The Select Objects dialog appears.

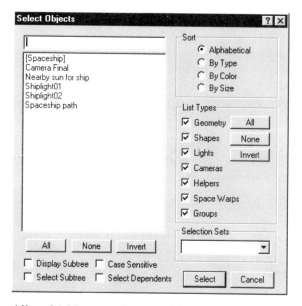

All unhidden, unfrozen objects appear on the list. To select a range of objects, select the first object, then hold down **<Shift>** and select the last object in the range. To select a series of nonconsecutive objects, hold down **<Ctrl>** while selecting objects.

You can also enter partial or full object names in the entry box above the list. As you enter an object name, all objects that start with the typed letters are highlighted.

Dialog Options

All   Highlights all objects for selection.

None   Unselects all objects.

Invert   Inverts the current selection. All selected objects are unselected, while all unselected objects become selected.

Display Subtree   Indents linked child objects below their parent objects on the list. If a parent object is hidden or frozen, it does not appear on the list, but its children will be indented regardless.

Toolbar

Select Subtree	This option causes all children of a selected parent object to also be selected. If a parent object is already selected when you turn on this checkbox, you must select the object again in order to select the children.
Case Sensitive	When **Case Sensitive** is checked, upper and lowercase letters typed in the entry box must match the case of an object name in order for it to be selected.
Select Dependents	Selects dependent objects such as instances, references and objects sharing a common modifier. If a dependent object is already selected when you turn on this checkbox, you must select the object again in order to select its dependents.
Sort	Choose to sort objects by name, type, color, or size. **Size** is the number of faces on the object. When **Display Subtree** is on, **Sort** is disabled.
List Types	Each checked object type is displayed on the list, while unchecked types are not. This option is useful when a scene contains many objects and you want to trim the list to make selection easier. This option does not affect the objects displayed in the scene.
Selection Sets	Use the pulldown arrow to access the list of selection sets. When a selection set is chosen, all the objects in the selection set are selected.
Select	Selects the chosen objects and exits the dialog.
Cancel	Exits the dialog without changing the selection.

Usage Notes

A Selection Floater can be displayed by choosing *Tools/Selection Floater* from the menu. The floater can stay on the screen as you work with objects, allowing you to more quickly select objects.

 **Select object**

Selects objects from the screen.

General Usage

Click **Select object** . Click on an object to select it.

You can select multiple objects by holding down the **<Ctrl>** key as you click on objects. To unselect an object, hold down the **<Alt>** key while clicking on the object.

To select objects with a bounding area, click and drag on the screen to create the bounding area.

Select object can also be used in the same manner to select sub-objects such as faces and vertices when at the sub-object level.

Usage Notes

When a selection bounding area is drawn and **Window Selection** is on, only objects completely inside the bounding area are selected.

When **Crossing Selection** is on, objects that are either inside the area or touching the boundary are selected.

The bounding area is set by choosing from the **Rectangular Selection Region** , **Circular Selection Region** and **Fence Selection Region** .

Selection Filter

Determines which object types can be selected.

General Usage

The **Selection Filter** pulldown is located to the left of the **Select by Name** button  on the Toolbar. Choose an object type from the pulldown. Selections made directly on the screen are limited to the object type chosen.

Pulldown Options

All   Selects any object type. This is the default setting.

Geometry   Selects only 3D geometry such as primitives, boolean objects and loft objects.

Shapes   Selects only shapes. Extruded, lathed or lofted shapes are considered geometry rather than shapes.

Lights   Selects only lights and light targets.

Cameras   Selects only cameras and camera targets.

Helpers   Selects only helper objects.

Warps   Selects only space warps.

Combos   Allows you to set up combinations of filters for your own use. When **Combos** is chosen, a dialog appears.

Select one or more object types at the left side of the dialog. Click **Add** to add the combination to the list. The combo is given a name based on the first letter of each object type chosen. For example, a combo of **Lights** and **Cameras** is named LC. Space warps are indicated by the letter W.

Click **OK** to accept the combos. The new combos appear on the **Selection Filter** pulldown list.

 Snapshot

Creates clones of an animated object's state at specified intervals during the animation.

General Usage

Create an object. Transform the object over time. This can be done by moving, rotating or scaling the object at various frames with the **Animate** button turned on.

From the **Array** flyout 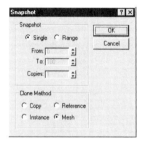, chose **Snapshot** . The Snapshot dialog appears.

You can take a snapshot of the object at the current frame, or you can make a series of snapshots over a range of frames. You can also choose whether the snapshot objects will be copies, instances, references or collapsed mesh objects.

For detailed information on using **Snapshot**, see the **Snapshot** entry in the *Menus* section of this book.

 # Spinner Snap Toggle

Turns spinner snap on and off.

General Usage

Click **Spinner Snap Toggle** to turn it on. Click an up or down arrow on a spinner once. The spinner value snaps up or down by the spinner snap increment.

Usage Notes

Spinners are the small up and down arrows that appear to the right of most number entries on panels and dialogs.

Spinner snap can be used to force all clicks on spinner arrows to cause value changes of a specific increment, such as 1.0.

The percent snap increment is set with the **Spinner Snap** value on the Preferences dialog, under the **General** tab. To access the Preferences dialog, right-click on **Spinner Snap Toggle** or choose *File/Preferences* from the menu.

# 3D Snap Toggle

Turns 3D snap on and off.

General Usage

Click **3D Snap** to turn it on. Create, modify or transform objects. The cursor snaps to grid lines or objects according to the settings on the Grid and Snap Settings dialog. When the cursor snaps to a grid or an object, a small box appears around the cursor.

Usage Notes

When **3D Snap** is on, the cursor can snap to any object in 3D space as well as to the construction grid.

All snap options are set on the Grid and Snap Settings dialog, under the **Snaps** tab. The size and color of the snap cursor is set under the **Options** tab. To access the Grid and Snap Settings dialog, right-click on **3D Snap** or choose *Views/Grid and Snap Settings* from the menu.

See the **Grid and Snap Settings** entry in the *Menus* section for information on setting snaps.

Time Configuration

Sets parameters for time-related functions.

General Usage

Click **Time Configuration** . The Time Configuration dialog appears.

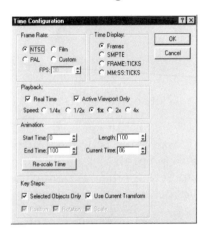

Dialog Options

Frame Rate Sets the frame rate for animation playback and for the settings under **Time Display**. The frame rate is also saved and used with any AVI files rendered from the scene. **NTSC**, the video standard for the United States and Canada, automatically sets the frames per second (FPS) to 30. **PAL**, the video standard for Europe, Australia and South America, sets FPS to 25. **Film** sets the FPS to 24. With the **Custom** option, you can change the FPS value to any number.

Time Display Determines how frame numbers will appear on the time slider. **Frames**, the default, displays frame numbers as sequential numbers. **SMPTE** uses the time code standard of Hours:Minutes:Seconds:Frames. **FRAME: TICKS** displays the frame number followed by the number of ticks. A *tick* is a time increment equal to 1/4800 of a second. **MM:SS:TICKS** displays minutes, seconds and ticks. For more information on time code and ticks, see the **Rendering Concepts** entry in the *Appendix.*

The settings under the **Playback** section control animation playback on the screen when **Play Animation** ▶ is clicked. These settings have no effect on rendered animation playback.

Real Time   When **Real Time** is on, animation attempts to play in viewports at the Frame Rate. If the screen cannot be refreshed at the frame rate and **Degradation Override** is on ⊞, the rendering method will drop down to lower levels. See the **Degradation Override** entry for information on setting these levels. If MAX still cannot keep up with the playback rate despite lower rendering methods, or if **Degradation Override** is off ▣, frames are skipped during playback to retain the playback rate.

When **Real Time** is off, every frame plays regardless of how long it takes to generate and display it.

Active Viewport Only   When this option is checked, animation plays back in the active viewport only. When this option is unchecked, animation is played in all viewports.

Speed   Sets the speed for playback in viewports as a fraction or multiple of the frame rate.

Start Time End Time   **Start Time** and **End Time** set the first and last frames that can be accessed with the time slider. Animation on frames before the **Start Time** and after the **End Time** remains part of the .*max* file, but is not played on the screen. These frames can still be rendered and altered in the Track View window.

Length   The length of time from the **Start Time** to **End Time**. **Length** changes automatically when **Start Time** and **End Time** are changed. If **Length** is changed, the **End Time** changes to reflect the new active length.

Current Time   The current time on the time slider.

Re-scale Time Scales all animation in the active time segment to a new time segment. When **Re-scale Time** is clicked, a dialog appears.

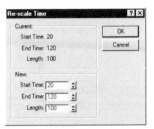

Enter a new **Start Time** and **End Time** for the animation. When you click **OK**, the distance between keys will be increased or decreased to fit the time entered. The **Start Time** and **End Time** on the Time Configuration dialog will also be changed accordingly.

Key Steps Determines how MAX will move from one key to another when the **Next Key** ▶| and **Previous Key** |◀ buttons are pressed. These buttons are only available when **Key Mode Toggle** ⌐ is on.

When **Selected Objects Only** is checked, the time slider moves between keys for the selected object only. When it is unchecked, the time slider moves between keys for all objects.

When **Use Current Transform** is checked, the time slider moves only to the type of keys set with the current transform. For example, if **Select and Move** ⊕ is currently on and **Use Current Transform** is checked, the time slider moves only to Position keys. When **Use Current Transform** is unchecked, the **Position**, **Rotation** and **Scale** checkboxes are enabled, and the time slider moves to keys for any checked transform regardless of the currently selected transform.

Toolbar

Opens a new Track View window.

General Usage

Click **Track View** 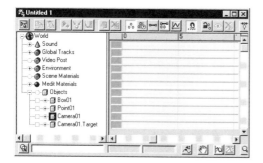. A new Track View window is opened and displayed on the screen.

The Track View window is used for controlling and adjusting animation and for setting up various aspects of the scene such as sound.

The default name of **Untitled-#** is automatically assigned to the Track View window, where **#** is the next sequential number for all the Track View windows you have created. The name is shown at the upper right of the Track View window. To change the name, access the name field and type in a new name.

When you are finished working with the Track View window, you can close it or minimize it. If you close it, you can later access it from the Track View menu. If you minimize it, you can choose it from its minimized position, usually at the bottom left of the screen. In order to delete a Track View window, you must first close it.

An entire book could be written on Track View itself. For this reason, only the most often used functions are described here. For more information on Track View and its more specialized uses, see your 3D Studio MAX documentation.

Working with the Hierarchy Listing

On the left side of the Track View window is the Hierarchy listing. Listed here are all aspects of the scene, including objects, environment settings such as fog and volume lights, and materials. To expand a hierarchy item, click a bounded plus sign ⊞ ⊕ next to the item.

The Track View window displays a *track* for each entry in the Hierarchy Listing. A track is a visual representation of the animation.

Toolbar

When you click the square bounded plus sign ⊞ next to **Objects**, a list of all objects in the scene appears. Click the round bounded plus sign ⊕ next to an object name to expand the tracks for that object.

Each type of animation has its own track name. For example, if you move an object at a frame other than zero with the **Animate** button on, you create a key on the **Position** track for that object. There are many different types of tracks that can be animated in the scene, and Track View displays each one.

Use the **Filters** ⊡ button on the Track View toolbar to limit the displayed tracks to just those you want to see. You can also right-click on a Hierarchy item to access a pop-up menu for further control of the listing.

Working with Track View can be confusing at first. See the **Sample Usage** below to get started in using Track View.

Editing Keys

Tracks can represent animation keys with lines, dots or curves. Keys are displayed only when you have expanded to an animated track at the lowest level of the Hierarchy. In other words, only an animated track with an empty circle ○ to the left of the track name can have keys displayed. For higher tracks on the Hierarchy listing (those with a plus sign bounded by a circle ⊕), lines are displayed to represent ranges of animation lower on the Hierarchy.

The picture below shows a Hierarchy listing for an object named **Donut01** expanded to its lowest level, displaying tracks for the donut's **Position**, **Rotation** and **Scale** tracks. Since the donut's rotation and scale have not been animated, there are no key dots displayed on the **Rotation** and **Scale** tracks. Instead, information about the donut's rotation and scale is displayed.

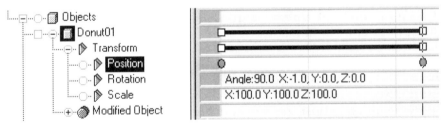

There are five buttons on the Track View toolbar that determine the mode for displaying and editing keys: **Edit Keys** ⚬⚬, **Edit Time** ⚬⊕, **Edit Ranges** ▭, **Position Ranges** ▭ and **Function Curves** ⩓. Only

one button can be on at any time. The Track View toolbar changes as each button is clicked to display functions for the selected mode. **Edit Keys** is on by default. **Edit Keys** and **Function Curves** are the two modes you will use most often.

Key Dots

Toolbar

When the **Edit Keys** button  is on, keys are represented with dots. The key dots of all expanded tracks are displayed on the Track View window.

To select one key, click on the key. A key turns white when selected. The key time (the frame on which the key sits) is displayed at the bottom center of the Track View window.

To select multiple keys, hold down the **<Ctrl>** key while clicking on keys. You can also select multiple keys by drawing a bounding box around them.

When **Move Keys** is on, selected keys can be moved by clicking and dragging on a selected key. To copy one or more selected keys, hold down the **<Shift>** key while moving the keys to another frame.

To see the values associated with a key dot, right-click on the key. A dialog appears that displays values for the key.

To add keys, click **Add Keys** and click on the window across from a track. A key is created with the frame's current values for that track.

To delete keys, select the keys and click **Delete Keys**.

Function Curves

When the **Function Curves** button is turned on, animation is represented as curved lines. To see the curves, highlight a track name at the lowest level of the Hierarchy and turn on **Function Curves**. Three lines appear, representing the animation for that track. If no lines appear when **Function Curves** is on and a track is highlighted, then **Function Curves** are not available for that track type.

The lines are red, green and blue. Each line represents a different component of the track. For example, when a **Position** track is highlighted, the red, green and blue lines represent the motion along the world coordinate system's X, Y and Z axes respectively. For a **Rotation** track, these lines represent the rotation around the X, Y and Z axes respectively.

Toolbar

Click on a line to see the keys on all lines. The keys are represented as black dots. The keys can be selected, moved and copied in the same manner as with **Edit Keys** mode.

To see information about a key, right-click on a key. A dialog displays with information about the key. At the bottom of the dialog are two curve buttons, **In** and **Out**. These buttons display the kind of curves coming into and going out of the currently selected key.

Although you set the exact location and value of each key yourself, 3D Studio MAX interpolates what will happen between each key. By default, MAX attempts to make a smooth curve between each key. This might not create the type of animation you want.

To change the shape of the curve between two keys, right-click on the key at the left end of the curve. Click and hold on the **Out** button to display the types of curves available, and select another type of curve. Select the key at the right end of the curve and change the **In** curve in the same fashion. You can also change curves for several selected keys at once.

Additional Tracks

You can add certain types of tracks to the Hierarchy list for specialty uses.

A **Visibility** track is used to make an object disappear or reappear in renderings. In order to create a **Visibility** track, you must be in **Edit Keys** mode. Highlight the object name on the Hierarchy listing and click **Add Visibility Track** ![icon] on the Track View toolbar. A track named **Visibility** is created under the object name. Click **Add Keys** ![icon] and click on the track across from the **Visibility** track name to add keys for disappearing and reappearing. Where the track is blue, the object is visible in renderings. Where the track is white, the object will not appear in renderings. To remove a **Visibility** track, highlight the track name and click **Delete Visibility Track** ![icon] on the Track View toolbar.

A **Note** track can be added while in any mode. **Note** tracks can be used to enter notes and information about frames or keys. Highlight any track and click **Add Note Track** ![icon]. A **Note** track appears under the highlighted track. Click **Add Keys** and click on the track across from the **Note** track name to add keys. Right-click on a key to access a dialog where notes can be entered. You can create as many note keys as you like.

Sound

A sound file can be set up to accompany animation, and to become part of an AVI file rendered with 3D Studio MAX. The sound file also plays when you click the **Play Animation** button ▶

To add a sound file, highlight the **Sound** track on the Hierarchy listing. Click **Properties** 🖳 on the Track View toolbar. The Sound Options dialog appears. Click **Choose Sound** and choose a sound file to accompany the animation. Close the Sound Options dialog.

The chosen sound file appears on the Track View window as a waveform. The peaks of the wave represent high sound volume. You can use the waveform to help you align animation keys to specific parts of the sound file. The sound file cannot be edited in any way in 3D Studio MAX.

Sample Usage

Once you have animated objects in the scene, you can use Track View to fine-tune the animation and make it look exactly the way you'd like. In this example you'll animate a box and adjust its motion with Track View.

1. In the lower left corner of the Front viewport, create a box with a length, width and height of about 50 units. Go to frame 50 and turn on the **Animate** button. In the Front viewport, move the box up by about 200 units. Go to frame 100. In the Front viewport, move the box to the right about 200 units.

2. Click **Play Animation** ▶ and watch the animation in the Front viewport.

The box moves in an arc. By default, frames between keyframes are interpolated as smooth arcs if possible. In this case, we'd like the box to move in a straight line to each key position. The motion can be adjusted with Track View.

3. Under the **Display** panel 🖳, turn on **Trajectory** under the Display Properties rollout.

This displays the trajectory for the box, which will make it easier to see how the changes in Track View affect the box's motion.

4. Click **Track View** 🖳 to open the Track View window. Move the Track View window down so you can see both the Track View window and the Front viewport.

5. Click **Filters** 🖳 and check the **Animated Tracks** option under the **Show Only** section. Click **OK** to exit the dialog.

Toolbar

6. On the Hierarchy list, right-click on **Objects**. Choose **Expand All** from the list.

The Hierarchy list expands to show the box's **Position** track, which is the only animated track in the scene.

7. Highlight **Position** on the Hierarchy list. Click **Function Curves** ⟨icon⟩. A curved representation of the animation appears.

8. Click on any of the curves to display key dots. Right-click on any of the dots at frame 50. A dialog appears with parameters for the **Position** key. Click and hold on the picture button below **In** ⟨icon⟩. Choose the second picture on the list ⟨icon⟩. Do the same for the button below **Out**.

The curves you just selected for the keys on frame 50 tell MAX to make the motion curve straight as it comes into and goes out of the key at frame 50. Note the changes to the trajectory in the Front viewport.

8. Click on a key dot at frame 0. Change the **Out** curve to the second curve on the picture list.

9. Click on a key dot at frame 100. Change the **In** curve to the second curve on the picture list.

Look at the trajectory in the Front viewport. The box now moves straight from one key position to another.

Usage Notes

The **Track View** button performs the same function as choosing *Track View/New Track View* from the menu.

Track View takes many months of use to master. Practice using the buttons described in this entry, and see your 3D Studio MAX documentation for more features.

Moves a camera or light, and its target if any, in the plane perpendicular to its direction.

General Usage

Activate a camera viewport or light viewport. Click **Truck Camera** or

Truck Light . Click and drag in the viewport to move the camera or light. The camera or light moves in the plane perpendicular to its direction.

Usage Notes

Every camera and light has a direction. **Truck Camera** and **Truck Light** use the plane perpendicular to this direction, and move the camera or light only in this plane.

When you click and drag in a camera or light view, **Truck Camera** and **Truck Light** actually move the camera or light in the direction opposite the drag. In this way, the camera's or light's view, rather than the camera or light itself, is moved in the direction of the drag.

Truck Camera is only available in an active camera viewport. It cannot be used to position the camera in other viewports. Likewise, **Truck Light** is only available when a light viewport is active, and cannot be used to adjust the light's position in other viewports.

With target cameras and spots, **Truck Camera** and **Truck Light** move the target with the camera or light.

2D Snap Toggle

Turns 2D snap on and off.

General Usage

From the **3D Snap** flyout ![icon], click **2D Snap** ![icon] to turn it on. Create, modify or transform objects. The cursor snaps to the construction plane or to objects on the construction plane according to the settings on the Grid and Snap Settings dialog. When the cursor snaps to a grid or an object, a small box appears around the cursor.

Usage Notes

When **2D Snap** is on, the cursor can snap only to the construction plane or to any object on the construction plane.

All snap options are set on the Grid and Snap Settings dialog, under the **Snaps** tab. The size and color of the snap cursor is set under the **Options** tab. To access the Grid and Snap Settings dialog, right-click on **3D Snap** or choose *Views/Grid and Snap Settings* from the menu.

See the **Grid and Snap Settings** entry in the *Menus* section for information on setting snaps.

2.5D Snap Toggle

Turns 2.5D snap on and off.

General Usage

From the **3D Snap** flyout , click **2.5D Snap** to turn it on. Create, modify or transform objects. The cursor snaps to any object, but when you click and drag to draw an object, the object is drawn on the construction plane.

Snaps occur according to the settings on the Grid and Snap Settings dialog. When the cursor snaps, a small box appears around the cursor.

Usage Notes

When **2.5D Snap** is on, the cursor can snap to any object in 3D space. However, when you click, the cursor actually snaps to the construction plane. The effect is similar to holding up a plate of glass in front of a 3D object and drawing the object on the glass.

All snap options are set on the Grid and Snap Settings dialog, under the **Snaps** tab. The size and color of the snap cursor is set under the **Options** tab. To access the Grid and Snap Settings dialog, right-click on **3D Snap** or choose *Views/Grid and Snap Settings* from the menu.

See the **Grid and Snap Settings** entry in the *Menus* section for information on setting snaps.

 **Undo**

Reverses the last operation.

General Usage

Perform an operation such as selecting or moving an object. Click **Undo** to undo the action. You can click **Undo** repeatedly to undo several consecutive operations.

You can also right-click on **Undo** button to display a list of recent actions from which you can choose several actions to undo at once.

Usage Notes

Most object creation and modification operations can be undone, but some cannot. See the **Undo [*last action*]** entry in the *Menus* section for lists of actions that can and cannot be undone.

By default, you can undo up to 20 operations. You can change this number by choosing *File/Preferences* from the menu. Click the **General** tab and change the **Undo Levels** value under the **Scene Undo** section. Lower **Undo Levels** take up less memory and improve system performance.

You can redo an undone action with the **Redo** button .

Removes a link between a selected object and its parent.

General Usage

Select one or more objects. Click **Unlink Selection** . The objects are unlinked from their parents.

Usage Notes

To quickly unlink all objects in a hierarchy, select all objects in the chain and click **Unlink Selection**. You can select all objects in a hierarchy by double-clicking on the root object.

■ Use Pivot Point Center

Uses an object's pivot point as the center for rotation and scaling.

General Usage

Choose a rotation or scale transform. If necessary, choose **Use Pivot Point Center** ■. Rotate or scale an object. The rotation or scale takes place about the object's pivot point.

Usage Notes

This button is a flyout with two other transform center selections, **Use Selection Center** ■ and **Use Transform Coordinate Center** ■.

When the **Animate** button is off and a single object is selected, the transform center defaults to **Use Pivot Point Center**, but can be changed. In addition, the transform center can be set differently for each transform. When you select a transform, the transform center selection changes to the last selection for that transform.

When the **Animate** button is on and an object is rotated or scaled, the transform center is forced to **Use Pivot Point Center**. To change the center about which rotation and scaling animation is performed, you can change the object's pivot point. For information on changing a pivot point, see the **Pivot** entry in the *Panels* section.

Use Selection Center

Uses the center of the object selection as the center for rotation and scaling.

General Usage

Choose a rotation or scale transform. From the **Use Pivot Point Center** flyout , choose **Use Selection Center** . Select two or more objects. With the **Animate** button off, rotate or scale the selection. The rotation or scale takes place about the object's pivot point.

Usage Notes

When the **Animate** button is off and a multiple objects are selected, the transform center defaults to **Use Selection Center**, but can be changed. In addition, the transform center can be set differently for each transform. When you select a transform, the transform center selection changes to the last selection for that transform.

When the **Animate** button is on and an object is rotated or scaled, the transform center is forced to **Use Pivot Point Center**. To set a different center for a selection, create a dummy object and link each object to the dummy. When you rotate the dummy with the **Animate** button on, all objects will rotate around the dummy's pivot point. For information on creating a dummy object, see the **Dummy** entry in the *Panels* section.

Toolbar

Use Transform Coordinate Center

Uses the origin of the current coordinate system as the center for rotation and scaling.

General Usage

Choose a rotation or scale transform. From the **Use Pivot Point Center** flyout, choose **Use Transform Coordinate Center**. Select two or more objects. With the **Animate** button off, rotate or scale the selection. The rotation or scale takes place about the coordinate system origin.

Sample Usage

Use Transform Coordinate Center is useful for rotating objects around a specified point. You can set the **Reference Coordinate System** to any object in the scene, and rotate objects around its pivot point. See the **Reference Coordinate System** entry for further information.

1. Create a box with a length, width and height of 50 units.

2. Create a point object about 100 units to the right of the box in the Top viewport. To find out how to create a point object, see the **Point** entry in the *Panels* section.

3. Click **Select and Rotate**. From the **Reference Coordinate System** pulldown, choose **Pick**. Click on the point object. The coordinate system changes to **Point01**.

4. Choose **Use Transform Coordinate Center**.

5. Select the box and rotate it. The box rotates around the point object.

Usage Notes

The transform center can be set differently for each transform. When you select a transform, the transform center selection changes to the last selection for that transform. In order to set the coordinate system center as the rotation or scale center as in the example above, you must be sure to select the transform first, then the **Reference Coordinate System**, then the transform center. Performing these steps in a different sequence might cause the transform center or **Reference Coordinate System** to change before you rotate or scale the object.

When the **Animate** button is on and an object is rotated or scaled, the transform center is forced to **Use Pivot Point Center**. To use a different transform center, create a dummy object where you want the center to be, and link objects to the dummy. See the **Dummy** entry in the *Panels* section for information on creating a dummy object.

Changes the current selection mode to crossing selection mode.

General Usage

Click the **Crossing Selection** button ⊡. The button changes to the

Window Selection button ⊡. Window selection is now enabled. Click
Window Selection to return to crossing selection mode.

When window selection mode is active, drawing a bounding area on
the screen for selection will select only objects contained completely
in the bounding box. Compare with crossing selection mode where
objects touching the box are also selected.

Toolbar

Zooms into or out of viewports with various controls.

Zoom Buttons

Several zoom buttons are available for controlling the views in viewports. These buttons are available only in orthographic viewports such as Top, Front, Left and Perspective. They are not available while camera or light views are selected.

 Zoom   Click **Zoom**, then click and drag in a viewport to zoom in or out of the viewport. By default, the zoom is centered on the viewport center. If the **Zoom About Mouse Point** checkbox on the **Viewports** tab of the Preference Settings dialog is checked, zooming in orthographic viewports is centered on the mouse point. This dialog can be accessed with the *File/Preferences* menu option.

Zoom All   Click **Zoom All** and click and drag in any viewport to zoom all viewports simultaneously. Camera and light views are not zoomed. To prevent Perspective views from zooming, hold down <**Ctrl**> while dragging.

Zoom Extents   Click **Zoom Extents** to zoom the current viewport to the extents of the scene. This option is useful for quickly restoring a view to show the entire scene.

Zoom Extents Selected   This button is available from the **Zoom Extents** flyout. Click **Zoom Extents Selected** to zoom the current viewport to the extents of selected objects.

Zoom Extents All   Click **Zoom Extents All** to zoom all viewports to the extents of the scene. Camera and light views will not be zoomed. To prevent Perspective views from zooming, hold down <**Ctrl**> while clicking this button.

Zoom Extents All Selected   This button is available from the **Zoom Extents All** flyout. Click **Zoom Extents All Selected** to zoom all viewports to the extents of selected objects. To prevent Perspective views from zooming, hold down <**Ctrl**> while clicking this button.

Toolbar

Panels

This section describes the panels in 3D Studio MAX. All panels are displayed in a column at the upper right or left side of the screen.

Panels are accessed by clicking one of six panel buttons. Entries in this section of the book can be found in different places, depending on the panel chosen.

Panel	Where to look for entry name
Create	Object Type rollout
Modify	Modifiers rollout
Hierarchy	Below object name
Motion	Below object name
Display	Below object name
Utilities	Utilities rollout

Create Panel

Under the **Create** panel [icon] are seven buttons.

[icon] **Geometry**

[icon] **Shapes**

[icon] **Lights**

[icon] **Cameras**

[icon] **Helpers**

[icon] **Space Warps**

[icon] **Systems**

The **Geometry** button [icon] is on by default. When any one of the seven buttons above is clicked, the panel changes to reflect the selection. Some selections have more than one class. Object classes can be selected from the pulldown menu above the Object Type rollout.

For information on a particular object type, see the entry corresponding to the object type listed on the Object Type rollout.

Geometry [icon] has nine classes of geometry for creating 3D objects.

- **Standard Primitives** are the standard building blocks used in 3D Studio MAX.
- **Extended Primitives** are slightly more complex building blocks.
- **Compound Objects** are objects created from two or more objects.
- **Loft Objects** are objects created by lofting one or more shapes along a path.
- **Particle Systems** generate streams of particles.
- **Patch Grids** are flat rectangles that can be deformed in a variety of ways.
- **NURBS** surfaces are smooth objects defined by curves rather than polygons.
- **Doors** and **Windows** provide a quick and easy way to generate objects commonly used in architectural models.

Shapes ⬚ has two classes of shapes. Shapes are two-dimensional objects that can be used to create 3D objects.

- **Splines** are shapes created from vertices, segments and splines. Vertices and segments can be edited to create sharp corners or smooth, rounded splines. The smoothness of round shapes is determined by the number of steps in each segment.
- **NURBS Curves** use NURBS technology to create smooth, round curves. NURBS curves created on this panel cannot be used to make NURBS objects, but can be used anywhere that spline shapes can, as in loft objects.

When **Start New Shape** is checked, each shape created is a separate object. When **Start New Shape** is unchecked, each shape created is part of the previously created shape.

Under **Lights** ⬚ and **Cameras** ⬚, the pulldown has only one choice, **Standard**.

Under **Helpers** ⬚ are four helper classes. Helper objects are non-rendering objects that can help you set up your scene.

- **Standard** helpers can be used with any model.
- **Atmospheric Apparatus** helpers are used to make atmospheric effects such as Combustion.
- **CameraMatch** is a class for helpers used with the **Camera Match** utility.
- **VRML 2.0** is a class for helpers used in setting up a VRML scene.

Under **Space Warps** ⬚ are three classes of space warps. Space warps deform objects or particle systems.

- **Geometric/Deformable** space warps deform geometry objects.
- **Particles & Dynamics** space warps deform particle systems, affecting the motion and direction of particles.
- **Modifier-Based** space warps are based on modifiers under the Modify panel.

Under **Systems** ⬚ is only one class, **Standard**. Systems are specialty objects for specific applications.

Modify Panel

The **Modify** panel ⬚ contains modifiers for altering objects. By default, ten modifiers are listed. The rest can be accessed by clicking the **More** button near the top of the panel. For information on a modifier, see the entry corresponding to the modifier name.

In order to modify objects, one or more objects must be selected. If no objects are selected, all options on the **Modify** panel are grayed out and cannot be selected.

When a modifier is applied to an object, the modifier is placed on a list called the *modifier stack*. The last modifier applied is placed on the top of the stack. The stack can be edited and manipulated with buttons on the Modifier Stack rollout of the **Modify** panel, such as **Edit Stack** 🖩 and **Remove modifier from the stack** 🖩. These buttons are described under the **Modify** entry in this section of the book.

After an object is created with one of the options under the **Create** panel, its creation parameters can be modified on the **Modify** panel. For most objects, the creation parameters are the same on both panels. For loft objects and NURBS surfaces, the **Modify** panel offers additional parameters. These parameters can be found under the object's entry. For example, parameters available under the **Modify** panel for loft objects can be found under the **Loft** entry.

See the **Modify** entry for more information on the **Modify** panel.

Hierarchy

The **Hierarchy** panel 🖩 has three buttons, **Pivot**, **IK** and **Link Info**. Information on these hierarchy types can be found under the **Pivot**, **IK** and **Link Info** entries.

Motion

The **Motion** panel 🖩 has two buttons, **Parameters** and **Trajectories**. Information on these hierarchy types can be found under the **Parameters** and **Trajectories** entries.

Display

The **Display** panel 🖩 has no further buttons. Information on the parameters on this panel can be found under the **Display** entry.

Utilities

The **Utilities** panel 🖩 contains various utilities for working with scenes. By default, eight utilities are listed. The rest can be accessed by clicking the **More** button near the top of the panel. For information on a utility, see the entry corresponding to the utility name.

A modifier that creates a smooth bump in a 3D object.

General Usage

On the **Modify** panel ![icon], click **More** if necessary, and choose **Affect Region**. Click **Sub-Object** to access the **Points** sub-object level, and move the tip and base points to the desired area. Adjust the **Falloff** value as necessary.

Sub-Object Level

By default, the tip and base points that control the effect are placed at the object's pivot point. At the **Points** sub-object level, you can move the tip or base points. These points can be animated.

Parameters

Falloff
Sets the radius, in units, of the area around the base point that will be affected by the modifier.

Ignore
Back Facing
When checked, only faces with normals pointing in the same general direction as the tip point arrow are affected.

Pinch
Affects the sharpness of the curve at the tip point. Higher values make a sharper point.

Bubble
Affects the slope of the curve below the tip. Higher values make a dome-type bubble, while lower values make a steeper slope.

Usage Notes

The 3D object to which the **Affect Region** modifier is applied must have sufficient detail to create the bump.

Anchor

Specifies a click-to-play trigger in a VRML scene.

General Usage

On the **Create** panel [icon], click **Helpers** [icon]. From the pulldown menu, choose **VRML 2.0**. Click **Anchor**. Click and drag to create the icon. Click **Pick Trigger Object** and select the trigger object. Choose **Hyperlink Jump** or **Set Camera** to specify the destination of the jump.

Anchor

Pick Trigger Object	To pick the trigger object, click **Pick Trigger Object** and click on the object on the screen. The object name appears next to **Trigger Object** below the button. You can also pick the object by clicking **Select by Name** [icon] or by pressing the **<H>** key.
Description	Enter a description for the trigger. This description will appear on the browser's status bar when the cursor is moved over the trigger object.
Hyperlink Jump	Instructs the anchor to jump to the specified URL.
URL	Specifies the destination for the jump.
Parameter	Specifies additional browser parameters.
Bookmarks	Selects a URL from a list of bookmarks. You can enter new bookmarks, or click **Import List** to import the list of bookmarks from your browser.
Set Camera	Instructs the anchor to jump to the specified camera.
Camera	Click the down arrow to select from a list of cameras in the scene.
Icon Size	Sets the diameter of the icon. The size of the icon does not affect the operation of the anchor.

Creates an arc shape.

General Usage

On the **Create** panel ⟨icon⟩, click **Shapes** ⟨icon⟩. Click **Arc**. To create the arc, click and drag on the screen, then move the cursor and click to set the arc. An arc has four Bezier vertices.

General

Steps	Sets the number of steps between vertices. A shape is made up of straight segments, or steps, between each vertex. When the shape is modified to make a 3D object, each step becomes a side or segment on the 3D object. A higher number of **Steps** makes a smoother shape.
Optimize	When checked, straight segments receive no steps, and curved segments receive the number of steps set with the **Steps** value.
Adaptive	When checked, the number of steps is automatically set by 3D Studio MAX. Straight segments receive no steps, while curved segments can receive up to 100 steps. Checking **Adaptive** disables **Steps** and **Optimize**.
Renderable	When checked, the shape appears in renderings. An imaginary circle is passed along the shape's length to skin the shape. The rendered object is similar to an object that would be created by lofting a circle along the shape. The imaginary circle and skin do not appear in viewports. If the shape is later used to make a 3D object, the circle and skin no longer appear in renderings.
Thickness	The diameter of the circle passed along the shape.
Generate Mapping Coordinates	Generates mapping coordinates for the imaginary skinned object. If the shape is later used to create a 3D object, these mapping coordinates are no longer valid.

Creation Method

The Creation Method rollout is available only on the **Create** panel.

End-End-Middle	When this option is on, the arc is created by first setting the two ends of the arc, then setting the midpoint of the arc. To create an arc in this way, click and drag and release the mouse to set the two ends of the arc, then move the mouse and click to set the midpoint.

Panels

Center-End-End	When this option is on, the arc is created by first setting the center and one end of the arc, then setting the other end of the arc. To create an arc in this way, click and drag and release the mouse to set the center and one end of the arc, then move the mouse and click to set the other end. When you first click and drag to set the center and first end, the center appears on the screen as a point. When you release the mouse, the center point disappears as it is not actually part of the arc.

Keyboard Entry

The Keyboard Entry rollout is available only on the **Create** panel.

X, Y, Z	Specifies the location of the center of the shape created when **Create** is clicked. The center is placed at the X, Y and Z location in the world coordinate system.
Radius From To	See these listings under Parameters below.
Create	Creates the shape with the parameters entered. The shape is placed on the current viewport's plane.

Parameters

Radius	Sets the radius of the arc.
From To	Sets the angle of the first and last points on the arc. Angles are calculated from the shape's local X axis. A point on the local X axis has an angle of 0, while a point on the local Y axis has an angle of 90 degrees. These angles can vary from 0 to 360.
Pie Slice	Creates a closed pie-shaped spline from the arc by connecting each end point with the center of the arc.
Reverse	Reverses the direction of the spline.

Usage Notes

An arc is considered to be a portion of a circle. The arc center is the center of an imaginary circle that would be created if the arc were extended all the way around to form a full circle. The **Radius** is the radius of this imaginary circle. The **From** and **To** angles refer to degrees around the imaginary circle.

ASCII Object Output

Outputs object information as an ASCII file.

General Usage

On the **Utilities** panel ⊤, choose **ASCII Object Output**. Click **Pick Object** and pick an object. Enter a filename in the file selector that appears. An ASCII file containing object information is saved with the extension *.asc.*

ASCII Object Output

Pick Object   Click **Pick Object** to pick the object and access the file selector for naming the ASCII file. Only one object may be picked.

Object Space   When checked, object information is listed using the object's local coordinates and orientation. When this checkbox is off, the world coordinate system is used to specify object information.

Shapes   This checkbox affects the listing of shapes only. When
as Beziers   checked, the shape is defined by vertex locations, with imaginary vertices generated along the shape frequently enough to adequately define the shape. When unchecked, a knot/vector format is used for each shape. If this information is later used to redraw the shape in another program, the information must be interpreted by a utility that reads knot/vector information.

Close   Closes the utility.

Usage Notes

This utility generates an ASCII file with information about an object such as object name, number and location of vertices, and smoothing information. The ASCII file can then be used by another program to regenerate the object.

If you want to use an object in another 3D program, an ASCII file is not the most efficient method of transferring the object information. Instead, choose *File/Export* from the menu and choose one of the file output types.

If a group is picked with **Pick Object**, information is generated for the first object in the group only.

Asset Manager

Displays thumbnails of *.max* files and bitmaps.

General Usage

On the **Utilities** panel , choose **Asset Manager**. The Asset Manager dialog appears. Access the directory containing the thumbnails you wish to view. Change the filter on the toolbar to display the type of file you wish to see. Thumbnails are displayed for all files of that type.

Drag and drop thumbnails into the Material Editor or other areas of MAX that use bitmaps. You can also drag and drop a *.max* file over a viewport to merge the file with the current scene.

Asset Manager

	View Selected Image	Displays the currently selected thumbnail at its full resolution. For *.avi* files, the Media Player is launched. This option is only available for bitmaps.
	Properties	Displays information about the selected thumbnail file, such as size and date saved.
	Cache Settings	A *cache* is an area on another part of the hard disk used to save information. The Asset Manager utility uses a cache to save thumbnails so they don't have to be generated every time you open the utility. When you click **Cache Settings**, a dialog appears where you can set the maximum size of the thumbnail cache and where it should be stored. Click the button next to the directory name ☐ to select a directory. The **Clear Cache** button deletes all thumbnails from the cache.
	Sort by Name Extension Size Date	Determines the order in which thumbnails are displayed.
	Display Small Medium Large Thumbnails	Determines the size of the displayed thumbnails.

Filter	Determines which thumbnails will be displayed from the current directory. If you choose **All Images**, then all bitmap thumbnails are displayed, but not *.max* files. In order to see *.max* files you must choose this file type specifically from the list.

Automatic Thumbnail Update When this button is on, thumbnails are created automatically for all files in the current directory that don't currently have thumbnails, or those files which have been updated since the Asset Manager's last access to this directory.

View All Images in Cache Displays all thumbnails in the cache regardless of the directory they come from.

Refresh Recreates thumbnails for the current directory.

Usage Notes

The Asset Manager dialog is a modeless dialog, meaning you can leave it on the screen while you perform other functions and access it whenever you need it. Thumbnails can be dragged and dropped into any area that accepts the type of file displayed. For example, bitmap thumbnails can be dragged into areas of the Material Editor where a bitmap is accepted.

If you drag and drop an old thumbnail for which there is no longer a corresponding file, the thumbnail itself will be used in place of the bitmap.

You can view the thumbnail at its full resolution by clicking **View Selected Image** or by double-clicking on the thumbnail itself. Another way to view a thumbnail is to right-click on it and choose **View** from the pop-up menu. You can also see a thumbnail's properties by choosing **Properties** from the pop-up menu. This is the same as clicking **Properties** on the Asset Manager toolbar.

To save time in generating thumbnails, turn off **Automatic Thumbnail Update** until you have accessed the desired directory, then turn it on to generate thumbnails just for that directory.

Assign Vertex Colors

Assigns vertex colors to selected objects.

General Usage

Set up a scene with objects and lights. In the Material Editor, create a **Blend** material. For **Material 1**, assign any **Bitmap** as a **Diffuse** map.

Turn on **Show Map in Viewport** ![icon] for the bitmap.

For **Material 2**, assign a **Vertex Colors** map as a **Diffuse** map. Leave the **Mix Amount** at 0.0. Assign the **Blend** material to one or more objects, and select the objects. The bitmap in **Material 1** appears in shaded viewports.

On the **Utilities** panel ![icon], choose **Assign Vertex Colors**. On the Assign Vertex Colors rollout, choose either the **Scene Lights** or **Diffuse** option. Click **Assign to Selected**. 3D Studio MAX takes a few moments to calculate the vertex colors. No change to the object color is visible in viewports.

On the **Display** panel ![icon], turn on **Vertex Colors** under the Display Properties rollout. Vertex colors are now displayed. Each vertex on selected objects is colored from the **Diffuse** map of the first material in the **Blend** material. If **Scene Lights** was selected on the **Utilities** panel when **Assign to Selected** was clicked, vertices were lit with lights in the scene. If **Diffuse** was selected, each vertex is lit. The areas in between vertices contain gradients from one vertex color to another.

Render the scene. Selected objects render with Material 1 from the Blend material.

Under the **Blend** material, change **Mix Amount** to 100.0. Render the scene. Selected objects render with vertex colors.

Change the lighting in the scene and render again. The colors on the object remain the same as they were before.

Usage Notes

The **Assign Vertex Colors** utility is primarily for programmers and game developers with special requirements. It assigns colors to vertices based on the current lighting setup. The colors are then glued to the vertices, and will not change even if the lighting in the scene or the material on the object is changed.

Using the **Assign Vertex Colors** utility on an object places a Vertex Colors modifier on the object's stack. You can remove the effects of this utility by removing the Vertex Colors modifier from the stack un-

der the **Modify** panel ![icon] with **Remove modifier from the stack** ![icon]. This modifier can only be added with the Assign Vertex Colors utility and not with a modifier on the **Modify** panel.

Vertex colors are stored on map channel 2. If you assign a **UVW Map** modifier to objects for which you have assigned vertex colors with this utility and turn on the **Channel 2** option for the modifier, objects may not render correctly.

Panels

AudioClip

Sets up a VRML link to a sound file.

General Usage

On the **Create** panel ![icon], click **Helpers** ![icon]. From the pulldown menu, choose **VRML 2.0**. Click **AudioClip**. Click and drag to create the icon. Enter the **URL** link to the sound file.

Audio Clip

URL	Specifies the URL for the sound. Only *.wav* or *.mid* files can be used.
Description	Enter a description for the sound. This description will appear on the browser's status bar when the sound is played.
Pitch	Sets the pitch, in octaves. A **Pitch** of 1 plays the sound file normally. A **Pitch** of 0.5 plays the sound at one octave below normal, while 2.0 plays the sound at one octave above normal.
Loop	When checked, the sound file loops after playing. If unchecked, the sound file plays once, then stops.
Start on World Load	When checked, the sound file plays immediately after the world is loaded.
Icon Size	Sets the diameter of the icon. The size of the icon does not affect the operation of the helper.

Usage Notes

AudioClip helpers can be selected by the **Sound**, **ProxSensor**, **TimeSensor** or **TouchSensor** helpers to play a sound.

Awning

Creates a window with hinged panels.

General Usage

On the **Create** panel , click **Geometry** . From the pulldown menu, choose **Windows**. Click **Awning**. Click and drag to create the width of the window, move the cursor and click to set the depth or height, then move the cursor and click again to set the remaining dimension.

Creation Method

Width/ Depth/ Height	The first click-and-drag sets the width of the window. The second click sets the depth while the third sets the height.
Width/ Height/ Depth	The first click-and-drag sets the width of the window. The second click sets the height while the third sets the depth.
Allow Non-vertical Jambs	When checked, jambs can be non-vertical. Turn this checkbox on to allow creation of tilted windows.

Parameters

Height Width Depth	Sets the dimensions of the window.
Frame	Sets the dimensions of the window frame. **Horiz. Width** and **Vert. Width** affect the size of the glazing in addition to the frame size. **Thickness** also affects the thickness of the rails in the window sashes.
Glazing	Sets the **Thickness** of the glass.
Rails and Panels	**Width** sets the width of the rails in the sashes. **Panel Count** sets the number of sashes.
Open Window	Opens window panels by the specified percentage. 100 percent is equal to 90 degrees.
Generate Mapping Coords	Generates planar mapping coordinates on the window.

Background

Sets up a sky and ground for a VRML world.

General Usage

On the **Create** panel ![icon], click **Helpers** ![icon]. From the pulldown menu, choose **VRML 2.0**. Click **Background**. Click and drag to create the icon. Specify colors and/or images to be used for the sky and ground.

Sky Colors

The parameters on the Sky Colors rollout create a gradient on a large sphere surrounding the entire scene.

Number of Colors	Specifies the number of colors used in the sky gradient.
Color	Sets colors 1, 2 and 3.
Angle	Sets the angle at which colors meet. The angle is calculated from the center of the scene pointing straight up.

Ground Colors

The parameters on the Ground Colors rollout create a gradient on the world ground plane.

Number of Colors	Specifies the number of colors used in the ground gradient.
Color	Sets colors 1, 2 and 3.
Angle	Sets the angle at which colors meet. The angle is calculated from the center of the scene pointing straight up.

Images

On the Images rollout, you can set up six images to be used as a background panorama in the scene. Enter the URL for each image. Only *.jpg* or *.gif* files can be used.

A modifier or space warp that bends an object.

General Usage

Create an object to bend. This object should have several segments so it can bend smoothly.

To apply **Bend** as a modifier, select the object. Click on the **Modify** panel ▦ and click **More** if necessary to access the full list of modifiers. Select **Bend**.

To apply **Bend** as a space warp, go to the **Create** panel ▦. Click **Space Warps** ▦ and select **Modifier-Based** from the pulldown menu. Click **Bend**. Click and drag on the screen to create the **Bend** space warp's length and width, then move the cursor and click to specify the space warp's height. Select the object, and click **Bind to Space Warp** ▦ on the Toolbar. Click and drag from the object to the space warp to bind it to the space warp. To modify the bend, select the bend space warp and go to the **Modify** panel.

Increase the **Angle** value to bend the object.

Sub-Object Levels

The **Modify** panel for the **Bend** modifier has two sub-object levels.

Gizmo	The controller for the bend. This controller can be transformed and animated to animate the bend.
Center	The center or base point of the bend.

Space Warp Parameters

Length **Width** **Height**	Sets the dimensions of the space warp object.
Decay	Sets the amount by which the bend will diminish over distance. When **Decay** is 0, the bend angle is the same over the entirety of the object. When **Decay** is higher than 0, the bend decays (becomes less) as the distance between the bound object and space warp increases.

Bend

Parameters

Angle   Sets the angle of the bend in degrees.

Direction   Rotates the bend around the **Bend Axis** by the specified number of degrees. If you cannot get the bend to go in the desired direction, try setting **Direction** to 90.

Bend Axis   Sets the axis for the bend. The **Bend Axis** uses the gizmo's or space warp's local axis as a reference, not the local axis of the object being bent.

Limit Effect   When checked, the bend is limited to parts of the object within the **Upper Limit** and **Lower Limit**. Other parts of the object will rotate in accordance with the bend, but will not bend. When **Limit Effect** is checked, the upper and lower limits appear as two rectangles spanning the length and width of the space warp or gizmo.

Upper Limit   Sets the upper limit for the bend. The limit is set in units above the center of the gizmo or space warp. This value must be a positive number. Only vertices that lie between the **Upper Limit** and **Lower Limit** are bent.

Lower Limit   Sets the lower limit for the bend. The limit is set in units below the center of the gizmo or space warp. This value must be a negative number. Only vertices that lie between the **Upper Limit** and **Lower Limit** are bent.

Bevel

A modifier that creates a 3D beveled object from a shape.

General Usage

Create a shape. Select the shape.

On the **Modify** panel 📷, click **More** if necessary, and choose **Bevel** from the modifier list. Set the **Height** for **Level 1** to extrude the shape. Set additional values under the Bevel Values rollout as desired.

Parameters

Start	Places a cap at the top of the extruded object.
End	Places a cap at the bottom of the extruded object.
Morph	Creates a cap with a repeatable number of vertices. Use this option on objects that will be used as morph targets.
Grid	Creates a cap with vertices and faces arranged in a grid. This type of cap works better with modifiers than the **Morph** type does.
Linear Sides	When this option is selected, **Bevel** makes straight edges.
Curved Sides	When this option is selected, **Bevel** makes curved edges. In order to make curved edges, **Segments** must also be greater than 1.
Segments	Sets the number of segments around the object in each bevel level.
Smooth Across Levels	When checked, all the object's faces are in the same smoothing group, making the object appear smooth regardless of the angle between levels.
Generate Mapping Coords	Generates cylindrical mapping coordinates on the object.
Keep Lines From Crossing	With nested shapes, the bevel settings might cause bevels to intersect. Checking **Keep Lines From Crossing** changes the bevels on the object so they are apart by at least the **Separation** value at all times, regardless of the settings on the Bevel Values rollout. This checkbox does not affect the displayed values on the Bevel Values rollout.

Bevel Values

Start Outline Changes the original shape, increasing its radius by the specified number of units. If the shape has nested splines, the radius of each nested spline is increased or decreased by this number. Setting **Start Outline** to 0 makes no change to the original shape.

Level 1 **Height** sets the height of the first bevel level. **Outline** sets the number of units by which the original shape's radius is increased at the first level of the bevel. If the shape has nested splines, the radius of each nested spline is increased or decreased by the **Outline** value at the first bevel level.

Level 2 To make a second and/or third bevel level, check **Level**
Level 3 **2** and/or **Level 3** and set the **Height** and **Outline** for each level.

Usage Notes

The **Bevel** modifier is available only when a shape is selected.

Bevel extrudes a 3D object from a shape and bevels it at the same time. If the shape consists of two or more nested splines, splines are beveled inward or outward toward the solid part of the object.

A modifier that creates a 3D object from a shape using another shape as a profile.

General Usage

Create a shape to extrude. Create a shape to use for the profile. Select the shape to extrude.

On the **Modify** panel ![icon], click **More** if necessary, and choose **Bevel Profile** from the modifier list. Click **Pick Profile** and pick the profile shape.

Parameters

Pick Profile	Click **Pick Profile** and pick the profile shape. One end of the profile is aligned with the outer edge of the shape, setting the size and shape of the extrusion. If the shape consists of nested splines, the profile shape affects each nested spline individually.
Generate Mapping Coords	Generates cylindrical mapping coordinates on the object.
Start	Places a cap at the top of the extruded object.
End	Places a cap at the bottom of the extruded object.
Morph	Creates a cap with a repeatable number of vertices. Use this option on objects that will be used as morph targets.
Grid	Creates a cap with vertices and faces arranged in a grid. This type of cap works better with modifiers than the **Morph** type does.
Keep Lines From Crossing	With nested shapes, the bevel settings might cause bevels to intersect. Checking **Keep Lines From Crossing** changes the bevels on the object so they are apart by at least the **Separation** value at all times.

Usage Notes

When **Bevel Profile** uses an instance of the profile shape to create the extruded object. This means you can later change the profile shape to change the extruded object.

BiFold

Creates a door with one or two hinged panels.

General Usage

On the **Create** panel ![icon], click **Geometry** ![icon]. From the pulldown menu, choose **Doors**. Click **BiFold**. Click and drag to create the width of the door, move the cursor and click to set the depth or height, then move the cursor and click again to set the remaining dimension.

Creation Method

Width/ Depth/ Height	The first click-and-drag sets the width of the door. The second click sets the depth while the third sets the height.
Width/ Height/ Depth	The first click-and-drag sets the width of the door. The second click sets the height while the third sets the depth.
Allow Non-vertical Jambs	When checked, jambs can be non-vertical. Turn this checkbox on to allow creation of tilted doors.

Parameters

Height Width Depth	Sets the dimensions of the door.
Double Doors	Creates two sets of hinged doors.
Flip Swing	Flips the doors to swing to the other side when **Open** is greater than 0.
Flip Hinge	Moves the door hinge to the other side of the door. This option is available only when **Double Doors** is unchecked.
Open	Opens door panels by the specified percentage. 100 percent is equal to 90 degrees.
Frame	Sets parameters for the frame. When **Create Frame** is unchecked, the frame is removed and the doors remain. **Width** and **Depth** set the width and depth of the frame. **Door Offset** moves the door or doors away from the default hinge location by the specified number of units.

| Generate Mapping Coords | Generates planar mapping coordinates on the door. |

Leaf Parameters

Thickness	Sets the thickness of panels. If **Glass** or **Beveled** panels are selected below, this value is the thickness of the panel frame.
Stiles/ Top Rail	Sets the width of panel frames, effectively setting the width of the panels themselves. The effect of this setting is visible only when **Glass** or **Bevel** panels are selected below.
Bottom Rail	Sets the height of panel frames, effectively setting the height of the panels themselves. The effect of this setting is visible only when **Glass** or **Bevel** panels are selected below.
# Panels Horiz # Panels Vert	Sets the number of horizontal and vertical panels in the doors.
Muntin	Sets the width of the separation between panels.
None	Leaves the panels as they are.
Glass	Creates a separate portion of the panel with its own **Thickness**. The thickness of the panel can be thicker than the door leaf. In order for the panel to look like glass, you must assign a glass material to the panel.
Bevel	Creates beveled panels.
Bevel Angle	The angle of the bevel between the outer surface of the door and the panel.
Thickness 1	The outer thickness of the panel.
Thickness 2	The thickness where the bevel begins.
Middle Thick	The thickness of the inner panel.
Width 1	The width where the bevel begins.
Width 2	The width of the inner panel.

Panels

Billboard

Causes linked objects to always face the VRML view.

General Usage

Create an object to face the VRML view.

On the **Create** panel ![icon], click **Helpers** ![icon]. From the pulldown menu, choose **VRML 2.0**. Click **Billboard**. Click and drag to create the icon.

Use **Select and Link** ![icon] to link the object to the icon.

Billboard

Screen Alignment	When checked, the icon and linked objects remain facing the viewer even when the view pitches and rolls.
Icon Size	Sets the diameter of the icon.

Usage Notes

The **Billboard** icon's local Y axis always faces the view. Before linking the object to the **Billboard** icon, align the icon's Z axis with the axis of the object that will face the view.

A particle system that creates whirling snow.

General Usage

On the **Create** panel 📷, click **Geometry** 🔾. Select **Particle Systems** from the pulldown menu. Click **Blizzard**. Click and drag on the screen to create the particle emitter.

Click **Play Animation** ▶ to see the default particle animation. Change parameters as desired.

Basic Parameters

See the Basic Parameters rollout under the **PArray** entry.

Particle Generation

Particle Quantity	Determines the particle quantity. When **Use Rate** is selected, the value below **Use Rate** sets the number of particles born on each frame. When **Use Total** is selected, the value below **Use Total** sets the total number of particles born between the **Emit Start** and **Emit Stop** frames. Note that if **Percentage of Particles** is less than 100, only a percentage of particles appear in viewports.
Speed	The speed of particles as they leave the emitter, in units per frame.
Variation	The amount of speed variation from one particle to another as a percentage of the **Speed** value. For example, if **Speed** is 20 and **Variation** is 10, particle speeds will vary by 10% of 20 = 2 units per frame, resulting in speeds varying from 18 to 22.

See the **Snow** entry for descriptions of the **Tumble** and **Tumble Rate** parameters.

See the **PArray** entry for descriptions of the remaining parameters, with the exception of one parameter.

Emitter Fit Planar	This option can be found near the bottom of the Particle Type rollout. When this option is checked, a planar map is applied to the emitter, and particles take on mapping based on the portion of the emitter from which they are born.

Bomb

A space warp that blows up an object and scatters its fragments.

General Usage

Create an object to blow up.

On the **Create** panel 🖼, click **Space Warps** 〰. Select **Geometric/ Deformable** from the pulldown menu. Click **Bomb**. Click and drag on the screen to create the **Bomb** space warp object. Move the bomb near the object so it points in the direction in which you would like the object to explode.

Select the object to be blown up, and click **Bind to Space Warp** 🖼 on the Toolbar. Click and drag from the object to the bomb to bind it to the space warp. Move the time slider beyond the **Detonation** frame number to see the object explode.

Bomb Parameters

Strength   Sets the power of the bomb. Higher values make a larger explosion.

Spin   The rate at which object fragments rotate, in revolutions per second. If **Chaos** is over 0.0, rotation speeds will vary.

Falloff   Sets the falloff distance for the bomb effect, in units. If **Falloff On** is on, any parts of the bound object that are a distance from the space warp greater than the **Falloff** distance are not affected by the explosion. However, all fragments are affected by the **Gravity** setting regardless of **Falloff**.

Falloff On   When this checkbox is on, the **Falloff** distance is in effect.

Min
Max   Sets the minimum and maximum number of faces per fragment. The number of faces on each fragment is a random number between **Min** and **Max**.

Gravity   Sets the acceleration of fragments in the negative Z direction to simulate gravity. When **Gravity** is 0.0, frragments fly into space and are not affected by gravity. When **Gravity** is greater than 0, fragments accelerate in the negative Z direction as they explode. **Gravity** can also be set to a negative value to accelerate fragments in the postive Z direction.

Gravity simply pushes particles along the Z direction. When the fragments strike another object such as a table or floor, they do not bounce off the obstacle as real particles would.

Chaos　　A value from 0 to 10 that causes random motion of fragments. When **Chaos** is 0.0, there is no random motion. Try low values such as 1.0 or 2.0.

Detonation　　The number of the frame on which the bound object begins to explode.

Seed　　Sets the base number for random calculations. Change this value to observe different bomb effects. This value cannot be animated.

Panels

Bones

Creates a system of linked bones.

General Usage

On the **Create** panel ![icon], click **Systems** ![icon]. Click **Bones**. Click and drag on the screen to create a series of bones. Right-click to end bones creation. **Bones** consist of links and nodes. The bones are automatically linked.

Bone Parameters

The settings in the **IK Controller** section automatically assign inverse kinematics to bones. This makes it possible to transform bones with inverse kinematics without turning on the **Inverse Kinematics on/off** toggle ![icon].

Assign to Children	When checked, inverse kinematics is automatically assigned to all bones in the chain except the first one.
Assign To Root	When checked, inverse kinematics is automatically assigned to the first bone in the chain. When unchecked, inverse kinematics motion does not affect the first bone in the chain, and the first bone can only be animated by moving or rotating it directly.
Create End Effector	An *end effector* is an object in a linked chain that dictates the motion of parent objects when IK is used. When **Create End Effector** is checked, the last node created in the bones chain is an end effector, meaning it can be moved and rotated directly. When **Create End Effector** is unchecked, bones can only be animated by being bound to an animated follow object. Check **Create End Effector** if you plan to animate the last node directly, or leave it unchecked if you want the bones' motion to come solely from a follow object. For information on using follow objects, see the **IK** entry.

The options under **Auto Boning** automatically create a chain of linked bones to match an existing chain.

Pick Root	Click **Pick Root** and click on the highest parent object in the existing chain. The bones chain is created based on the picked chain.

| Auto Link | When checked, bones are linked to the corresponding objects in the existing chain. Note that bones are always linked to each other, but are only linked to the picked chain if this checkbox is checked. |

| Copy Joint Parameters | Copies the Rotational Joints and Sliding Joints parameters from the picked chain to the new bones. These parameters are set on the **Hierarchy** panel ⬚ under **IK**. For information on IK, see the **IK** entry in this section of the book. |

| Match Alignment | When checked, each bone's local coordinate system is aligned with the local coordinate system from the corresponding object in the linked chain. When unchecked, each bone's local coordinate system is aligned with the world coordinate system. |

Usage Notes

Bones are non-rendering objects. **Bones** objects are intended to be used with a third-party plug-in that attaches the bones to a mesh and deforms the mesh with the bones' animation. Character Studio and Bones Pro MAX are examples of two of these plug-ins.

It can be difficult to understand the usage of bones before you understand how to use inverse kinematics. For general information on inverse kinematics, see the **IK** entry in this section of the book, and see the **Inverse Kinematics on/off toggle** entry in the *Toolbar* section. Further bones parameters can be set on the **Motion** panel ⬚ under **Parameters**. See the 3D Studio MAX online documentation for more information.

Boolean

Calculates the subtraction, union or intersection of two intersecting 3D objects.

General Usage

Select the first object for the boolean operation. Under the **Create** panel , click **Geometry** . Choose **Compound Objects** from the pulldown menu. Click **Boolean**.

Under the **Operation** section, choose the type of operation. Click **Pick Operand B** and click on the second object for the operation. The boolean operation is calculated and the boolean result appears.

Pick Boolean

Pick Operand B	To pick the second operand for the boolean operation, click **Pick Operand B** and click on the object on the screen. You can also pick the object by choosing **Select by Name** from the Toolbar or by pressing the **\<H\>** key. If you want to perform a boolean operation on several objects, click the **Boolean** button in between clicking the **Pick Operand B** button.
Reference Move Copy Instance	Determines how operand B is used for the boolean operation. Choosing **Move** uses the object itself. Choosing **Reference**, **Copy** or **Instance** uses one of these clone types as the operand, leaving the original object in the scene. For information on these clone types, see the **Clone** entry in the *Menus* section.

Parameters

Operands	Lists the boolean operands. To change an operands creation parameters, go to the **Modify** panel and select the operand from this list. The operand can now be accessed on the modifier stack.
Operand A/B Name	Displays each operand name. You can enter a new name for each operand.
Extract Operand	Creates an **Instance** or **Copy** of the operand currently highlighted on the Operands list. This option is only available on the **Modify** panel.
Operation	Sets the operation type. **Union** combines the two operands. **Intersection** results in the intersection of the two operands. **Subtraction (A-B)** subtracts operand B from

operand A, and **Subtraction (B-A)** subtracts operand A from operand B. Operands must intersect in order for the boolean operation to succeed.

Display   An object resulting from a boolean operation is called the result. Choose to display the **Result** or **Operands**. **Display Hidden Operands** displays operands as wireframes in shaded viewports when **Result** is selected.

Update   Determines when the boolean result is updated on the screen. **Always**, the default, immediately updates the result whenever a change is made. **When Selected** updates the result when it is selected. **When Rendering** updates the result when the scene is rendered. **Manually** updates the result when the **Update** button is checked.

Optimize Result removes coplanar faces that can sometimes create visible edges in a rendering.

Sub-Object Level

On the **Modify** panel, a boolean object has one **Sub-Object** level, **Operands**. This sub-object level can be used to select and transform either operand.

Usage Notes

When you perform a boolean operation on two objects, they lose their original mapping coordinates, but faces retain their material IDs and smoothing groups. New faces created during a subtraction operation take on the same material IDs as the object being subtracted. In other words, if you create a hole in a box by subtracting a cylinder from the box, the faces on the hole will have the same material ID as the faces on the cylinder.

To edit an operand used in a Boolean operation, go to the **Modify** panel and highlight the operand on the **Operands** list under the Parameters rollout. You can then choose the operand's creation level from the bottom of the modifier stack. The operand's creation parameters will appear on the panel where they can be modified.

Boolean operations are computation-intensive, and sometimes don't work as expected. If your boolean operation doesn't seem to work, try changing the amount of detail on one or both operands, or moving one or both objects slightly. Always save your file before performing a boolean operation.

Panels

Box

Creates a box of any dimensions, including a cube.

General Usage

On the **Create** panel [icon], click **Geometry** [icon]. If necessary, select **Standard Primitives** from the pulldown menu. Click **Box**. Click and drag on the screen to set the length and width, then move the mouse and click again to set the height.

The orientation of the box's local X, Y and Z axes is determined by the viewport in which you begin drawing. The object's local X and Y axes are set on the viewport's drawing plane.

Creation Method

Cube   Creates a cube. When this option is selected, it affects only boxes created by dragging in a viewport. Only one dimension size is needed to create the cube. Keyboard entry overrides this setting.

Box   When this option is selected, the sides of the box can be different sizes.

Keyboard Entry

X, Y, Z   Sets the location of the box base center at the X, Y and Z location in the world coordinate system.

Length   See these listings under Parameters below.
Width
Height

Create   Creates the object using the parameters entered.

Parameters

Length   The dimensions of the box in units along its local X, Y
Width   and Z axes.
Height

Length Segs The number of divisions along the local X, Y and Z axes.
Width Segs
Height Segs

Generate   Generates box mapping coordinates on the box.
Mapping
Coords

BoxGizmo

Creates a box to contain an atmospheric effect.

General Usage

On the **Create** panel ⟨⟩, click **Helpers** ⟨⟩. Select **Atmospheric Apparatus** from the pulldown menu. Click **BoxGizmo**. Click and drag on the screen to set the length and width of the box, then move the mouse and click again to set the height.

Box Gizmo Parameters

Length
Width
Height   Sets the length, width and height of the box.

Seed   Atmospheric effects rely on a seed number for calculating their effects. For example, a combustion effect with identical parameters in two equal-sized **BoxGizmos** will produce exactly the same combustion effect if both have the same seed. If each has a different seed, the combustion effect will be different. If you have more than one **BoxGizmo** in your scene and you want to be sure they have different effects, make sure the **Seed** is set differently for each **BoxGizmo**.

New Seed   Randomly generates a new **Seed**.

Usage Notes

Atmospheric effects are created with the help of **BozGizmo**, **SphereGizmo** and **CylGizmo** objects. These gizmos are used to set the boundaries and the **Seed** for each of these effects. For information on atmospheric effects that use gizmos, see the **Environment** entry in the *Menus* section.

Camera Map

Generates mapping coordinates on an object based on a camera view.

General Usage

Create a scene with at least one object and one camera. Apply a material with a **Diffuse** map with **Texture** coordinates to the object. Set up the same map as the background image using **Screen** coordinates for the map. Position the object so it is viewed by the camera, and change one viewport to the camera view.

Select the object. On the **Modify** panel 🗿, click **More** if necessary, and choose **Camera Map** under the World Space Modifiers list, or **Camera Map** under the Object Space Modifiers list. Click **Pick Camera** and pick the camera.

When the camera view is rendered, the object's **Diffuse** map matches the background image. If the object's material has 100% **Self-Illumination**, the object is invisible against the background until the material is changed.

Camera Mapping

Pick Camera   To pick the camera for which to match mapping coordinates, click **Pick Camera** and pick the camera from the scene.

Usage Notes

When the **Camera Map** World Space Modifier is applied to an object, the object's mapping coordinates will shift when the object or camera is moved, so the mapping coordinates always match the camera view.

When the **Camera Map** Object Space Modifier is applied, the object's mapping coordinates match the camera view only on the frame that was current when the camera was picked with the **Camera Map** modifier. If the camera or object is animated on another frame, there will be a visible difference between the object's mapping coordinates and the background image.

You can use **Pick Camera** to pick another camera for a previously assigned **Camera Map** modifier, or to pick the same camera again after it has been moved.

The **Camera Map** modifier works properly on only one object at a time. If the **Camera Map** modifier is applied to a selection of objects, only the first object in the selection will be mapped correctly.

For information on setting up a background image, see the **Environment** entry in the *Menus* section.

Matches a camera's position to a scanned photograph.

General Usage

Scan in a photograph of a land area or structure for which you know at least five XYZ coordinates. In 3D Studio MAX, ensure that your current units setup matches the units for which the landscape or structure was measured. To change the current units setup, choose *Views/ Units Setup* from the menu to access the Units Setup dialog.

Set the rendering resolution to the same resolution as the bitmap. To do this, choose *Rendering/Render* from the menu. Under the **Output Size** section of the Render Scene dialog, change the **Width** and **Height** values to match the bitmap.

Set up the bitmap as a rendering background. For information on this procedure, see the **Environment** entry in the *Menus* section of this book.

On the **Create** panel, click **Helpers**. From the pulldown menu, choose **CameraMatch**. Click **CamPoint**. Create at least five **CamPoint** helpers and position each one so it corresponds with one of the real-life coordinates from the photograph.

Set up the bitmap as a background for one viewport. To do this, choose *Views/Background Image* from the menu. On the Viewport Background dialog, turn on the **Use Environment Background** checkbox.

On the **Utilities** panel, choose **Camera Match**. Select a helper from the list and click **Assign Position.** On the photograph in the view, visually locate the part of the photo where the selected helper should sit. Click on this location in the viewport.

Select the next helper from the list, click **Assign Position**, and click on the viewport at the appropriate location to assign the next point. Continue for each helper until a position has been assigned for each helper.

Click **Create Camera** to create a camera that matches the perspective of the helpers to the photograph. Change the viewport with the background to the camera view. The scene should now match the photograph bitmap.

CamPoint Info

The list at the top of the CamPoint Info rollout displays all **CamPoint** helpers currently in the scene.

Input Screen Coordinates Displays the screen coordinates most recently selected when the **Assign Position** button is on.

Use This Point When this checkbox is on, this helper point will be used to calculate the camera position. If the **Current Camera Error** is high, this checkbox can be turned off for one or more helper points to troubleshoot the problem.

Assign Position Click this button to assign a position on the bitmap background to the selected helper.

Camera Match

Create Camera Click to create a new camera. A free camera is created with its position, orientation and FOV based on the assigned positions and **CamPoint** helpers. If not enough helpers have been assigned positions, the camera will not be created.

Modify Camera To use an existing free camera, select the camera and click **Modify Camera**. The camera's position, orientation and FOV are changed based on the assigned positions and **CamPoint** helpers. This option is only available for free cameras.

Iterations A number related to the number of calculations used to estimate the camera location, orientation and FOV. The higher the **Iterations**, the more accurate the solution, and the longer it takes to calculate. If the helper and assigned position information is reasonably accurate, an **Iterations** value of 100 to 500 will yield a good result.

Freeze FOV If you know the FOV of the real-life camera used to take the original scanned photograph, create a free camera in MAX and set the FOV accordingly. Turn on the **Freeze FOV** checkbox, select the camera, and use **Modify Camera** to reorient the camera without changing the set FOV.

Current Camera Error	Displays the error factor in the calculation used to orient the camera. Values of 0 to 1.5 are common. If this value is above 10, one or more helpers are inaccurate.
Close	Closes the utility.

Usage Notes

You must use at least five **CamPoint** helpers in order to get results with the **Camera Match** utility. For ease of use, rename each helper to correspond to its real-world coordinate.

If possible, use XYZ positions spaced far apart in the photograph. Points spread far apart will yield the most accurate results. You can also maximize the camera view with the **Min/Max Toggle** before clicking **Assign Position** to accurately choose points on the bitmap background.

When you click **Create Camera** or **Modify Camera**, the camera is created or modified and the camera error is calculated. If the **Current Camera Error** is over 10, at least one **CamPoint** helper or selected screen position is far from its correct position. If you have more than five **CamPoint** helpers on the list, try turning off the **Use This Point** checkbox for one or more helpers to fix the problem. You must leave at least five helpers selected in order to position the camera.

CamPoint

Creates a helper object for the **Camera Match** utility.

General Usage

On the **Create** panel ✍, click **Helpers** 🖾. Select **CameraMatch** from the pulldown menu. Click **CamPoint**. Click on the screen to place the **CamPoint** helper.

To place the **CamPoint** exactly, use the parameters under the Keyboard Entry rollout. To place a **CamPoint** after creating it, click **Select and Move** ✛ and choose *Tools/Transform Type-In* from the menu. On the Move Transform Type-In dialog, enter the exact coordinates for the **CamPoint** under **Absolute:World**.

Keyboard Entry

X, Y, Z   Sets the location of the helper object at the X, Y and Z location in the world coordinate system.

Create   Creates the helper object at the specified location.

Parameters

Show Axis Tripod   When checked, the tripod axis is displayed.

Axis Length   Sets the size of each axis on the tripod axis that appears when the helper object is unselected.

Usage Notes

CamPoint helpers are used to place a camera with the **Camera Match** utility. For information on how **CamPoints** are used, see the **Camera Match** entry in this section of the book.

Cap Holes

Caps holes in an object.

General Usage

Select an object with missing faces. On the **Modify** panel ⬛, click **More** if necessary, and choose **Cap Holes**.

Camera Mapping

Smooth New Faces   Applies the same smoothing group to all new faces.

Smooth With Old Faces   Applies the same smoothing group to faces at the edge of the hole. When checked, new faces inherit the material ID of old edge faces. When unchecked, new faces are assigned their own material ID.

All New Edges Visible   When checked, all new face edges are visible in the viewport.

Usage Notes

For information on smoothing groups, see the **Smooth** entry in this section of the book.

For information on material IDs, see the **Multi-Sub/Object** entry in the *Mat Editor* section.

Capsule

Creates a cylinder with rounded caps.

General Usage

On the **Create** panel [icon], click **Geometry** [icon]. Choose **Extended Primitives** from the pulldown menu. Click **Capsule**. Click and drag to set the radius of the capsule. Move the cursor and click to set the capsule height.

Creation Method

Edge   Draws the capsule starting from the outer edge of the capsule hemisphere.

Center   Draws the capsule starting from the center of the capsule hemisphere.

Keyboard Entry

X, Y, Z   Sets the location of the capsule base center along the X, Y and Z axes of the coordinate system.

Radius   See these listings under Parameters below.
Height
Overall
Centers

Create   Creates the capsule using the parameters entered.

Parameters

Radius   Radius of the end hemispheres and the cylindrical center portion of the capsule.

Height   When the **Overall** option is on, this value refers to the height of the entire capsule. When the **Centers** option is on, this value refers the height of just the straight part of the capsule.

The **Height** value is always forced to at least twice the **Radius** value. When the **Overall** radio button is active, the height of the straight portion of the capsule is equal to the **Height** minus twice the **Radius**. To make a perfect sphere, turn on the **Overall** radio button and make the **Height** value exactly twice the **Radius** value.

Overall	When this option is on, the **Height** value refers to the height of the entire capsule, including the hemispheres. See **Height** above.
Centers	When this option is on, the **Height** value sets just the height of the straight part of the capsule. See **Height** above.
Sides	Sets the number of sides on the capsule.
Height Segs	Sets the number of segments along the length of the capsule.
Smooth	When the **Smooth** checkbox is on, the surface renders as a smooth, rounded object. When **Smooth** is off, the surface renders as faceted.
Slice On	Slices the object around its local Z axis according to the **Slice From** and **Slice To** angles. The slice function starts at the **Slice From** angle and moves counterclockwise to the **Slice To** angle, cutting out the portion of the object in this area. The sliced area is capped with two flat surfaces. Slice angles are measured from the object's local X axis, which is considered to be zero degrees.
Generate Mapping Coords	Generates mapping coordinates on the object. Cylindrical mapping coordinates are applied to the cylindrical midsection of the capsule while planar mapping coordinates are applied to the end hemispheres. Sliced areas receive planar mapping.

Panels

Casement

Creates a window with one or two hinged sashes.

General Usage

On the **Create** panel , click **Geometry** . From the pulldown menu, choose **Windows**. Click **Casement**. Click and drag to create the width of the window, move the cursor and click to set the depth or height, then move the cursor and click again to set the remaining dimension.

Creation Method

See the Creation Method rollout under the **Awning** entry.

Parameters

Height **Width** **Depth**	Sets the dimensions of the window.
Frame	Sets the dimensions of the window frame. **Horiz. Width** and **Vert. Width** affect the size of the glazing in addition to the frame size. **Thickness** also affects the thickness of the rails in the window sashes.
Glazing	Sets the **Thickness** of the glass.
Panel Width	Sets the width of the rails in the sashes. The window can have **One** or **Two** sashes. If the window has two sashes, the sashes open from the center when **Open** is above 0.
Open	Opens window panels by the specified percentage. 100 percent is equal to 90 degrees.
Flip Swing	Flips the window swing to the other side of the window frame. The effect of **Flip Swing** is visible only when **Open** is above 0.
Generate Mapping Coords	Generates planar mapping coordinates on the window.

Creates an extruded C-shaped object.

General Usage

On the **Create** panel ⬚, click **Geometry** ⬚. Select **Extended Primitives** from the pulldown menu. Click **C-Ext**.

Click and drag to set the overall size of the object. Move the cursor and click to set the height, then move the cursor and click again to set the width of all three sections of the object.

The top of the C-extension, as viewed from the drawing viewport, is called the *front*. The side is called the *side*, and the bottom is referred to as the *back*. When you create a C-extension by drawing on the screen, each section is drawn with the same width. If you want varying widths on your C-extension, you must type in the values under the Parameters rollout after creating the object. You can also give each section a different number of segments.

Creation Method

Corners   Draws the C-extension starting at a front corner.

Center   Draws the C-extension from the center of the object.

Keyboard Entry

X, Y, Z   Sets the location of the C-extension base corner at the X, Y and Z location in the world coordinate system.

Back Length   See these listings under Parameters below.
Side Length
Front Length
Back Width
Side Width
Front Width
Height

Create   Creates the C-extension using the parameters entered.

Parameters

Back Length   Length of the bottom of the C-extension as viewed from the current viewport.

Side Length   Length of the side of the C-extension as viewed from the current viewport.

Front Length	Length of the top of the C-extension as viewed from the current viewport.
Back Width	Width of the C-extension back.
Side Width	Width of the C-extension side.
Front Width	Width of the C-extension front.
Height	Height of the C-extension. Each section of the C-extension shares the same height.
Back Segs	Number of segments along the bottom of the C-extension.
Side Segs	Number of segments along the side of the C-extension.
Front Segs	Number of segments along the top of the C-extension.
Width Segs	Number of segments along the width of the C-extension.
Height Segs	Number of segments along the height of the C-extension.
Generate Mapping Coords	Generates mapping coordinates on the C-extension.

ChamferBox

Creates a box with beveled or rounded edges.

General Usage

On the **Create** panel , click **Geometry** . Select **Extended Primitives** from the pulldown menu. Click **ChamferBox**. Click and drag to set the length and width of the box, then move the mouse and click again to set the height. Move the mouse and click again to set the size of the chamfer edges.

The orientation of the box's local X, Y and Z axes is determined by the viewport in which you begin drawing. The object's local X and Y axes are set on the viewport's drawing plane.

Creation Method

Cube	Creates a cube. When this option is selected, it affects only boxes created by dragging in a viewport. Only one dimension size is needed to create the cube. Keyboard entry overrides this setting.
Box	When this option is selected, the sides of the box can be different sizes.

Keyboard Entry

X, Y, Z	Sets the location of the chamfer box base center at the X, Y and Z location in the world coordinate system.
Length **Width** **Height** **Fillet**	See these listings under Parameters below.
Create	Creates the object using the parameters entered.

Parameters

Length **Width** **Height**	Sets the size of the dimensions of the box along its local X, Y and Z axes.
Fillet	Sets the amount of each box side to be replaced with a fillet. For example, a **Fillet** value of 2 replaces 2 units of each box's side with a fillet. A value equal to one-half of a box size fillets the entire side. Values beyond one-half of the box side have no further effect on the object.

Length Segs **Width Segs** **Height Segs**	Sets the number of divisions along the local X, Y and Z axes.
Fillet Segs	Sets the number of segments in the fillet. Leaving this value at 1 creates a sharp, beveled edge on the object. Higher values round out the fillet.
Smooth	When the **Smooth** checkbox is on, the surface renders as a smooth, rounded object. When **Smooth** is off, the surface of the box renders as faceted.
Generate Mapping Coords	Generates box mapping coordinates on the box.

ChamferCyl

Creates a cylinder with beveled or rounded edges.

General Usage

On the **Create** panel ![icon], click **Geometry** ![icon]. Select **Extended Primitives** from the pulldown menu. Click **ChamferCyl**. Click and drag to set the radius of the cylinder, then move the mouse and click again to set the height. Move the mouse and click again to set the size of the chamfer edges.

Creation Method

Edge   Draws the cylinder starting at the outer edge of the cylinder base.

Center   Draws the cylinder starting at the center of the cylinder base.

Keyboard Entry

X, Y, Z   Sets the location of the chamfer cylinder base center at the X, Y and Z location in the world coordinate system.

Radius
Height   See these listings under Parameters below.
Fillet

Create   Creates the chamfer cylinder using the parameters entered.

Parameters

Radius   Sets the radius of the cylinder.

Height   Sets the height of the cylinder.

Fillet   Sets the amount of each end to be replaced with a fillet. For example, a **Fillet** value of 2 replaces 2 units of each end with a fillet. The **Fillet** value cannot exceed one-half of the **Radius** or one-half of the **Height**, whichever is smaller.

Height Segs   Sets the number of segments along the height of the cylinder.

Panels *(side tab)*

Fillet Segs	Sets the number of segments in the fillet. Leaving this value at 1 creates a sharp, beveled edge on the object. Higher values round out the fillet.
Sides	Sets the number of sides on the cylinder.
Cap Segs	Sets the number of circular segments on the cylinder cap.
Smooth	When the **Smooth** checkbox is on, the surface renders as a smooth, rounded object. When **Smooth** is off, the surface renders as faceted.
Slice On	Slices the cylinder around its local Z axis according to the **Slice From** and **Slice To** angles. The slice function starts at the **Slice From** angle and moves counterclockwise to the **Slice To** angle, cutting out the portion of the cylinder in this area. The sliced area is capped with two flat surfaces. Slice angles are measured from the object's local X axis, which is considered to be zero degrees.
Generate Mapping Coords	Generates cylindrical mapping coordinates on the cylindrical portion of the object, and planar mapping coordinates on caps and fillets. Sliced areas receive planar mapping.

Circle

Creates a circle shape.

General Usage

On the **Create** panel 🖐, click **Shapes** 🔷. Click **Circle**. Click and drag on the screen to create the circle. A circle has four Bezier vertices.

General

See the General rollout under the **Arc** entry.

Creation Method

Edge   When **Edge** is selected, the circle is created starting from the circle edge when you click and drag on the screen. To keep a vertex at the first location clicked, drag up, down, left or right rather than diagonally.

Center   When **Center** is selected, the circle is created from the center outward when you click and drag on the screen.

Keyboard Entry

X, Y, Z   Specifies the location of the center of the shape created when **Create** is clicked. The center is placed at the X, Y and Z location in the world coordinate system.

Radius   Sets the radius of the circle.

Create   Creates the shape with the parameters entered.

Parameters

Radius   Specifies the radius of the circle.

Collapse

Collapses one or more objects to an editable mesh, or performs a boolean operation.

General Usage

Select one or more objects. On the **Utilities** panel ![T icon], choose **Collapse**. Set parameters as desired, and click **Collapse Selected**.

Collapse

Selected Object	Displays the name of the selected object. If more than one object is selected, the name Multiple Selected is displayed.
Collapse Selected	Performs the collapse operation according to the settings below.
Modifier Stack Result	When this option is selected, objects are collapsed in the same manner as the **Collapse All** button on the Edit Modifier Stack dialog. This dialog can be accessed by clicking **Edit Stack** ![icon] on the **Modify** panel ![icon].
	In most cases, the object will be collapsed to an Editable Mesh. If the object has an **Edit Patch** modifier, it will be collapsed to a patch object. If the object is a spline with an **Edit Spline** modifier, it will be collapsed to an Editable Spline.
	When this option is chosen, all other options are disabled. Each object is collapsed separately and remains separate.
Mesh	When this option is selected, each selected object is collapsed to an Editable Mesh regardless of the original object type or the modifiers on the stack. When **Multiple Objects** is selected, each object is collapsed to an individual Editable Mesh. When the **Single Object** option is selected, all selected objects are collapsed to one Editable Mesh containing all the objects.
Multiple Objects	Collapses each object individually to an Editable Mesh.
Single Object	Collapses all selected objects to one Editable Mesh containing all objects.

Panels

Boolean	Performs a boolean operation with all selected objects. Click **Union**, **Intersection** or **Subtraction** to choose the type of boolean desired. Note that some or all selected objects must intersect in order to perform a boolean operation.
	If the boolean operation fails for any object, the object is skipped and the boolean operation moves on to the next selected object. The boolean result is an Editable Mesh and cannot be used as an animated boolean.
	For more information on boolean operations, see the **Boolean** entry in the *Panels* section.
Union	Combines all objects and removes intersecting geometry.
Intersection	Removes all geometry except intersecting parts.
Subtraction	Starts with the first selected object and subtracts each subsequently selected object from the first one.
Close	Closes the **Collapse** utility.

Usage Notes

The **Collapse** utility is useful for making several objects into one mesh quickly. The **Mesh** option is handy for turning an object into an Editable Mesh regardless of its origins or modifiers.

Color Clipboard

Sets up and displays a floating color clipboard for use with the Material Editor.

General Usage

Access the Material Editor. On the **Utilities** panel ⊤, choose **Color Clipboard**. Drag a color from the Material Editor to one of the large color swatches under the Color Clipboard rollout. Change to another material or map, and drag the color from the clipboard to a color swatch on the Material Editor. Choose to Swap or Copy the color, or to cancel the operation.

To use the floating color clipboard, click **New Floater** to open the floater. Change the color swatches by clicking on each one and changing the color on the Color Selector, or by dragging colors from the Material Editor and other areas of 3D Studio MAX.

Color Clipboard

New Floater   Displays a new color clipboard floater. On the floater, click **Open** to open a previously saved color clipboard (*.ccb*) file. You can also click **Save** or **Save As** to save the current colors in a color clipboard (*.ccb*) file. See Usage Notes below for information on *.ccb* files.

Close   Closes the utility.

Usage Notes

Colors can be dragged to and from the floating color clipboard from any area of MAX that utilizes color swatches. Color swatches are used for the background color, fog color, light colors, and many other areas of MAX.

A *.ccb* file is an ASCII file containing 12 lines of three numbers each. The numbers represent the R, G and B values of each color respectively. You can create or change a *.ccb* file with any word processing program or text editor.

Creates a compass rose.

General Usage

On the **Create** panel ⚞⚟, click **Helpers** ⚞⚟. If necessary, choose **Standard** from the pulldown menu. Click **Compass**. Click and drag on the screen to create the compass rose.

Parameters

Show
Compass
Rose   When checked, the compass rose is displayed.

Radius   Sets the radius of the compass rose.

Usage Notes

A compass can be placed and rotated to indicate direction in a scene. A compass created with the **Compass** option is for reference only, and is not used with any other feature of 3D Studio MAX.

A compass is automatically created when a **Sunlight** object is created, and is used to indicate the direction of sunlight. A compass created here cannot be used with a **Sunlight** system. For information on the **Sunlight** system, see the **Sunlight** entry.

Cone

Creates a cone.

General Usage

On the **Create** panel ![icon], click **Geometry** ![icon]. If necessary, select **Standard Primitives** from the pulldown menu. Click **Cone**. Click and drag to set the cone bottom radius, then move the cursor and click to set the height. Move the cursor and click again to set the cone top radius.

Creation Method

Edge   Draws the cone starting at the outer edge of the cone base.

Center   Draws the cone starting at the center of the cone base.

Keyboard Entry

X, Y, Z   Sets the location of the cone base center at the X, Y and Z location in the world coordinate system.

Radius 1   See these listings under Parameters below.
Radius 2
Height

Create   Creates the object using the parameters entered.

Parameters

Radius 1   The radius of the cone base and cone top.
Radius 2

Height   The height of the cone.

Height Segs The number of segments along the height of the cone.

Cap Segs   The number of top and bottom circular segments.

Sides   Sets the number of sides on the cone.

Smooth   See these parameters under the **Capsule** entry.
Slice On
Slice From
Slice To

Generate   Generates cylindrical mapping coordinates on the cone.
Mapping   Sliced areas receive planar mapping.
Coords

Conforms one object's vertices to another object.

General Usage

Select the object that will be conformed to another object. On the **Create** panel ![icon], click **Geometry** ![icon]. Choose **Compound Objects** from the pulldown menu. Click **Conform**. Click **Pick Wrap-To Object** and click on the object to which the first object is to be wrapped.

Pick Wrap-To Object

Pick Wrap-To Object	Click **Pick Wrap-To Object** and pick the object to which to conform the currently selected object.
Reference Copy Move Instance	Uses a **Reference**, **Copy** or **Instance** of the picked object. **Move** uses the object itself. For information on **Reference**, **Copy** and **Instance**, see the **Clone** entry in the *Menus* section.

Parameters

Objects	Lists the two objects that comprise the compound object. To access one of the objects' modifier stack, highlight the object, then access the object on the modifier stack.
Wrapper Name	Displays the wrapper object name. You can change the name here if you like.
Wrap-To Object Name	Displays the wrap-to object name. Change the name if desired.

The next six options determine how the wrapper is projected onto the wrap-to object. To conform the object, the wrapper's vertices are projected in the specified direction. If a projected vertex doesn't hit the wrap-to object, it is not conformed to the object.

Use Active Viewport	Projects wrapper vertices in the direction pointing into the active viewport. If you change viewports, click **Recalculate Projection** to recalculate the wrap operation.
Use Any Object's Z Axis	Projects wrapper vertices in the direction of a selected object's local Z axis. To select the object, click **Pick Z-Axis Object**.
Along Vertex Normals	Projects wrapper vertices in the direction opposite their normals.

Towards Wrapper/ Wrap-To Center/ Pivot	Projects vertices toward the center of the wrapper or wrap-to object's center or pivot point. **Conform** uses the position of the object's pivot point before the **Conform** operation was begun.
Default Projection Distance	The distance a vertex moves if it is not projected onto the wrap-to object.
Standoff Distance	Each vertex projected onto the wrap-to object sits above the object by the number of units specified by **Standoff Distance**. When **Standoff Distance** is 0, all wrapped vertices sit directly on the wrap-to object.
Use Selected Vertices	When checked, only selected vertices on the wrapper are projected to the wrap-to object.
Update	An object resulting from a compound operation is called the *result*. These options determine when the result is updated on the screen. **Always**, the default, immediately updates the result whenever a change is made. **When Rendering** updates the result when the scene is rendered. **Manually** updates the result when the **Update** button is checked.
Hide Wrap-To Object	When checked, the wrap-to object is hidden.
Display	Choose to display the **Result** or **Operands**.

Sub-Object Level

On the **Modify** panel, the compound object has one **Sub-Object** level, **Operands**. This sub-object level can be used to select and transform either the wrapper or the wrap-to object.

Usage Notes

Conform (Compound Object) projects the vertices of a wrapper object in the specified direction and wraps them around another object.

Conform (Compound Object) does not change the number of vertices in the wrapper object. For this reason, it can be used to create morph targets using two copies of an object. One copy of the wrapper can be left in its original shape, while the other copy can be conformed to the wrap-to object by using the **Along Vertex Normals** option or any of the **Towards** options. One copy can then be morphed to the other. For information on how to perform morphing, see the **Morph** entry.

A space warp that flattens one object against another.

General Usage

Create two objects, one to be flattened and another against which the first object will be flattened. On the **Create** panel ⬡, click **Space Warps** ⬭. Select **Geometric/Deformable** from the pulldown menu. Click **Conform**. Click and drag on the screen to create the conform space warp object.

Note the direction of the large arrow on the conform space warp object. Rotate the space warp object until the arrow points in the direction in which the first object will be moved in order to flatten against the second object. Click **Pick Object** and pick the object against which the first object will be flattened.

Select the first object and click **Bind to Space Warp** ⬡ on the Toolbar. Click and drag from the object to the conform space warp to bind it to the space warp. The first object is flattened against the second object.

Parameters

Pick Object   Selects the object against which the bound object will be flattened. Click **Pick Object** and pick the object from the scene. Only one object can be selected.

Default Projection Distance   The distance the bound object's vertices will move toward the flattening object if it they cannot be flattened. To conformthe bound object, the **Conform** space warp projects the vertices of the object in the direction of the arrow until they strike the flattening object that was selected with **Pick Object**. If a vertex does not strike the flattening object, it is not flattened but is moved by the **Default Projection Distance** in the direction of the space warp arrow.

Standoff Distance   The bound object does not have to lie flat against the flattening object. **Standoff Distance** places the bound object a specified distance away from the flattening object, in the direction opposite the arrow on the conform space object. **Standoff Distance** is expressed in units.

Panels

Use Selected Vertices	Conforms selected vertices only. To use this option, you must apply a **Mesh Select** modifier to the bound object and select the desired vertices. The selection of vertices can take place either before or after the object is bound to the space warp.
Icon Size	Sets the size of the icon in square units. The size of the icon has no relationship to the conform effect.

Usage Notes

The **Conform** space warp object can be animated to change the flattening direction. This is useful for flattening an object moving around a sphere or other round surface.

Conform is ideal for animating amorphous blobs over curved or bumpy surfaces. For example, you could animate a sphere passing over a terrain model, then use **Conform** to flatten the sphere against the terrain. The sphere will then move over the terrain in the manner of the famous man-eating blob in film *The Thing*.

When the bound object is moved in such a way that some vertices will no longer strike the flattening object, the vertices no longer conform to the flattening object. Instead, these vertices are moved in the direction of the space warp arrow by the **Default Projection Distance**. With this in mind, you can animate an object so it starts out unflattened, then flattens as it passes over the flattening object.

Once a bound object is flattened, it cannot be moved away from the flattening surface with **Select and Move** ⊕. To move a flattened object toward or away from the flattening object, change the **Standoff Distance**.

Connect

Connects two objects that have missing faces by building faces between them.

General Usage

Create two objects. Delete one or more faces from each object. Rotate one or both objects so the areas with missing faces are facing each other.

Select one object. On the **Create** panel ![icon], click **Geometry** ![icon]. Choose **Compound Objects** from the pulldown menu. Click **Connect**. Click **Pick Operand** and click on the second object. The two objects are connected by a bridge of faces.

Pick Operand

Pick Operand	To pick an operand, click **Pick Operand** then pick the object from the screen. You can also pick the object by clicking **Select by Name** ![icon] or by pressing the <H> key.
Reference	Use a reference of the picked operand to create the surface.
Move	Use the picked operand itself to create the surface.
Copy	Use a copy of the picked operand to create the surface.
Instance	Use an instance of the picked operand to create the surface.

Parameters

Operands	The first operand and all picked operands are listed here.
Delete Operand	To remove an operand from the **Connect** effect, highlight the operand on the **Operands** list and click **Delete Operand**.
Segments	Sets the number of segments (cross sections) on the bridge between operands.
Tension	Sets the curvature on the bridge. At 0, there is no curvature. Higher values attempt to make a smooth transition from the normals at each end of the bridge, but can cause pinching effects. Lower values cause curvature in the opposite direction. This value takes effect only when **Segments** is higher than 0.

Bridge   Smooths the bridge between two operands.

Ends   Smooths the ends, where the bridge meets the oper-
                    and. If this checkbox is unchecked, a new material ID
                    number is assigned to the bridge. When it is checked,
                    the material ID for the bridge is taken from one of the
                    operands. For information on material IDs, see the
                    Multi/Sub-Object entry in the *Mat Editor* section.

Usage Notes

To connect several objects to each other, you can click **Pick Operand**
and pick an object, click **Pick Operand** and pick an object, over and
over as many times as necessary. This creates one surface connecting
all the objects.

You can also connect a mesh to a shape. In order to do this, the shape'
must be oriented so its negative local Z axis points in the direction of
the mesh.

Creates a NURBS CV curve.

General Usage

On the **Create** panel , click **Shapes** . Select **NURBS Curves** from the pulldown menu. Click **CV Curve**. Click on the screen to set the first control point, then move the cursor and click as many times as desired to create more control points. Right-click to end creation of the curve.

Keyboard Entry

On the Keyboard Entry rollout, a CV curve is created one control point at a time with the **Add Point** button. The **Close** and **Finish** buttons end creation of the curve.

The Keyboard Entry rollout is available only on the **Create** panel.

X, Y, Z	Specifies the location of the control point created when **Add Point** is clicked. The control point is placed along the X, Y and Z axes in the world coordinate system.
Weight	Each control vertex has a weight. At its default position, the **Weight** is 1.0. Increasing the **Weight** pulls the curve toward the control vertex, while decreasing the **Weight** pushes the curve away from the control vertex.
Add Point	Adds a control point at the location specified by X, Y and Z. **Add Point** defines control points used to control the CV curve, not actual points on the curve.
Close	Closes the curve by creating a segment that connects the last point with the first point.
Finish	Completes the curve without closing it.

Curve Approximation

CV curves are approximated with steps between control points, just as standard splines are.

For information on the parameters on this rollout, see the General rollout under the **Arc** entry.

Modify Panel

On the **Modify** panel, a NURBS curve can be used as a parent NURBS object for more NURBS curves. A NURBS curve cannot be used as a parent NURBS object to create NURBS surfaces.

For a description of the parameters that appear on the **Modify** panel, see the **Point Curve** entry.

Usage Notes

A **CV Curve** is a spline surrounded by control vertices (CVs). Control vertices may or may not lie on the object's surface. The control vertices are used to change the shape of the curve. NURBS curves are convenient for making smooth curves quickly.

NURBS curves created on the **Create** panel cannot be used to make NURBS surfaces. These curves can only be used as you would any ordinary MAX shape. For example, you might apply a **Lathe** or **Extrude** modifier to a NURBS curve to make a 3D object.

Panels

Creates a NURBS CV surface.

General Usage

On the **Create** panel ⬚, click **Geometry** ⬚. Select **NURBS Surfaces** from the pulldown menu. Click **CV Surf**. Click and drag on the screen to create the CV surface.

Keyboard Entry

X, Y, Z	Sets the location of the CV surface center at the X, Y and Z location in the world coordinate system.
Length **Width** **Length CVs** **Width CVs**	See these listings under Create Parameters below.
Create	Creates the CV surface using the parameters entered.

Create Parameters

Length **Width**	Sets the dimensions of the CV surface along its local X and Y axes.
Length CVs **Width CVs**	Sets the number of CV points along the surface's local X and Y axes.
Generate Mapping Coords	Generates planar mapping coordinates on the CV surface.

Modify Panel

More NURBS surfaces can be created and joined to the **CV Surf** to create a more complex NURBS surface. The options for creating these surfaces appear under the **Modify** panel when a **CV Surf** is selected.

For a description of the parameters that appear on the **Modify** panel, see the **Point Surf** entry.

Usage Notes

A **CV Surf** is a flat surface surrounded by control vertices (CVs). Control vertices may or may not lie on the object's surface. The control vertices are part of a lattice, or framework, around the CV surface. The surface can be deformed by moving the control vertices. For information on working with NURBS surfaces, see the **Point Surf** entry.

CylGizmo

Creates a cylinder to contain an atmospheric effect.

General Usage

On the **Create** panel ⟨icon⟩, click **Helpers** ⟨icon⟩. Select **Atmospheric Apparatus** from the pulldown menu. Click **CylGizmo**. Click and drag on the screen to set the radius of the cylinder, then move the cursor and click again to set the height.

Cylinder Gizmo Parameters

Radius Sets the radius and height of the **CylGizmo**.
Height

Seed   Atmospheric effects rely on a seed number for calculating their effects. For example, a combustion effect with identical parameters in two equal-sized **CylGizmos** will produce different versions of combustion if each has a different seed. Two **CylGizmos** with the same seed will produce exactly the same combustion effect. If you have more than one **CylGizmo** in your scene and you want to be sure they have different effects, make sure the **Seed** is set differently for each **CylGizmo**.

New Seed Randomly generates a new **Seed**.

Usage Notes

Atmospheric effects are created with the help of **BoxGizmo**, **SphereGizmo** and **CylGizmo** objects. These gizmos are used to set the boundaries and the **Seed** for each of these effects. For information on atmospheric effects that use gizmos, see the **Environment** entry in the *Menus* section.

Creates a cylinder.

General Usage

On the **Create** panel ⬚, click **Geometry** ⬚. If necessary, select **Standard Primitives** from the pulldown menu. Click **Cylinder**. Click and drag to set the radius of the cylinder, then move the cursor and click to set the height.

Creation Method

Edge	Draws the cylinder starting at the outer edge of the cylinder base.
Center	Draws the cylinder starting at the center of the cylinder base.

Keyboard Entry

X, Y, Z	Sets the location of the cylinder base center at the X, Y and Z location in the world coordinate system.
Radius **Height**	See these listings under Parameters below.
Create	Creates the object using the parameters entered.

Parameters

Radius	Sets the radius of the cylinder.
Height	Sets the height of the cylinder.
Height Segments	Sets the number of segments along the height of the cylinder.
Cap Segments	Sets the number of circular segments on the cylinder cap.
Sides	Sets the number of sides on the cylinder.
Smooth **Slice On** **Slice From** **Slice To**	See these parameters under the **Capsule** entry.
Generate Mapping Coords	Generates cylindrical mapping coordinates on the cylinder. Sliced areas receive planar mapping.

Deflector

A space warp that deflects particles with a flat plane.

General Usage

Create a particle system object. On the **Create** panel ![icon], click **Space Warps** ![icon]. Select **Particles & Dynamics** from the pulldown menu. Click **Deflector**. Click and drag on the screen to create the space warp object. Move the space warp so it sits in the path of oncoming particles.

Select the particle system object and click **Bind to Space Warp** ![icon] on the Toolbar. Click and drag from the particle system object to the space warp to bind it to the space warp.

Parameters

Bounce
: Sets the speed at which particles bounce off the deflector. When **Bounce** is 0, particles don't bounce off the deflector. When **Bounce** is 1.0, particles bounce at the same speed they have when they hit the deflector. Values between 0 and 1.0 make particles move at a fraction of the hit speed after bouncing. Values over 1.0 make particles move faster than the hit speed after bouncing.

Width
: Sets the width of the space warp object. Since particles bounce only when they hit the space warp object, this parameter can affect the bouncing of particles as well.

Length
: Sets the length of the space warp object. Since particles bounce only when they hit the space warp object, this parameter can affect the bouncing of particles as well.

DeleteMesh

Deletes selected faces, vertices or edges from a 3D object.

General Usage

Select an object. On the **Modify** panel ![icon], click **MeshSelect** or **Edit Mesh**. Select faces, vertices or edges. On the **Modify** panel, click **More** if necessary, and choose **DeleteMesh**. The selected faces, vertices or edges are deleted.

The **DeleteMesh** modifier has no parameters.

Usage Notes

In order to delete faces, edges or vertices, they must first be selected with the **Mesh Select** or **Edit Mesh** modifiers. For information on using the **Mesh Select** and **Edit Mesh** modifiers, see these entries in this section.

When faces, edges or vertices are selected with one of these modifiers and you leave the **Sub-Object** button when choosing DeleteMesh, the selected faces, edges or vertices are deleted. If you don't use **Mesh Select** or **Edit Mesh** before **DeleteMesh** or you turn off the **Sub-Object** level before choosing **DeleteMesh**, the entire object is deleted from the screen. The object still exists in the scene, however, and can be recovered by removing the **DeleteMesh** modifier with the **Remove modifier from the stack** button ![icon].

DeleteSpline

A modifier that deletes selected splines, segments or vertices.

General Usage

On the **Create** panel ⟨icon⟩, click **Shapes** ⟨icon⟩. Uncheck **Start New Shape**. Make one or more shapes as part of the same shape.

On the **Modify** panel ⟨icon⟩, click **More** to access the list of modifiers. Choose **SplineSelect**. Choose **Spline** as the **Sub-Object** level. Select one or more splines.

On the **Modify** panel, click **More** to access the list of modifiers. Choose **DeleteSpline**. The selected splines are deleted.

Usage Notes

The **DeleteSpline** modifier is available only when a shape is selected. **DeleteSpline** has no parameters.

DeleteSpline can also delete a vertex or segment selection passed up the stack.

If you do not select splines before choosing **DeleteSpline**, the entire shape is deleted.

Splines can also be deleted at the **Spline Sub-Object** level of the **Edit Spline** modifier or an **Editable Spline**'s **Modify** panel. The **DeleteSpline** modifier, however, uses less memory and takes less time than these other methods.

Displace

A modifier or space warp that deforms an object according to a map's grayscale values.

General Usage

In the Material Editor, set up a map to use as a displacement map. Create an object or particle system to displace. Objects should have several segments in order to deform smoothly.

To apply **Displace** as a modifier, select the object and go to the **Modify** panel ![icon]. Click **More** if necessary, and choose **Displace**.

To use **Displace** as a space warp, go to the **Create** panel ![icon] and click on **Space Warps** ![icon]. Select **Geometric/Deformable** from the pulldown menu. Click **Displace**. Click and drag on the screen to create the displace space warp object. Select the object to be displaced, and use **Bind to Space Warp** ![icon] to bind it to the space warp. Position the space warp over or near the object to align mapping as desired.

Under the **Map** section at the bottom of the Parameters rollout, select the type of mapping to apply to the space warp or object.

To assign a bitmap for displacement, click the button labeled **None** under **Bitmap**. A file selector dialog appears where you can choose the displacement bitmap. To assign another kind of map for displacement, click on the button labeled **None** under **Map** to access the Material/Map Browser. Under **Browse From**, turn on **Mtl Editor** to view the maps currently in the sample slots, and select the displacement map from the list. You can also drag the map directly from the Material Editor sample slot to the **None** button under **Map**.

Parameters

Strength   Sets the intensity of the displacement. The whitest areas of the map (pixels with RGB color 255,255,255) will cause displacement by this number of units. Each remaining pixel will displace the geometry by a percentage of this number depending on how bright its color is. Strength can be a negative number to displace the object away from the space warp in the opposite direction.

Decay   Sets the amount by which the displacement strength will diminish over distance. When **Decay** is 0, strength is the same over the entirety of the bound object. When **Decay** is higher than 0, the strength decays as the distance between the bound object and space warp increases. Try fractional values such as 0.005 or 0.2 for best results.

Luminance Center
Shifts the entire bound object based on the **Center** value.

Center
A value from 0 to 1 that sets the center for the displacement when **Luminance Center** is checked. When **Center** is 0.5, parts of the map with a luminance of 128 are not displaced, while pixels with a luminance above 128 are displaced higher, and those with a luminance below 128 are displaced lower. Change this value to shift the displacement.

Bitmap
Click the button under **Bitmap** to assign a bitmap as the displacement map. A file selector appears where you can select a bitmap. To remove the assignment, click **Remove Bitmap**.

Map
Click the button under **Map** to assign any map as the displacement map. The Material/Map Browser appears, where you can select a map. You can also drag a map directly from the Material Editor to this button to assign the map. To remove the assignment, click **Remove Map**.

Blur
A value from 0 to 10 that softens the effects of sharp transitions in the displacement map. Increase **Blur** to 1.0 or 2.0 to keep displacement angles from becoming too harsh.

Planar
Cylindrical
Spherical
Shrink Wrap
Sets the type of mapping coordinates to use for the displacement. For information on mapping coordinates, see the **Mapping Coordinates** entry in the *Mat Editor* section.

Length
Width
Height
Sets the dimensions of the space warp, which sets the size of the mapping coordinates gizmo. The **Height** value has no effect on the size of the gizmo if **Planar** is chosen as the mapping coordinates type.

U Tile
V Tile
W Tile
Tiles the displacement effect in the U, V and/or W directions.

Modifier Parameters

The **Displace** modifier has additional parameters.

Sub-Object	The **Displace** modifier has one sub-object, **Gizmo**. The gizmo takes the shape of the current displacement mapping coordinates. The gizmo can be moved, rotated or scaled to change the displacement mapping.
Use Existing Mapping	When checked, the object uses existing mapping coordinates for displacement.
Apply Mapping	When checked, the displacement mapping coordinates set on this panel are used for material assignment.
X, Y, Z	Aligns the **Displace** gizmo so it is perpendicular to the object's local **X**, **Y** or **Z** axis.
Fit	Fits the gizmo to the object.
Bitmap Fit	Fits the gizmo to a bitmap's aspect ratio. Click **Bitmap Fit** and select the bitmap from the file selector.
View Align	Aligns the gizmo with the current viewport.
Reset	Resets the gizmo to its default position and size.
Center	Centers the gizmo on the object.
Normal Align	Aligns the gizmo with a face normal. Click **Normal Align**, then click and drag on the object to select a face normal.
Region Fit	Fits the gizmo to a custom area. Click **Region Fit** and click and drag on the screen to set the gizmo dimensions.
Acquire	Acquires a displacement gizmo from another object. Click **Acquire** and select another object. The gizmo from the selected object's latest displacement modifier is acquired. If an object does not have a displacement modifier with a gizmo, it cannot be selected. You are prompted with two options, **Acquire Relative** and **Acquire Absolute**. **Acquire Relative** positions and orients the gizmo with the object's local axes in the same way it was aligned on the selected object. **Acquire Absolute** does not change the gizmo's location or orientation.

Panels

Usage Notes

The results of the **Displace** space warp or modifier can be likened to those of a **Bump** map in the Material Editor. Both give an object bumps, but with **Displace**, the bitmap changes the actual geometry of the object. A **Bump** map does not change the geometry, and its effects appear only in renderings.

A displacement map raises the geometry at light areas and depresses or pushes in the geometry at dark areas of the map. The map can be a bitmap created with a paint program or other means, or can be any other kind of map supported by the Material Editor.

Use the **Displace** space warp when displacing particles, a large number of objects, or an object relative to its position in world space. Use the **Displace** modifier when you want to displace one or two objects only.

The **Displace** space warp relies on mapping coordinates to define the size, shape and direction of the warp. The size of the displacement mapping coordinates is set by the **Displace** icon itself. The size of the mapping coordinates can be changed by changing the **Length**, **Width** and **Height** values on the Parameters rollout. For information on mapping coordinates, see the **Mapping Coordinates** entry in the *Mat Editor* section.

Controls the display of objects in the scene.

General Usage

Go to the **Display** panel 🖼️. Choose options as desired.

Display Color

Wireframe Shaded	Sets the display color for wireframe and shaded objects. **Object Color** uses the object color, while **Material Color** uses the **Diffuse** color of any assigned materials. If no materials are assigned to an object, the object color is used.

Hide by Category

The checkboxes on this rollout can hide all objects of a specified type. Objects hidden with this rollout can only be unhidden by unchecking the appropriate checkbox on this rollout.

All	Checks all options.
None	Unchecks all options.
Inverts	Inverts the checked and unchecked options.

Hide

Hide Selected / Unselected	Hides all selected or unselected objects.
Hide by Name	Displays a list where you can choose the objects to hide.
Hide by Hit	To hide objects by picking them from the screen, click **Hide by Hit** and pick objects to hide. When finished picking objects, click **Hide by Hit** again to turn it off.
Unhide All	Unhides all hidden objects.
Unhide by Name	Displays a list where you can choose objects to unhide.

Freeze

Freezing an object causes it to remain onscreen, but prevents it from being selected or altered in any way. A frozen object turns dark gray on the screen regardless of its object or material color.

Freeze / **Selected** **Unselected**	Freezes all selected or unselected objects.
Freeze **by Name**	Displays a list where you can choose the objects to freeze.
Freeze by Hit	To freeze objects by picking them from the screen, click **Freeze by Hit** and pick objects to freeze. When finished picking objects, click **Freeze by Hit** again to turn it off.
Unfreeze All	Unfreezes all frozen objects.
Unfreeze **by Name**	Displays a list where you can choose objects to unfreeze.
Unfreeze **by Hit**	To unfreeze objects by picking them from the screen, click **Unfreeze by Hit** and pick objects to unfreeze. When finished picking objects, click **Unfreeze by Hit** again to turn it off.

Display Properties

See the **Display Floater** entry in the *Menus* section for a description of the first five parameters on this rollout.

Vertex Colors	Displays the object's vertex colors. When **Shaded** is turned on, vertex colors appear in shaded viewports. For information on vertex colors, see the **Assign Vertex Colors** entry.

Link Display

Display Links	Displays links between selected objects. Links displayed with this option look similar to **Bones** system objects. Displayed links extend between objects' pivot points.
Link **Replaces** **Object**	Replaces linked objects with a link display. Checking this checkbox automatically turns on the **Display Links** checkbox.

Usage Notes

You can also use the Display Floater to hide and freeze objects. To access the Display Floater, choose *Display Floater* from the *Tools* menu.

Donut

Creates two circle shapes, one inside the other.

General Usage

On the **Create** panel ![icon], click **Shapes** ![icon]. Click **Donut**. Click and drag on the screen to create one circle. Move the cursor inside or outside the first circle and click again to create the second circle. Each circle has four Bezier vertices.

General

See the General rollout under the **Arc** entry.

Creation Method

The Creation Method rollout is available only on the **Create** panel.

Edge   When **Edge** is turned on, the first circle is created starting from a circle edge when you click and drag on the screen. To keep a vertex at the first location clicked, drag up, down, left or right rather than diagonally.

Center   When **Center** is on, circles are created from the center outward when you click and drag on the screen.

Keyboard Entry

The Keyboard Entry rollout is available only on the **Create** panel.

X, Y, Z   Specifies the location of the center of the donut created when **Create** is clicked. The center is placed at the X, Y and Z location in the world coordinate system.

Radius 1   See these listings under Parameters below.
Radius 2

Create   Creates the shape with the parameters entered. The shape is placed on the current viewport's plane.

Parameters

Radius 1   Specifies the radius of one circle.

Radius 2   Specifies the radius of the second circle. **Radius 2** may be either smaller or larger than **Radius 1**.

Dummy

Creates a dummy object in the scene.

General Usage

On the **Create** panel ![icon], click **Helpers** ![icon]. Click **Dummy**. Click and drag in a viewport to create the dummy object. Link objects to the dummy object as desired.

Usage Notes

A dummy object is a non-rendering cube object that can be used to control other objects. One or more objects can be linked to the dummy, then the dummy can be transformed (moved, rotated, scaled). All linked objects will be transformed along with the dummy object.

Dummy objects are useful for rotating an object from a pivot point other than the object's pivot point. For example, a moon orbiting a planet can be made to orbit the planet by creating a dummy at the planet's center, linking the moon to the dummy and rotating the dummy.

Dynamics

A utility that applies real-world physics to objects, simulating collisions and other effects such as gravity.

General Usage

Set up an animated scene with objects that touch or pass through one another.

On the **Utilities** panel , click **Dynamics**. On the Dynamics rollout, click **New** to start a new simulation. Click **Edit Object List** to select the objects to be affected by dynamics.

On the Timing & Simulation rollout, set the range of frames to be affected by dynamics with **Start Time** and **End Time**. On the Dynamics rollout, click **Solve** to calculate collisions and create new key frames based on these collisions.

The **Dynamics** setup is called a *simulation*.

Dynamics

Simulation Name	Displays the name of the current simulation. The simulation name can be changed here.
New	Creates a new simulation with default parameters and a new **Simulation Name**. Change the simulation name if desired.
Remove	Removes the currently displayed simulation from the scene. This will make the *.max* file smaller, but will not remove keys created with this simulation.
Edit Object List	Displays a dialog where you can choose the objects to include in the simulation.
Edit Object	Displays a dialog where you can set various parameters for object collision and effects. For information on this dialog, see the **Edit Object** section later in this entry.
Select Objects in Sim	Selects all objects chosen for the simulation with **Edit Object List**. This button enables you to quickly select all objects in the simulation.
Effects by Object	When **Effects by Object** is chosen, space warps are applied to objects only when they are selected with the **Assign Object Effects** button on the **Edit Object** dialog.

Global Effects	Assigns effects to all objects in the simulation. Click **Assign Global Effects** to display a dialog where effects can be selected.
Collisions by Object	When **Collisions by Object** is chosen, collisions take place between objects only when colliding objects are selected with the **Assign Object Collisions** button on the **Edit Object** dialog.
Global Collisions	Enables you to choose objects, all of which will collide with one another. Click **Assign Global Collisions** to display a dialog where objects can be selected.
Solve	Click **Solve** to generate keys based on the collision and effects parameters set with this utility. This operation cannot be undone, so save your work before clicking **Solve**. Check **Update Display w/Solve** before clicking **Solve** to see each key frame as it is generated.

Timing & Simulation

Start Time End Time	Sets the starting and ending frames for which keys will be generated when **Solve** is clicked. To create keys for collision, the **Start Time** must be after the frame at which animation starts, but before the first collision. For example, if animation takes place over frames 0 through 50 and the first collision is at frame 10, the **Start Time** must be any frame from 1 to 9 in order for the collision keys to generate properly.
Calc Intervals Per Frame	Sets the number of times the solution is calculated for each frame. For example, if this value is 5, the solution will be calculated every 1/5th of a frame. If fast-moving objects don't collide properly, this value must be increased. Note that regardless of the number of times the solution is calculated per frame, keys are set at the **Keys Every N Frames** interval.
Keys Every N Frames	Sets the interval for key generation. For example, setting this value to 5 generates a key every 5 frames.
Time Scale	Adjusts the timing of the simulation. Higher values increase speed while lower values reduce speed. Use this value to fine-tune the simulation.
Use IK Joint Limits	Uses joint limits set on the **Hierarchy** panel under **IK**. See the **IK** entry for information on limiting joints.

Panels

Use IK Joint Damping   Uses the joint damping set on the **Hierarchy** panel under **IK**. See the **IK** entry for information on damping.

Density   Sets the air resistance. The default value of 0 simulates a vacuum with no air resistance. A value of 100 simulates air. Air resistance causes objects to slow down after a period of motion, and also causes objects to tumble as they move.

Edit Object

Clicking **Edit Object** on the Dynamics rollout accesses the Edit Object dialog, where you can set various parameters for each object individually.

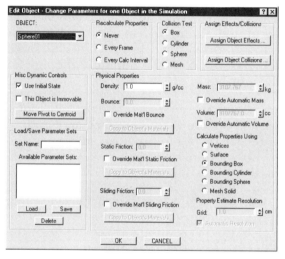

OBJECT   From the pulldown menu, select the object for which you want to set parameters. In order to set parameters for an object, it must first be put on the object list by clicking **Edit Object List** on the Dynamics rollout.

Use Initial State   When checked, any motion of the object at the Start Time frame will be maintained when the simulation starts. When unchecked, the object is considered to be stationary at the start of the simulation.

This Object is Immovable   When checked, the object will not be affected by collisions or effects, and will remain stationary throughout the entire simulation.

Move Pivot to Centroid   When this button is turned on, the object's pivot point is moved to its physical center. The pivot point is not moved until you click **OK** to exit the dialog.

Panels

Load/Save Parameter Sets	You can save the settings on the Edit Object dialog for use for another object or simulation. To save the settings, enter a **Set Name** and click **Save**. To load a previously saved parameter set, highlight the set name on the **Available Parameter Sets** list and click **Load**. To remove a set from the list, highlight the set and click **Remove**.
Recalculate Properties	If an object is deformed by an animated space warp or modifier, its **Volume**, and thus its **Mass**, will change over time. **Never** uses the **Volume** and **Mass** values set on this dialog, and doesn't recalculate the **Volume** regardless of the changes the object experiences over the course of the animation. **Every Frame** recalculates these values on every frame of the simulation, taking into account the new shape of the object. **Every Calc Interval** recalculates these values at intervals set by the **Calc Intervals Per Frame** setting on the Timing & Simulation rollout. For an object that is not deformed by an animated space warp or modifier, select **Never** to minimize the time to solve the simulation.
Collision Test	Sets the type of bounding area used to detect collisions. **Box**, **Cylinder** and **Sphere** use a bounding area around the object in the shape of the primitive. **Mesh** uses the surface of the object. Choosing **Mesh** makes more accurate collision calculations, but takes longer to solve.
Density	The **Density** of an object affects its inertia. Objects with a higher **Density** are harder to move when struck by another object. If two objects of equal size collide at the same speed, the one with the lower **Density** will bounce off and move faster than the one with the higher **Density**. **Density** is set in grams per cubic centimeter (g/cc).
Assign Object Effects	Displays a dialog where you can choose space warps to affect this object in the simulation.
Assign Object Collisions	Displays a dialog where you can specify objects that will collide with this object in the simulation.
Bounce	A value from 0 to 1 that sets the bounciness of the object. A value of 0, which creates little bounce, is suitable for a bowling ball or cannonball. A value of 1 creates a high bounce like that of a rubber ball. This parameter is available only when **Override Mat'l Bounce** is checked.

Override Mat'l Bounce	When checked, the **Bounce Coefficient** setting for the object's material is overridden by the **Bounce** parameter above. When unchecked, the **Bounce Coefficient** setting from the material is used.
Static Friction	A value from 0 to 1 that sets the object's friction when sitting still. Higher friction makes it harder to move the object. A value of 0 creates little friction, while a value of 1 creates an enormous amount of friction. Use low values such as 0.1 or 0.3 for realistic results. This parameter is available only when **Override Mat'l Static Friction** is checked.
Override Mat'l Static Friction	When checked, the **Static Friction** setting for the object's material is overridden by the **Static Friction** parameter above. When unchecked, the **Static Friction** setting from the material is used.
Sliding Friction	A value from 0 to 1 that sets the object's friction when moving. As a start, set **Sliding Friction** to half the **Static Friction** value. This parameter is available only when **Override Mat'l Sliding Friction** is checked.
Override Mat'l Sliding Friction	When checked, the **Sliding Friction** setting for the object's material is overridden by the **Sliding Friction** parameter above. When unchecked, the **Sliding Friction** setting from the material is used.
Copy to Object's Materials	Copies the parameter to the object's material.
Mass	Specifies the mass for the object. *Mass* is a technical term for what we commonly call weight. In physics, mass is density times volume. The **Mass** value on this dialog is automatically calculated from the **Density** and **Volume** when **Override Automatic Mass** is unchecked. When **Override Automatic Mass** is checked, the **Mass** value can be entered directly, and is never recalculated regardless of the **Recalculate Properties** setting. **Mass** is expressed in kilograms (kg).
Volume	Specifies the volume of the object. When **Override Automatic Volume** is checked, the **Volume** can be entered directly, and is never recalculated regardless of the **Recalculate Properties** setting.

When **Override Automatic Volume** is unchecked, the volume is calculated with one of the methods under the **Calculate Properties Using** section. **Volume** is expressed in cubic centimeters (cc).

Calculate Properties Using

These settings determine how **Volume** is calculated when **Override Automatic Volume** is unchecked.

Vertices assigns a volume of 1cc to each vertex. **Surface** assumes the surface area has a thickness of 1 centimeter.

Bounding Box, **Bounding Cylinder** and **Bounding Sphere** approximate the volume with a primitive surrounding the object.

Mesh Solid calculates the volume based on the object's geometry. This is the most accurate method, but also takes the longest to solve.

Grid

When **Mesh Solid** is selected as the method for determining the **Volume**, a **Grid** is used to calculate the center of mass for the object. The smaller the **Grid**, the more accurate the calculation. When **Automatic Resolution** is checked, the **Grid** value is automatically calculated based on the object's size and detail. When **Automatic Resolution** is unchecked, the **Grid** value can be entered directly.

Usage Notes

There are many motion subtleties that can be applied with **Dynamics**. See the 3D Studio MAX manual for further information on setting **Dynamics** parameters.

Sample Usage

This simple exercise will get you started with **Dynamics**.

1. Create a sphere and a box some distance apart in the scene. Go to frame 25 and turn on the **Animate** button. Move the sphere so it sits on or inside the box. Go to frame 50 and move the sphere past the box. Turn off the **Animate** button.

The sphere now passes through the box as it moves from frames 0 to 50.

2. Go to the **Utilities** panel ⊤ and click **Dynamics**. On the Dynamics rollout, click **New** to start a new simulation.

Panels

3. Click **Edit Object List** and select both the box and the sphere to be included in the simulation.

4. Turn on **Global Collisions**. Click **Assign Global Collisions** and select both the box and the sphere for global collision. On the Timing and Simulation rollout, change **Keys Every N Frames** to 5. Leave **Start Time** and **End Time** at their default values of 10 and 100.

5. On the Dynamics rollout, click **Solve.** Wait a few moments while the collisions are calculated and keys are created.

6. When the calculation is complete, click **Play Animation** .

The sphere collides with the box, causing it to move. The sphere continues to move after it hits the box, but with a different speed and/or direction.

Try creating a **Gravity** space warp and selecting it with the **Assign Global Effects** button to make it affect all objects in the scene.

By default, a box-shaped bounding area is used to approximate both objects, making the collision inaccurate. To make the sphere hit the box more accurately, you can use a spherical bounding area around the sphere. Click **Edit Objects**. On the Edit Object dialog, select the sphere as the **OBJECT**. Under Collision Test, select **Sphere**. Click **OK**, and click **Solve** to generate new keys.

Panels

Edit Mesh / Editable Mesh

Edits vertices, faces or edges of a 3D object, and attaches objects.

General Usage

Select an object. On the **Modify** panel , click **More** if necessary, and choose **Edit Mesh**. By default, the **Vertex** sub-object level is selected. Select vertices and modify them, or choose **Face** or **Edge** from the **Sub-Object** pulldown and select faces or edges to modify. When you have finished working with vertices, faces and edges, you can turn off the **Sub-Object** button before applying another modifier, which will apply the next modifier to the entire object. You can also leave the **Sub-Object** button on to apply the next modifier to the selected vertices, faces or edges only.

You can access all the options under **Edit Mesh**, and more, by changing the object to an **Editable Mesh** instead of applying an **Edit Mesh** modifier. On the **Modify** panel, click **Edit Stack** . If the pop-up menu appears, choose **Editable Mesh**. If the Edit Modifier Stack dialog appears, click **Collapse All**, then click **OK**. The object is changed to an **Editable Mesh**. The **Sub-Object** levels for the object now include **Vertex**, **Face** and **Edge**, just as the **Edit Mesh** modifier does. However, there are more options under **Editable Mesh**. The disadvantage is that once an object has been converted to an **Editable Mesh**, you can no longer access its creation parameters on the modifier stack.

Sub-Object Vertex

Copy
: Copies a named selection set of vertices so the selection can used with another **Mesh Select** modifier or with an **Edit Mesh** modifier elsewhere on the stack. When you click **Copy**, the list of selection sets of vertices for this object at this level of stack appears. Choose a selection set.

This option is available only when the object being edited has been collapsed to an **Editable Mesh**.

Paste
: Pastes a selection set of faces, vertices or edges from another **Mesh Select** or **Edit Mesh** modifier. **Paste** is enabled only when a **Copy** has been performed on a **Mesh Select** or **Edit Mesh** modifier elsewhere on the stack for the same object. Vertex selections can be pasted to faces or edges, and vice versa. You can also paste to the same modifier and sub-object for which a **Copy** was performed. In this case, you will be prompted for a new name for the selection set.

This option is available only when the object being edited has been collapsed to an **Editable Mesh**.

Target
When this button is clicked, vertices can be picked up and moved to other vertices. When a vertex is placed on or near another vertex, the two vertices are welded together. Vertices must be apart by no more than the **Target Thresh** distance in order to weld. **Target Thresh** is expressed in units. Click **Target** to turn it off when finished welding vertices.

Selected
Welds selected vertices that are closer to each other than the unit distance specified by **Weld Thresh**.

Break
Replaces each selected vertex with two or more vertices so each vertex is at the corner of only one face. This option is available only when the object being edited has been collapsed to an **Editable Mesh**.

Affect Region
When checked, transforming vertices also affects the vertices around them, deforming them slightly to make a smoother transformation.

Ignore Backfacing
When checked, vertices affected by transformations around them do not include vertices whose normals point in the direction opposite the average direction of selected vertices.

Edge Distance
When checked, the **Iterations** value is used to limit the vertices that are affected by the transformation. **Iterations** specifies the maximum number of edges between a selected vertex and a vertex affected by the transformation.

Edit Curve
When **Edit Curve** is clicked, a dialog appears where you can control which surrounding vertices are affected by the transformation.

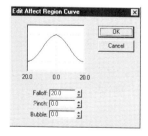

The curve represents how strongly nearby vertices are affected by the transformation. The distance set by

Panels

Falloff is graphed against the height of the curve. Where the curve is high, vertices surrounding transformed vertices are strongly affected by the transformation. **Pinch** affects the sharpness of the top of the curve, while **Bubble** affects the roundness of the curve. Try setting **Falloff** to about ¼ the diameter of the object for very smooth transformations.

Construction Plane
Aligns selected vertices to the active grid. If the home grid and an orthographic view are active, selected vertices are aligned to the viewport plane. This option is available only for **Editable Mesh** objects.

Viewport
Aligns selected vertices to the active viewport plane. This option is available only for **Editable Mesh** objects.

Create
Creates vertices as part of the object. Click **Create** and click on the screen once or more to place each new vertex. Vertices are created on the construction plane. To finish creating vertices, click **Create** again, or click a selection tool such as **Select object** . Vertices created in this way are usually used to create faces with the **Build** option under the **Face Sub-Object** level.

Delete
Deletes selected vertices.

Collapse
Collapses all selected vertices into one vertex at the center of the selection.

Detach
Detaches selected vertices, and the faces they bound, as a separate object.

Hide
Causes selected vertices to be hidden from view, where they can no longer be selected. Faces and edges of hidden vertices are still visible. Vertices remain hidden until **Unhide All** is clicked, either on this **Edit Mesh** modifier or on a later one. This option can be used to temporarily hide vertices that you wish to avoid selecting.

Unhide All
Unhides all vertices hidden at this level or lower levels of the stack.

Remove Isolated
Removes any vertices that are not used as part of a face.

Vertex Color
Displays and sets the color of selected vertices. Click the color swatch to set the color with the Color Selector.

Select Click to select vertices with the color below.

Color Click the color swatch choose a color with the Color Selector.

Range Sets a range for each color component. For example, if R, G and B are set to 10, vertices with colors with R, G and B values that are within 10 points of the color in the color swatch will be selected.

Sub-Object Face

Selection Faces can be selected by **Face** ◁, **Polygon** ▢ or **Element** ▢. *Polygons* make up the surface of the mesh. A *face* is a triangular polygon, or triangular portion of a polygon with more than three sides. For example, a sphere has many four-sided polygons which are made up of two faces each. An *element* is several polygons that make up a particular section of a mesh, such as the lid of a teapot. If you choose **Face**, dotted lines appear across polygons to show the separation of faces.

By Vertex When checked, faces can be selected by selecting vertices. All faces surrounding selected vertices are selected.

Ignore Backfaces When a bounding box is drawn, this option causes only faces facing away from you in the active viewport to be selected, rather than both faces facing toward and away from you. This checkbox has no effect when **By Vertex** is checked. This option is available only for **Editable Mesh** objects.

Ignore Visible Edges When unchecked, face selection is limited to the faces that are selected directly, either by clicking on them or drawing a bounding box around them.

Planar Thresh The **Planar Thresh** value is used to select faces around a directly selected face when **Ignore Visible Edges** is off. When **Planar Thresh** is above 0, all coplanar faces that share edges with the directly selected face are also selected. As **Planar Thresh** is increased, angled faces around the directly selected face are also selected. This option is available only for **Editable Mesh** objects.

Copy Copies a named selection set of faces so the selection can used with another **Mesh Select** modifier or with an

Panels

Edit Mesh modifier elsewhere on the stack. When you click **Copy**, the list of selection sets of faces for this object at this level of stack appears. Choose a selection set. This option is available only when the object being edited has been collapsed to an **Editable Mesh**.

Paste
See **Paste** under the **Sub-Object Vertex** section above.

Extrude
Extruding faces pulls them out from the mesh without affecting the surrounding faces. To extrude faces interactively, select faces to extrude and click **Extrude**. Click and drag on the selected faces to extrude them. The amount of extrusion in units automatically appears as the **Amount** value. Click **Extrude** to turn it off when finished extruding.

Amount
Displays the amount of the current extrusion in units. To extrude faces by a known amount, select the faces and enter the extrusion distance as the **Amount**. An **Amount** of 0 creates no extrusion.

Tessellate
When **Tessellate** is clicked, faces are added to the mesh. Faces are added with either the **Edge** or **Face-Center** method depending on which is chosen below. Note that a face is always triangular, and a polygon may be made up of two or more faces.

Edge
Edge tessellation places a vertex in the middle of each face edge and creates three lines connecting these vertices, dividing each face into four faces.

Face-Center
Face-Center places a vertex at the center of each face and creates a line from this vertex to each of the three original vertices, dividing each face into three faces.

Tension
Sets the tension for **Edge** tessellation. Each new vertex created at the middle of a face edge can be pushed toward or away from the center of the object with the **Tension** value. Positive values push the vertex away from the center, while negative values push it toward the center of the object. If you want faces to be created without any visible change to the object, set **Tension** to 0.

Explode
Separates selected faces into separate objects or elements depending on which is chosen below. When you click **Explode** and **Objects** is selected below, you are prompted for a base name for the new objects that will be created from the selected faces.

Angle Thresh   When **Explode** is clicked, faces at an angle to one another of less than the **Angle Thresh** remain connected. Where two faces meet with an angle greater than **Angle Thresh**, they are broken apart.

Objects   When **Objects** is selected, faces are detached to separate objects when **Explode** is clicked. The new objects can no longer be edited with this **Edit Mesh** modifier as they are now separate objects. To edit the object, turn off the **Sub-Object** level and select the new object.

Elements   When **Elements** is selected, faces are detached to separate elements when **Explode** is clicked. The elements remain part of the currently selected object, and can be further modified with the Edit Mesh options. The main advantage of separate elements is that they can easily be

selected when the **Element** button ⬚ is turned on under the **Selection** section of the Edit Face rollout.

Construction Plane   Aligns selected faces to the active grid. If the home grid is active and an orthographic view is active, selected faces are aligned to the viewport plane. This option is available only when the object being edited has been collapsed to an **Editable Mesh**.

Viewport   Aligns selected faces to the active viewport plane. This option is available only when the object being edited has been collapsed to an **Editable Mesh**.

Detach   Detaches faces to a separate object. When **Detach** is clicked, you are prompted for a name for the new object that will be created from the selected faces. The new object can no longer be edited with this Edit Mesh modifier as it is now a separate object. To edit the object, turn off the **Sub-Object** level and select the new object.

Collapse   Removes selected faces from the mesh, and collapses the surrounding faces to fill in the empty space on the mesh.

Make Planar   Adjusts all selected faces so they lie on the same plane. The plane is determined by averaging the angle of all selected faces.

Build Face   The **Build Face** option allows you to connect three vertices to build a custom face. Click the **Build Face** button, then click on three vertices to build a face. You may continue to click on sets of three vertices to build more faces until the **Build Face** button is turned off. In order to use vertices with **Build Face**, the vertices must already be part of the currently selected object. To create new vertices, go to the **Vertex Sub-Object** level and use the **Create** option.

Delete	Deletes selected faces.
Hide	Hides selected faces. The faces can be unhidden only by clicking **Unhide All** on an **Edit Mesh** modifier.
Unhide All	Unhides all hidden faces.
Material ID	Displays or sets the material ID for selected faces. If selected faces contain more than one material ID, Material ID will be blank. To set the material ID, type in the new material ID number for selected faces. For more information on material IDs, see the **Multi/Sub-Object** entry in the *Mat Editor* section of this book.
Select by ID	Selects faces by material ID. Click **Select by ID**. A dialog appears.

Enter the material ID for which you want to select faces. Check **Clear Selection** to clear the previous selection of faces, or uncheck **Clear Selection** to add the material ID face selection to the current selection. Click **OK** to select faces by material ID.

Smoothing Groups

Sets the smoothing groups for selected faces. When a face is selected, or more than one face from the same smoothing group is selected, the number of the faces' smoothing group is depressed. If faces from more than one smoothing group are selected, the numbers of the smoothing groups are blanked out.

To change the smoothing group number of a face or faces, click the depressed number to remove the faces from the smoothing group, then click another number to assign that smoothing group number to the faces. You can also depress additional numbers to assign the faces to more than one smoothing group.

Panels

For information on smoothing, see the **Smooth** entry in this section of the book.

Auto Smooth   Automatically assigns smoothing groups to selected faces based on the **Threshold**. When faces are at an angle to one another that is less than the **Threshold** angle, they are assigned to the same smoothing group. Faces at angles greater than the **Threshold** angle are assigned different smoothing groups.

Clear All   Clears selected faces from all smoothing groups. This option is available only for **Editable Mesh** objects.

Select by Smooth Group   When **Select by Smooth Group** is clicked, a dialog appears.

Depress the numbers of smoothing groups you wish to select. When **Clear Selection** is unchecked, the new selection will be added to the existing selection. When **Clear Selection** is checked, the existing selection is cleared before the new selection is made. Click **OK** to select faces by smoothing groups.

Flip   Flips the normals of selected faces. For more information on normals, see the **Normal** entry in the *Panels* section of this book.

Unify   Attempts to unify normals on a mesh. When a mesh is imported from another program, it sometimes arrives in 3D Studio MAX with random face normals pointing in the wrong direction. This makes the mesh appear to have holes in it when it appears in a rendering. To correct this problem, try **Unify**. If this doesn't work, apply a two-sided material to the object. See the **Standard** entry in the *Mat Editor* section of this book, under **2-Sided**, for information on making a two-sided material.

Flip Normal Mode   To flip normals interactively, click **Flip Normal Mode**. Click on a face to reverse its normal.

Show Normals   Displays normals on selected faces as small blue and white lines. The white area of the line indicates the head of the normal arrow. For more information on normals, see the **Normal** entry in this section of the book.

Scale Scales the normal lines. Changing this value has no effect on the normals themselves, and merely serves to make normals easier to see.

Select by Normals If **Front Facing** is selected, clicking **Select by Normals** selects all faces with normals pointing toward you in the active viewport. If **Back Facing** is selected, clicking **Select by Normals** selects all faces facing away from you in the active viewport.

Vertex Color Displays and sets the vertex color for vertices on selected faces.

Sub-Object Edge

An edge one side of a face. Each face has three edges. Edges are usually shared between two faces.

Copy Copies a named selection set of edges so the selection can used with another **Mesh Select** modifier or with an **Edit Mesh** modifier elsewhere on the stack. When you click **Copy**, the list of selection sets of edges for this object at this level of stack appears. Choose a selection set. This option is available only when the object being edited has been collapsed to an **Editable Mesh**.

Paste See **Paste** under the **Sub-Object Vertex** section above.

Extrude Extruding edges pulls them out from the mesh without affecting the surrounding faces. To extrude edges interactively, select edges and click **Extrude**. Click and drag on the selected edges to extrude them. The amount of extrusion in units automatically appears as the **Amount** value. Right-click when finished extruding.

Amount Displays the amount of the current extrusion in units. To extrude edges by a known amount, select the edges and enter the extrusion distance as the **Amount**. An **Amount** of 0 creates no extrusion.

Visible Makes selected edges visible.

Invisible Makes selected edges invisible. Invisible edges can still be selected. When selected, invisible edges appear as dotted lines.

Auto Edge	Automatically makes some selected edges invisible based on the **Angle Thresh** value.
Divide	Places a new vertex at the center of each selected edge, dividing each edge into two edges.
Turn	Turns a selected edge within its face.
Delete	Deletes selected edges.
Collapse	Collapses a selected edge to one vertex.
Create Shape	Creates a 2D shape from selected edges. Each edge becomes a segment in the new shape. Click **Create Shape**. A dialog appears.

Enter a name for the new shape object next to **Curve Name**. Choose **Smooth** or **Linear** as the shape type. Check **Ignore Hidden Edges** if you don't want hidden edges to be used to create the shape. Click **OK** to create the shape. The shape is a separate object and must be selected and edited separately. This option is available only when the object being edited has been collapsed to an **Editable Mesh**.

Select Open Edges	Selects all edges that bound only one face. Open edges exist at the boundaries of missing faces, so this option can be used to show you where faces are missing. This option is available only when the object being edited has been collapsed to an **Editable Mesh**.

Object Level

Attach	Attaches objects to the currently selected object. To attach objects, click **Attach**, then click on an object to attach to the current object. Attached objects become elements, which can be selected, modified and detached under the **Face Sub-Object** level.
Attach Multiple	Displays a dialog where multiple objects can be selected for attachment. This option is available only when the object being edited has been collapsed to an **Editable Mesh**.

Panels

Usage Notes

Edit Mesh is primarily for editing vertices, faces and edges. It can also be used to select vertices, faces or edges to which to apply the next modifier on the stack. However, the **Mesh Select** modifier also performs the task of selecting vertices, faces or edges for the next modifier. **Mesh Select** uses less memory and works faster than **Edit Mesh**. Wherever possible, use **Mesh Select** instead of **Edit Mesh**.

You can also use **Edit Mesh** to delete vertices, faces and edges, but the **DeleteMesh** modifier performs this task faster. To use **DeleteMesh**, first use **Mesh Select** to select the vertices, faces or edges you want to delete, then apply the **DeleteMesh** modifier.

Changes made to vertices, faces and edges with **Edit Mesh** often affect the mesh to the point where later changes to lower levels of the stack, such as the creation level, can yield unpredictable results. For example, suppose you create a cylinder and move its vertices around with **Edit Mesh**. If you later return to the **Cylinder** level of the stack and change the number of segments on the cylinder, the moved vertices will now be at different locations in the mesh. The mesh will look quite different, and probably be wrong for your purposes.

When you attempt to return to a level on the stack below an **Edit Mesh** modifier that has changed the object considerably, a warning message appears.

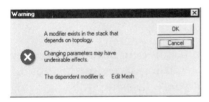

You are allowed to return to lower in the stack and make changes. However, changes to the number of faces, vertices or edges will change the shape of the mesh and may ruin the model. Be sure to save your work before making changes after the warning message appears.

When **Edit Mesh** is at a **Sub-Object** level, as it is by default, an asterisk (*) appears next to **Edit Mesh** in the modifier stack. When a modifier is applied just after *Edit Mesh*, an asterisk appears next to the modifier to indicate that it is applied only to faces, vertices or edges selected with **Edit Mesh**. To return to application of modifiers to the entire object, apply a **Mesh Select** modifier and turn off **Sub-Object**.

Edits an object with patches.

General Usage

Select a patch grid or other object. On the **Modify** panel , click **Edit Patch**. Move patch vertices to deform the grid.

When the **Edit Patch** modifier is applied to an object that is not a patch grid, the object is converted to editable patches called a *surface*. The surface looks exactly like the original object, except that some faces may be different shapes. A *lattice* is also created from the object's overall shape. The lattice is made up of vertices, edges and patches. This lattice may sit above the surface, or sit directly on it. When you choose a **Sub-Object** level of the **Edit Patch** modifier and work with vertices, edges or patches, you are working with the lattice rather than the surface. The transformations are then passed on from the lattice to the surface to deform the surface.

Quad Patch or **Tri Patch** grids are built as a surface and lattice, so the **Edit Patch** modifier uses the vertices and patches that already exist.

Patch vertices have Bezier handles which can be moved separately to deform the surface around the vertex.

Sub-Object Vertex

At the **Vertex Sub-Object** level, you can right-click on a selected vertex to choose a vertex type from the pop-up menu.

Lock Handles When two or more patch or lattice vertices with handles are selected, turning on **Lock Handles** causes handles on all vertices to move when a handle on one vertex is moved. When **All** is on, all handles on selected vertices will move when one handle is moved. When **Alike** is on, only like-sided handles will move together.

Weld Welds selected vertices on different patches that are closer to each other than the unit distance specified by **Weld Threshold**.

Delete Deletes selected vertices and their surrounding patches.

Lattice When checked, the lattice is displayed. When unchecked, only the vertices are displayed.

Surface When checked, the surface is displayed.

Vertices   When checked, vertices can be selected and transformed. Either **Vertices** or **Vectors**, or both options, must be checked.

Vectors   When checked, vertex handles can be transformed. In order to transform a vertex handle, the vertex must be selected. Vertices can only be selected when **Vertices** is checked.

Sub-Object Edge

Subdivide   Click **Subdivide** to divide the patches of selected edges into more patches. When **Propogate** is checked, all contiguous patches are subdivided as well.

Add Tri   Adds a patch with triangular polygons to each selected open edge.

Add Quad   Adds a patch with four-sided polygons to each selected open edge.

Lattice   When checked, the lattice is displayed.

Surface   When checked, the surface is displayed.

Sub-Object Patch

Detach   Detaches selected patches to a separate object. You are prompted for a new object name.

Reorient   When checked, the new detached objects is positioned and oriented so it has the same position and orientation relative to the world coordinate system that the original patches had to the object's local coordinate system.

Copy   When checked, the selected patches are not deleted, and a copy of the patches is used to make the new object. When unchecked, selected patches are deleted when the new object is made.

Subdivide   Click **Subdivide** to divide selected patches into more patches. When **Propogate** is checked, all contiguous patches are subdivided as well.

Lattice   When checked, the lattice is displayed.

Surface   When checked, the surface is displayed.

ID   Sets the material ID for selected patches.

Select by ID   **Select by ID**   Selects patches by material ID. Click **Select by ID**. A dialog appears.

Enter the material ID for which you want to select patches. Check **Clear Selection** to clear the previous selection of patches, or uncheck **Clear Selection** to add the new selection to the current selection. Click **OK** to select patches by material ID.

For information on material IDs, see the **Multi/Sub-Object** entry in the *Mat Editor* section of this book.

Object Level

Lattice   When checked, the lattice is displayed.

Surface   When checked, the surface is displayed.

Steps   Sets the level of detail on the surface when **Fixed** is chosen under the **Mesh Parameters** section below.

Attach   Attaches other objects to the currently selected object. To attach objects, click **Attach**, then click on an object to attach to the current object. Attached objects become patch objects, each with its own lattice.

Reorient   When checked, an attached object is reoriented so its local axes' position and orientation matches that of the current object.

Viewports   Two sets of surface parameters can be set with this
Renderer   rollout. Choose **Viewports** to set the parameters for viewports, or **Renderer** to set them for the renderer.

The settings in the **Mesh Parameters** section determine how the surface will be generated.

Fixed   Calculates the face number and distribution using the **Steps** value.

Parametric   Generates faces based on an equation using the **U Steps** and **V Steps** values.

Spatial   Uses the **Edge** value to specify the maximum edge length for faces, in units. Lower values make more faces.

Panels

Curvature	Uses the curvature of the surface to determine how faces will be generated. **Distance** is the percentage by which the generated surface is allowed to deviate from the original object's surface. The percentage is calculated from the diagonal measurement of the surface's bounding box. **Angle** is the maximum angle between faces on the surface. Lower values increase the number of faces.
View-Dependent	When checked, the renderer takes the object's distance into account when calculating the detail on the object. Objects far from the view receive fewer faces than those close to the view.

Usage Notes

Both the **Edit Patch** modifier and **FFD** modifiers such as **FFD 4x4x4** and **FFD(Box)** use lattices to deform objects.

Panels

Edit Spline / Editable Spline

Edits vertices, segments and splines on a 2D shape.

General Usage

Select a 2D shape. On the **Modify** panel 🔧, click **Edit Spline**. By default, the **Vertex Sub-Object** level is selected. Select vertices and modify them, or choose **Segment** or **Spline** from the **Sub-Object** pulldown and select segments or splines to modify.

You can access all the options under **Edit Spline**, and more, by changing the object to an **Editable Spline** instead of applying an **Edit Spline** modifier. On the **Modify** panel, click **Edit Stack** 📋. If the pop-up menu appers, choose **Editable Spline**. If the Edit Modifier Stack dialog appears, click **Collapse All**, then click **OK**. The object is changed to an **Editable Spline**. The **Sub-Object** levels for the object now include **Vertex**, **Segment** and **Spline**, just as the **Edit Spline** modifier does. However, there are more options under **Editable Spline**. The disadvantage is that once an object has been converted to an **Editable Spline**, you can no longer edit its creation parameters on the modifier stack.

Sub-Object Vertex

At the **Vertex Sub-Object** level, you can right-click on selected vertices and choose a vertex type from the pop-up menu that appears.

Copy	Copies a named selection set of vertices so the selection can used with another **Edit Spline** modifier elsewhere on the stack. When you click **Copy**, the list of selection sets of vertices for this object at this level of stack appears. Choose a selection set. This option is available only for **Editable Spline** objects.
Paste	Pastes a selection set of vertices, segments or splines from another **Edit Spline** modifier. **Paste** is enabled only when a **Copy** has been performed on an **Edit Spline** modifier elsewhere on the stack for the same object. Vertex selections can be pasted to segments or splines, and vice versa. You can also paste to the same modifier and sub-object for which a **Copy** was performed. In this case, you will be prompted for a new name for the selection set. This option is available only for **Editable Spline** objects.
Connect	Connects two vertices. Each vertex must be at the end of an open spline. Click **Connect** and click and drag from one vertex to the other. Right-click when finished.

Break	Breaks a spline at the selected vertex or vertices. At each selected vertex, two vertices are placed. The vertices can then be moved independently.
Refine	Inserts a vertex. Click **Refine** and click on a segment to insert a vertex. Continue clicking on the spline to insert more vertices. Right-click when finished.
Insert	Inserts a vertex and allows you to move it when placing the vertex. Click Insert, then click on a segment to place a vertex. Move the cursor and click to place the vertex. Move the cursor again to place another vertex. Right-click to finish placing vertices. Right-click again to turn off **Insert**.
Make First	Makes the currently selected vertex the first vertex. The first vertex is indicated by a small box around the vertex. When lofting, the first vertices of shapes are lined up to set the flow of segments along the loft. If first vertices are not aligned, the mesh will twist along the path. The first vertex is assigned automatically when a spline is created, but can be changed with this button. **Make First** is only effective when only one vertex is selected. On a closed spline, the first vertex can be any vertex. On an open spline, the first vertex must be one of the vertices at either end of the spline.
Weld	Glues selected vertices together to make one vertex. In order for two vertices to be welded together, they must both be selected, and must be apart from each other by no more than the number of units specified by **Weld Threshold**.
Delete	Deletes selected vertices.
Lock Handles	When two or more vertices with handles are selected, turning on **Lock Handles** causes handles on all vertices to move when a handle on one vertex is moved. When **All** is on, all handles on selected vertices will move when one handle is moved. When **Alike** is on, only like-sided handles will move together. For example, if the handle on the right side of a Bezier Corner vertex is moved, only handles on the right side of other selected vertices will move. On Bezier vertices, both handles move in unison regardless of whether **Alike** or **All** is selected.

Sub-Object Segment

At the **Segment Sub-Object** level, you can right-click on selected segments and choose a segment type from the pop-up menu that appears.

Copy	Copies a named selection set of segments so the selection can used with another **Edit Spline** modifier elsewhere on the stack. When you click **Copy**, the list of selection sets of segments for this object at this level of stack appears. Choose a selection set. This option is available only for **Editable Spline** objects.
Paste	See **Paste** under **Sub-Object Vertex** above.
Break	Splits the segment. To break a segment, click **Break** and click on the segment. Two vertices are placed at the break point, and the spline is open at that point. The two vertices can then be moved under the **Vertex Sub-Object** level.
Refine	Inserts a vertex. Click **Refine** and click on a segment to insert a vertex. Continue clicking on the spline to insert more vertices. Right-click when finished.
Detach	Detaches selected segments to a separate object. To detach, select one or more segments. Click **Detach**. You will be prompted for a name for the new object. Enter a new name and click **OK**. The new object can no longer be edited with the current modifier as it is now a separate object.
Same Shape	When checked, the detached segment remains part of the same shape.
Reorient	When checked, the detached segment is positioned and oriented so it has the same position and orentation relative to the world coordinate system that the original segment had to the shape's local coordinate system.
Copy	When checked, the segment is copied when detached. When unchecked, the segment is removed from the shape when detached.
Divide	Divides the selected segment into the number of segments specified by **Divisions**.
Insert	Inserts vertices on the segment.
Delete	Deletes selected segments.

Sub-Object Spline

Panels

Copy	Copies a named selection set of splines so the selection can used with another **Edit Spline** modifier elsewhere on the stack. When you click **Copy**, the list of selection sets of segments for this object at this level of stack appears. Choose a selection set. This option is available only for **Editable Spline** objects.
Paste	See **Paste** under Sub-Object Vertex above.
Close	If the selected spline is open, clicking **Close** creates a segment attaching each end of the spline.
Outline	Creates an outline of the selected spline. Outlines can be created interactively or with an exact number of units. To create an outline interactively, click **Outline**, select a spline, and click and drag to create the outline. To make an outline by an exact number of units, enter a value for **Outline Width**. The outline is created and **Outline Width** returns to 0. If **Center** is unchecked, the original shape is retained and one outline is created. If **Center** is checked, two outlines are created, one on the inside and one on the outside of the original shape, and the original shape disappears.
Boolean	Performs a boolean operation on two splines. Choose **Union** ⬡, **Subtraction** ⬡ or **Intersection** ⬡. Select the first spline for the boolean operation. Click **Boolean**, and click on the second spline. The splines must be part of the same shape.
Mirror	Mirrors selected splines. The mirror axis is set by the **Mirror Vertically** ⬡, **Mirror Horizontally** ⬡ and **Mirror Both** ⬡ buttons. If **Copy** is checked, a copy of the spline is mirrored. To mirror a spline, select the spline and click **Mirror**.
Detach	Detaches the selected spline to a separate object. When Detach is clicked, you are prompted for a name for the new object.

Reorient	When checked, the detached spline is positioned and oriented so it has the same position and orentation relative to the world coordinate system that the original spline had to the shape's local coordinate system.
Copy	When checked, the spline is copied when detached. When unchecked, the spline is removed from the shape when detached.
Reverse	Reverses the vertex assignment on the selected spline. If the spline is open, the first vertex will be moved to the opposite end of the spline.
Insert	Inserts vertices on the spline. Click **Insert** and click on the spline to insert vertices.
Delete	Deletes selected splines.
Vertex Count	Displays the number of vertices in the currently selected splines.

Object Level

Attach	Attaches shapes to the currently selected shape. To attach shapes, click **Attach**, then click on a shape to attach to the current shape.
Attach Multiple	Displays a dialog where multiple shapes can be selected for attachment. This option is available only when the object being edited has been collapsed to an **Editable Spline**.
Reorient	When checked, an attached shape is reoriented so its local axes' position and orientation matches that of the current shape.
Create Line	Click **Create Line** to draw a line as part of the current shape.

For the remaining parameters on this rollout, see the General rollout under the **Arc** entry.

Usage Notes

The **Lathe** and **Extrude** modifiers modify the entire shape even if vertices or segments have been selected at a **Sub-Object** level of the **Edit Spline** modifier.

Ellipse

Creates an ellipse shape.

General Usage

On the **Create** panel [icon], click **Shapes** [icon]. Click **Ellipse**. Click and drag on the screen to create the ellipse.

General

See the General rollout under the **Arc** entry.

Creation Method

The Creation Method rollout is available only under the **Create** panel.

Edge   When **Edge** is turned on, the ellipse is created starting from the ellipse edge when you click and drag on the screen.

Center   When **Center** is on, the ellipse is created from the center outward when you click and drag on the screen.

Keyboard Entry

The Keyboard Entry rollout is available only under the **Create** panel.

X, Y, Z   Specifies the location of the center of the shape created when **Create** is clicked. The center is placed at the X, Y and Z location in the world coordinate system.

Length   See these listings under Parameters below.
Width

Create   Creates the shape with the parameters entered. The shape is placed on the current viewport's plane.

Parameters

Length   Sets the length of the diameter of the ellipse along its local Y axis.

Width   Sets the width of the diameter of the ellipse along its local X axis.

Extrudes a 2D shape to make a 3D object.

General Usage

Create a 2D shape. Select the shape. On the **Modify** panel 🔧, click **Extrude**. Change the **Amount** parameter to extrude the shape.

Parameters

Amount	Sets the amount by which the shape is extruded, in units. **Amount** can be a positive or negative value.
Segments	Sets the number of segments along the length of the extruded object.
Capping	**Cap Start** and **Cap End** cap the two ends of the extruded object. **Grid** creates caps with faces arranged in a grid. **Morph** creates cap faces in a pattern that can be repeated from one extruded object to another. Use **Morph** if you plan to morph the extruded object to another extruded object.
Output	**Patch** makes the extruded object into a patch type object that can be modified with the **Edit Patch** modifier. **Mesh** makes the object a mesh type object that can be edited with the **Edit Mesh** modifier. **NURBS** makes the extruded object a NURBS type object. If you choose NURBS and collapse the object's stack, the **Modify** panel display will change to allow the editing of a NURBS surface. For information on NURBS surfaces, see the **Point Surface** entry in this section of the book.
Generate Mapping Coords	Generates cylindrical mapping coordinates along the length of the extruded object, and planar coordinates on caps.
Generate Material IDs	Assigns different material IDs to the extruded object's sides and caps.

Usage Notes

The **Extrude** modifier is available only when a 2D shape is selected.

If an **Extrude** modifier is applied to a 2D shape and **Amount** is left at 0, the object is still considered a 3D object.

Panels

Face Extrude

Extrudes selected faces on a 3D object.

General Usage

Select an object. On the **Modify** panel , click **MeshSelect** or **Edit Mesh**. Select faces on the object. On the **Modify** panel, click **More** if necessary, and choose **Face Extrude**. Change the **Amount** to extrude the selected faces.

Sub-Object Level

The **Face Extrude** modifier has one **Sub-Object** level, **Extrude Center**. When **Extrude From Center** is checked, you can move this center to change the direction of the extrusion.

Parameters

Amount	Extrudes faces from their original positions by this distance in units.
Scale	Scales the extruded faces.
Extrude From Center	When checked, the **Extrude Center** sub-object is used as the center of the extrusion, rather than the object center.

Usage Notes

In order to extrude faces, they must first be selected with the **Mesh Select** or **Edit Mesh** modifiers. For information on using the **Mesh Select** and **Edit Mesh** modifiers, see these entries in this section.

FFD 2x2x2/3x3x3/4x4x4

A modifier that deforms an object using a box-shaped lattice and control points.

General Usage

Create an object with several segments along its length, width and height.

Go to the **Modify** panel ![icon]. Click **More** and choose **FFD 2x2x2**, **FFD 3x3x3** or **FFD 4x4x4**. A lattice appears around the object. Depending on the FFD modifier chosen, the lattice has 2x2x2, 3x3x3 or 4x4x4 control points.

Click **Sub-Object** to access the **Points** sub-object level. Move one or more control points to deform the object.

Sub-Object Level

These modifiers have one **Sub-Object** level, **Points**. Access the sub-object level to move points and deform the object.

FFD Parameters

See the Parameters rollout under the **FFD(Box)** entry.

Usage Notes

These modifiers work similarly to the **FFD(Box)** modifier. The choice of which modifier to use is a matter of personal taste. All perform the same function, but with different numbers of control points. The **FFD 2x2x2**, **FFD 3x3x3** and **FFD 4x4x4** modifiers are simply quicker ways to set up an **FFD(Box)** modifier.

FFD(Box)

A space warp or modifier that deforms an object using a box-shaped lattice and control points.

General Usage

Create an object such as a box or similarly shaped object. Give the box several segments along its length, height and width.

To assign **FFD(Box)** as a modifier, select the object and go to the **Modify** panel 🔧. Click **More** and choose **FFD(Box)**. A box lattice appears around the object.

To assign **FFD(Box)** as a space warp, go to the **Create** panel 🖰 and click **Space Warps** 〰. Select **Geometric/Deformable** from the pulldown menu. Click **FFD (Box)**.

Click and drag on the screen to create the FFD box lattice length and width, then move the cursor and click to set the height. If necessary, move the FFD lattice so it surrounds or is near the object.

For the space warp, select the bound object, and click **Bind to Space Warp** 🔗 on the Toolbar. Click and drag from the object to the lattice to bind it to the lattice. Select the FFD lattice and go to the **Modify** panel to access parameters for the FFD lattice.

Click **Sub-Object** to access the **Points** sub-object level. Move one or more control points to deform the bound object.

Sub-Object Level

The **FFD(Box)** modifier and space warp have one **Sub-Object** level, **Points**. Access the sub-object level to move points and deform the object.

FFD Parameters

Length **Width** **Height**	Sets the dimensions of the FFD lattice in units.
Set Number **of Points**	Click this button to set the number of points along the length, width and height of the lattice.
Lattice	When checked, all control points and connecting lines in the lattice are displayed in their current positions. When unchecked, only control points are displayed.

Source Volume	When checked, lattice control points are displayed in their original positions rather than moved positions. If **Lattice** is checked, connecting lines on the original lattice appear also. This option is useful for checking the position of the original lattice in relationship to the object. You can still click and drag on control points to deform the bound object, but the moved positions of the control points will not display until this option is unchecked.
Only in Volume	When this option is selected, only the bound object vertices that lie inside the original lattice are deformed. If a control point lies inside the bound object when the lattice is in its original state but is then moved outside the bound object, it will still not affect the bound object.
All Vertices	When this option is selected, all vertices on the bound object are deformed regardless of whether they lie inside or outside the original lattice.
Falloff	Controls how much vertices outside the lattice are affected by movement of control points. Depending on which part of the lattice the control point is from, the lattice's **Length**, **Width** or **Height** is multiplied by the **Falloff** value. This distance is compared with the distance from the original control point position and bound object vertices. Any vertices that are farther away than this distance are not affected by the movement of the control point.
	When **Falloff** is 0, all nearby vertices outside the lattice are affected by the movement of a control point. When **Falloff** is 1.0, vertices that sit away from the original control point position by a distance greater than the **Length**, **Width** or **Height** (depending on where the control point is from) are not affected. When **Falloff** is 2.0, vertices that sit away from the original control point position by a distance greater than twice the **Length**, **Width** or **Height** are not affected.
Tension	A value from 0 to 50 that sets the sharpness or roundness of the deformation. Low values create rounder curves, while high values make sharper curves.
Continuity	A value from 0 to 50 that affects the lean of the deformation. Lower values lean the deformation more toward the original bound object, while higher values push the deformation toward the control point.

Selection   These three buttons aid in the selection of control points. When **All X** is on and a control point is selected, all control points along the lattice's local X axis for that point are also selected. **All Y** and **All Z** work similarly. All three buttons can be turned off or on at any time.

About   Displays information about the author of **FFD (Box)**.

Usage Notes

The abbreviation FFD stands for *free form deformation*. This type of space warp or modifier allows you to freely deform an object simply by moving control points on the lattice.

Although the **FFD (Box)** space warp or modifier works best with box-shaped objects, it can be used with any kind of deformable object. Whatever object you use, be sure to give it several segments along each of its dimensions or the object will not deform.

If you change the number of points in the lattice with the **Set Number of Points** option after the object has been deformed, the points will return to their default positions. Be sure you have the correct number of points before beginning deformation.

Control points can be animated to animate the deformation of the bound object.

The **FFD(Cyl)** space warp or modifier can also be used to deform objects, as can the **FFD 2x2x2**, **FFD 3x3x3** and **FFD 4x4x4** modifiers.

Panels

A space warp or modifier that deforms an object using a cylindrical lattice and control points.

General Usage

Create an object such as a cylinder or similarly shaped object. Give the cylinder several segments along its height.

To assign **FFD(Cyl)** as a modifier, select the object and go to the **Modify** panel ![icon]. Click **More** and choose **FFD(Cyl)**. A cylindrical lattice appears around the object.

To assign **FFD(Cyl)** as a space warp, go to the **Create** panel ![icon] and click **Space Warps** ![icon]. Select **Geometric/Deformable** from the pulldown menu. Click **FFD (Cyl)**.

Click and drag on the screen to create the FFD cylinder lattice radius, then move the cursor and click to set the height. If necessary, move the FFD lattice so it surrounds or is near the object.

For the space warp, select the bound object, and click **Bind to Space Warp** ![icon] on the Toolbar. Click and drag from the object to the lattice to bind it to the lattice. Select the FFD lattice and go to the **Modify** panel to access parameters for the FFD lattice.

Click **Sub-Object** to access the **Points** sub-object level. Move one or more control points to deform the bound object.

Sub-Object Level

The **FFD(Cyl)** modifier and space warp have one **Sub-Object** level, **Points**. Access the sub-object level to move points and deform the object.

FFD Parameters

Radius   Sets the radius of the FFD lattice in units.

Height   Sets the height of the FFD lattice in units.

See the **FFD(Box)** entry for the remaining parameters on this rollout. Also see the **Usage Notes** under the **FFD(Box)** entry for information on FFD space warps and modifiers.

Fillet/Chamfer

A modifier that creates rounded or beveled corners on a shape.

General Usage

Create a shape with one or more Corner or Bezier Corner vertices, such as a rectangle.

On the **Modify** panel ![icon], click **More** to access the modifier list, and choose **Fillet/Chamfer**. The **Vertex Sub-Object** level is on by default. Select one or more vertices. Change **Radius** to fillet the selected vertices, or change **Distance** to chamfer the vertices.

Sub-Object Level

The **Fillet/Chamfer** modifier has one **Sub-Object** level, **Vertex**. All parameters are accessed under this sub-object level.

The **Radius** and **Distance** values replace each selected vertex with two or more vertices to create an arc or straight edge where each vertex once was.

Radius
: Sets the radius of the fillet. A *fillet* is an arc between two vertices. The size of the arc is set by making a full circle with the **Radius** value as its radius, and using the part of the circle that fits between the two segments around the vertex as the arc.

Distance
: A *chamfer* is a straight segment between two vertices. Each selected vertex is replaced with two new vertices and a straight segment between them. **Distance** sets the distance, in units, between each new vertex and the original vertex location.

  When **Radius** or **Distance** is changed, its new value is applied to selected vertices. When vertices are selected after **Radius** or **Distance** has been changed, the value is not applied to vertices automatically. Click **Apply** to apply the **Radius** or **Distance** value to vertices that have just been selected.

Usage Notes

The **Fillet/Chamfer** modifier is available only when a shape is selected.

The **Fillet/Chamfer** modifier works only on Corner or Bezier Corner vertices. In addition, segments on either side of the vertex must be straight in order for **Fillet/Chamfer** to work on the vertex.

Creates a window that does not open.

General Usage

On the **Create** panel ![icon], click **Geometry** ![icon]. From the pulldown menu, choose **Windows**. Click **Fixed**. Click and drag to create the width of the window, move the cursor and click to set the depth or height, then move the cursor and click again to set the remaining dimension.

Creation Method

See the Creation Method rollout under the **Awning** entry.

Parameters

Height **Width** **Depth**	Sets the overall dimensions of the window.
Frame	Sets the dimensions of the window frame. **Horiz. Width** and **Vert. Width** affect the size of the glazing in addition to the frame size. **Thickness** also affects the thickness of the rails in the window sashes.
Glazing	Sets the **Thickness** of the glass.
Width	Sets the width of the rails.
# Panels Horiz **# Panels Vert**	Sets the number of horizontal and vertical panels.
Chamfered **Profile**	When checked, rails have chamfered edges similar to those of a wooden window.
Generate **Mapping** **Coords**	Generates mapping coordinates on the window.

Fog

Creates fog in a VRML world.

General Usage

On the **Create** panel ⊡, click **Helpers** ⊡. From the pulldown menu, choose **VRML 2.0**. Click **Fog**. Click and drag to create the icon. Set the **Visibility Range** and other parameters as desired.

Fog

Type   Choose from two types of fog, **Exponential** and **Linear**. **Linear** fog increases as a direct function of the distance from the view. **Exponential** fog uses an exponential equation to calculate the increase of fog at any distance from the viewer. **Exponential** fog looks more realistic than **Linear** fog, but takes longer to calculate.

Color   Sets the color of the fog.

Visibility Range   Sets the distance from the viewer at which objects become obscured by fog. When this value is 0, there is no fog effect.

Icon Size   Sets the diameter of the icon. The size of the icon does not affect the fog.

Usage Notes

The scene background is not affected by fog. For the most realistic looking scene, make the fog color the same as the background color.

A utility that causes a moving object to bank and follow.

General Usage

Animate an object's position. Select the object. Under the **Utilities** panel ⬚, choose **Follow/Bank**. Set parameters as desired and click **Apply Follow**.

Follow/Bank

Selected Object	Displays the name of the currently selected object.
Apply Follow	When this button is clicked, the object is reoriented so its local X axis points along the path at all times. In addition, any parameters set below are applied to the object. All rotation keys for the object are deleted and regenerated according to the values below when **Apply Follow** is clicked.
Bank	When checked, the object will tilt from side to side as it follows the path after **Apply Follow** is clicked.
Bank Amount	Adjusts the amount of banking. The effects of this value vary greatly depending on the change in steepness on the path curve. Values under 5 work best. To adjust banking accurately, move the time slider to a frame where banking occurs, and observe the object as you change the **Bank Amount**. Click **Apply Follow** to see the effect of each change. This option is available only when **Bank** is on.
Smoothness	When a path has many sharp curves, banking might cause the object to move wildly from side to side. Increasing the **Smoothness** value smooths out the object's banking. Values of 2 or 3 work well. This option is available only when **Bank** is on.
Allow Upside Down	Objects moving down a vertical path sometimes flip from one side to the other. When this option is checked, flipping on a vertical path can occur. When this option is unchecked, flipping is prevented.
Start End	Sets the first frame and last frames over which to apply banking.

Panels

| Samples | Sets the number of keyframes to be generated over the range. This number does not include the key at frame 0, so setting **Samples** to 5 will generate a total of 6 keyframes including the key at frame 0. |

Usage Notes

The **Follow/Bank** utility is useful for applying follow and banking properties to an object's motion without using the **Path** controller.

Each time you change the values on this panel, you must click **Apply Follow** again to regenerate keys.

This utility replaces all existing rotation keys for the object, and cannot be undone.

Free

Creates a free camera.

General Usage

On the **Create** panel , click **Cameras** . Click **Free**. Click and drag in a viewport to place the camera. Move the camera or change parameters as desired.

Parameters

See the Parameters rollout under the **Target** entry.

Target Distance   Sets the distance from the camera to its imaginary target. This target distance determines the length of the cone, and is also used with camera viewport control buttons such as **Orbit Camera** .

Usage Notes

After a camera is set up, you can change a viewport to the camera view by activating the viewport and pressing the **<C>** key. If more than one camera is in the scene, you will be prompted to choose a camera for the view.

A **Target** camera is similar to a **Free** camera. The only difference is that **Target** cameras have a target that can be moved and animated as a separate object. The direction of a **Free** camera is controlled by rotating the camera itself, while the direction of a **Target** camera is controlled by the position of the target. The decision of which type of camera to use depends on how you plan to animate the camera, or which type of camera you find easier to work with.

Free Direct

Creates a free direct light.

General Usage

On the **Create** panel ⬚, click **Lights** ⬚. Click **Free Direct**. Click and drag in a viewport to place the light. Move the light or change parameters as desired.

General Parameters

See the General Parameters rollout under the **Target Spot** entry.

Directional Parameters

See the Directional Parameters rollout under the **Target Direct** entry.

Target Distance Sets the distance from the direct light to its imaginary target. This target distance determines the length of the cone, and is also used with light viewport control buttons such as **Orbit Light** ⬚.

Shadow Parameters

See the Shadow Parameters rollout under the **Target Spot** entry.

Usage Notes

Direct lights create parallel shadows, and so are excellent for simulating daylight.

A **Target Direct** light is similar to a **Free Direct** light. The only difference is that **Target Direct** lights have a target that can be moved and animated as a separate object. The direction of a **Free Direct** is controlled by rotating the light itself, while the direction of a **Target Direct** is controlled by the position of the target. The decision of which type of light to use depends on how you plan to animate the light, or which type of light you find easier to work with.

Creates a target spotlight.

General Usage

On the **Create** panel , click **Lights** . Click **Target Spot**. Click and drag in a viewport to place the light and target. Move the light or target, or change parameters as desired.

General Parameters

See the General Parameters rollout under the **Target Spot** entry.

Spotlight Parameters

See the Spotlight Parameters rollout under the **Target Spot** entry.

Target Distance	Sets the distance from the direct light to its imaginary target. This target distance determines the length of the cone, and is also used with light viewport control buttons such as **Orbit Light** .

Shadow Parameters

See the Shadow Parameters rollout under the **Target Spot** entry.

Usage Notes

Free Spot lights are useful for illuminating specific areas of the scene. A **Free Spot** is also ideal for use as a volume light for making a visible light cone generated by an overhead light fixture.

A **Target Spot** light is similar to a **Free Spot** light. The only difference is that **Target Spot** lights have a target that can be moved and animated as a separate object. The direction of a **Free Spot** is controlled by rotating the light itself, while the direction of a **Target Spot** is controlled by the position of the target. The decision of which type of light to use depends on how you plan to animate the light, or which type of light you find easier to work with.

Gengon

Creates an extruded polygon with sides of equal size.

General Usage

On the **Create** panel ✍, click **Geometry** 🔯. Select **Extended Primitives** from the pulldown menu. Click **Gengon**. Click and drag to set the size of the gengon base. Move the cursor and click to set the height, then move the cursor and click again to set the size of the fillet.

The orientation of the object's local X, Y and Z axes is determined by the viewport in which you begin drawing. The object's local X and Y axes are set on the viewport's drawing plane.

Creation Method

Edge   Draws the gengon starting at the outer edge of the gengon base.

Center   Draws the gengon starting at the center of the gengon base.

Keyboard Entry

X, Y, Z   Sets the location of the gengon base center at the X, Y and Z location in the world coordinate system.

Sides   See these listings under Parameters below.
Radius
Height
Fillet

Create   Creates the object using the parameters entered.

Parameters

Sides   Sets the number of sides on the gengon.

Radius   Sets the radius of the gengon.

Fillet   Sets the amount of each side to be replaced with a fillet. For example, a **Fillet** value of 20 replaces 20 units of each side with a fillet. The **Fillet** value cannot exceed one-half the size of a side.

Height   Sets the height of the gengon.

Side Segs	Sets the number of segments on each side. The number of side segments does not affect the number of fillet segments.
Height Segs	Sets the number of segments along the height of the gengon.
Fillet Segs	Sets the number of segments in each fillet. Leaving this value at 1 creates sharp, beveled edges on the object. Higher values round out the fillets.
Smooth	When **Smooth** is checked, the surface renders as a smooth, rounded object. When **Smooth** is off, the surface renders as faceted.
Generate Mapping Coords	Generates cylindrical mapping coordinates on the cylindrical portion of the object, and planar mapping coordinates on the top and bottom caps.

Panels

GeoSphere

Creates a sphere with triangular polygons.

General Usage

On the **Create** panel ⌖, click **Geometry** 🗔. If necessary, select **Standard Primitives** from the pulldown menu. Click **GeoSphere**. Click and drag to set the radius of the sphere.

The orientation of the object's local X, Y and Z axes is determined by the viewport in which you begin drawing. The object's local X and Y axes are set on the viewport's drawing plane.

Creation Method

Diameter   Draws the geosphere starting at a point on the edge of the geosphere.

Center   Draws the geosphere from the center outward.

Keyboard Entry

X, Y, Z   Sets the location of the geosphere center at the X, Y and Z location in the world coordinate system.

Radius   Sets the radius of the geosphere.

Create   Creates the object using the parameters entered.

Parameters

Radius   Sets the radius of the geosphere.

Segments   A geosphere is created from a base object, either a tetrahedron, octahedron or icosahedron, as set below. The **Segments** value determines the number of segments added to each side of the base object. The number of faces in the object is the number of sides in the base object times the **Segments** squared.

Tetra   Creates a sphere based on a tetrahedron, a 3D object with 4 equal sides. Each side is segmented according to the **Segments** value. When **Tetra** is checked, the number of faces in the geosphere is 4 times the number of segments squared, and the geosphere can be divided into 4 equal sections.

Octa
A sphere based on an octahedron, a 3D object with 8 equal sides. Each side is segmented according to the **Segments** value. When **Octa** is checked, the number of faces in the geosphere is 8 times the number of segments squared, and the geosphere can be divided into 8 equal sections.

Icosa
A sphere based on an icosahedron, a 3D object with 20 equal sides. Each side is segmented according to the **Segments** value. When **Icosa** is checked, the number of faces in the geosphere is 20 times the number of segments squared. This type of geosphere can be divided into equal sections in many ways based on multiples and divisions of 20, such as 2, 4, 5 or 10 equal sections.

Hemisphere
When checked, the sphere is chopped in half to create a hemisphere. When a hemisphere is created, the geometry in the geosphere is adjusted accordingly. An **Icosa** hemisphere has more faces than a full geosphere.

The only option for a geosphere is to cut the sphere exactly in half. To chop a sphere with greater or larger fractions, create a sphere with the **Sphere** option. For more information on this procedure, see the **Sphere** listing in the *Panels* section of this book.

Base to Pivot
When this option is checked, the geosphere's pivot point is placed at the base of the sphere. When this option is unchecked, the pivot point is placed at the sphere's center.

Generate Mapping Coords
Generates spherical mapping coordinates on the geosphere.

Usage Notes

The **GeoSphere** option creates a sphere with triangular polygons. A sphere created with the **Sphere** option is built with longitude and latitude lines. Either type of sphere will work for most applications. The difference between the two types of spheres is only important when the shape of the polygons that make up the object is important.

Panels

Gravity

A space warp that pushes particles or objects in a specified direction, similar to a gravitational pull.

General Usage

Create a particle system, or create an object and use it in a **Dynamics** simulation.

On the **Create** panel ![icon], click **Space Warps** ![icon]. Select **Particles & Dynamics** from the pulldown menu. Click **Gravity**. Click and drag on the screen to create the gravity space warp icon. Move or rotate the space warp so the arrow points in the direction in which you wish the gravity to act.

To use the space warp with a particle system, bind the particle system to the space warp with **Bind to Space Warp** ![icon] on the Toolbar. With a **Dynamics** simulation, assign the **Gravity** space warp as an effect.

Select the gravity space warp icon. Click **Modify** ![icon] to access the parameters. Change the parameters as desired.

Parameters

Strength   The strength of the gravity. When **Strength** is 0, gravity is not in effect. Higher values make gravity stronger. Negative values push objects and particles in the direction opposite the space warp arrow.

Decay   Sets the amount by which the gravity strength will diminish over distance. When **Decay** is 0, strength is the same over the entirety of the bound object. When **Decay** is higher than 0, the strength decays as the distance between the bound object and space warp increases.

Planar   Causes the gravity effect to always be perpendicular to the plane of the gravity space warp.

Spherical   Causes the gravity effect to be spherical, where gravity emanates from the center of the space warp in all directions.

Icon Size   Sets the overall size of the gravity space warp icon in square units. The size of the icon has no relationship to the gravity effect. The only purpose for changing the **Icon Size** is to make it easier for you to see and position the icon.

Creates a grid helper object.

General Usage

On the **Create** panel 🖼, click **Helpers** 🖳. Click **Grid**. Click and drag on the screen to create the grid. Rotate the grid if necessary.

Parameters

Length **Width**	Sets the dimensions of the grid object. The grid extends infinitely into space regardless of the grid object dimensions.
Grid	Sets the size of the smallest grid square.
Active Color	Sets the color of the grid.
Display	Sets the grid plane in the world coordinate system. Select the plane before drawing the grid.

Usage Notes

A grid helper object is useful for temporarily changing the construction plane. Create a grid, and rotate or align it as desired. With the grid object selected, choose *Views/Grids/Activate Grid Object* from the menu. Any objects drawn while the grid is active will be drawn on the grid. To return to the home grid, choose *Views/Grids/Activate Home Grid* from the menu.

Panels

Hedra

Creates a multi-sided 3D object.

General Usage

On the **Create** panel [image], click **Geometry** [image]. Choose **Extended Primitives** from the pulldown list. Click **Hedra**. Click and drag on the screen to create the object.

Parameters

Family	Choose the type of multi-sided object to create. The easiest way to find out what kind of object each type creates is to try each type.
Family Parameters	**P** and **Q** change the size of facets depending on a number of factors. Increase and decrease **P** and **Q** to observe different effects.
Axis Scaling	A hedra has **P**, **Q** and **R** axes. The relationship between these axes differs depending on many different factors. Increase and decrease these values to observe different effects. **Reset** sets the axis scales to their default values.
Vertices	Determines how vertices are distributed over the object.
Radius	Sets the radius of the hedra object.

General Usage

You might wonder what a hedra can be used for. Hedras make dandy knick knacks, and look especially nice as Christmas tree ornaments. The 3D Studio MAX icon is a hedra.

Creates a coil shape.

General Usage

On the **Create** panel ⟮⟯, click **Shapes** ⟮⟯. Click **Helix**. Click and drag to create the radius of the top of the helix, then release the mouse. Move the cursor and click to set the helix height, then move the cursor and click again to set the radius for the bottom of the helix.

General

See the General rollout under the **Arc** entry.

Creation Method

The Creation Method rollout is available only on the **Create** panel.

Panels

Edge   When Edge is turned on, the helix circle is created starting from the circle edge when you click and drag on the screen. To keep a vertex at the first location clicked, drag up, down, left or right rather than diagonally.

Center   When **Center** is on, the helix circle is created from the center outward when you click and drag on the screen.

Keyboard Entry

The Keyboard Entry rollout is available only on the **Create** panel.

X, Y, Z   Specifies the location of the center of the shape created when **Create** is clicked. The center is placed at the X, Y and Z location in the world coordinate system.

Radius 1   See these listings under Parameters below.
Radius 2
Height

Create   Creates the shape with the parameters entered.

Parameters

Radius 1   Specifies the radius at the top of the helix.

Radius 2   Specifies the radius at the bottom of the helix.

Height   Specifies the height of the helix.

Turns   Specifies the number of times the helix makes a complete 360-degree turn.

Bias	Sets the distribution of turns along the helix. As **Bias** moves closer to 1.0, turns are pushed toward the bottom of the helix. As **Bias** moves closer to -1.0, turns are pushed to the top of the helix.
CW **CCW**	Determines whether the helix turns clockwise (**CW**) or counterclockwise (**CCW**).

Usage Notes

A helix is a spiral or coil shape. A helix is particularly useful as a loft path for making springs and spirals.

IFL Manager

Generates an *.ifl* file listing based on a selected file.

General Usage

On the **Utilities** panel ![icon], choose **IFL Manager**. Click **Select** and select a name from the file selector that appears. Click **Create** and enter a filename to create the *.ifl* file.

IFL Manager

Working File Prefix	Displays the prefix for the selected files. This prefix does not necessarily become the *.ifl* filename. A different *.ifl* filename can be entered when **Create** is clicked.
Start	Sets the starting number for the *.ifl* file. The value displayed is based on the first sequentially numbered file found with the selected prefix.
End	Sets the ending number for the *.ifl* file. The value displayed is based on the last sequentially numbered file found with the selected prefix.
Every nth	Sets the increment for the file listing. For example, if 2 is entered, then every other file in the sequence will be listed in the *.ifl* file.
Multiplier	Sets the number of times in a row each file will be listed in the *.ifl* file. For example, a **Multiplier** of 3 will list each file three times in a row before listing the next file.
Select	Selects a file as a base for the **IFL Manager** utility. When a file is selected, the utility searches for all numbered files with the same prefix and extension, and uses the numbers of files found to display the **Start** and **End** values.
Create	Creates an *.ifl* file based on the file selected and values entered. Click **Create**, enter a filename, and click **Save** to save the file. This option is only available when a file has been selected with the **Select** button, and more than one sequentially numbered file has been found. The created file has the extension *.ifl*.
Edit	Selects a previously created *.ifl* file for editing. The selected file is displayed in the Notepad text editor.
Close	Closes the utility.

Usage Notes

An .*ifl* file is an ASCII listing of bitmap filenames. An .*ifl* file can be used in the Material Editor or in any place in MAX where a bitmap can be chosen. The listed files are used in sequential order.

Although an .*ifl* file can be any list of files, it is usually a list of sequentially numbered bitmaps, such as those created when an animation is rendered to a single-image file type. For example, rendering an animation to the single-image file type .*bmp* creates a series of sequentially numbered .*bmp* files. These files could then be listed in an .*ifl* file and used in the Material Editor as a **Diffuse** map, as one example. The **IFL Manager** utility is intended to simplify the task of creating an .*ifl* file from sequentially numbered files.

An .*ifl* file can be created that lists files in descending number order rather than ascending order. To do this, set the **Start** number higher than the **End** number. Files will be listed starting with the **Start** number and descending to the **End** number.

Sets limits and other parameters for inverse kinematics.

General Usage

Create a series of objects and link them with **Select and Link** 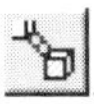 .

On the **Hiearchy** panel  , click **IK**. On this panel, you can "bind" an object to its original position or to another object in the scene. When an object is bound, it attempts to stay put while other links in the chain move and rotate. An object to which the object is bound is called the *follow object*.

There are three ways to use inverse kinematics with a series of linked objects. All three use the settings under the Rotational Joints and Sliding Joints rollouts.

To use the IK settings interactively, turn on the **Inverse Kinematics on/off toggle**  . Move or rotate either the last child object in the chain or the follow object. Objects higher in the chain will move or rotate accordingly. This is called *Interactive IK*.

If one or more objects in the chain are bound to a follow object, you can leave the **Inverse Kinematics on/off toggle** off and animate the follow object. You can then select an object in the chain and click **Apply IK** on the Inverse Kinematics rollout to calculate the IK solution (motion and rotation keys). This is called *Applied IK*.

A third method uses non-rendering bones. Turn on the **Animate** button and move bones as desired. With this method, the inverse kinematics is built into the bones, so it is not necessary to turn on the **Inverse Kinematics on/off toggle** or use the **Apply IK** button. This method is called *New IK*. For information on creating bones, see the **Bones** entry.

Inverse Kinematics

Apply IK	Calculates the IK solution for the chain.
Apply Only To Keys	When checked, keys are set for objects in the chain only on frames where a key exists for the follow object. When unchecked, keys are set at all frames in the range.
Update Viewports	When checked, key frames are displayed as keys are generated.
Clear Keys	When checked, all position and rotation keys in the **Start** to **End** range are removed from the chain before

Panels (side tab)

Panels

new keys are generated. If unchecked, old keys remain on frames where new keys are not generated.

Start   Sets the range of frames over which new keys are gener-
End   ated.

Object Parameters

Terminator   Makes the currently selected object the terminator in the chain. Any IK motion will not affect this object, nor will it affect any of the objects above it on the chain.

Bind Position   When checked, the selected object is bound to its original position, or to the follow object's position if one is picked with the **Bind** button below.

R Relative   When turned on and a follow object is selected with the **Bind** button, the follow object's transforms are mimicked by the bound object. When this option is not on, the bound object's pivot point attempts to match the follow object's pivot point position. This option has no effect if a follow object has not been selected with the **Bind** button.

Axis   The object can be bound along any or all of the world **X**, **Y** and **Z** axes.

Weight   Each object can be bound to only one follow object, but different objects in the chain may be bound to different objects. When there is more than one follow object assigned in a chain, **Weight** sets the priority for each. Priority is given to the binding with the highest **Weight**. If there is only one follow object assigned to the chain, **Weight** has no effect.

Bind   When checked, the selected object is bound to its origi-
Orientation   nal orientation, or to the follow object's orientation if one is picked with the **Bind** button below. The options for **Bind Orientation** work similarly to those for **Bind Position**, described above.

Bind   Binds the object to a follow object. Click **Bind**, and click and drag from the object to the follow object. The name of the follow object appears above the **Bind** button. There is no immediate change in the scene. To see the object jump to the follow object's position or orientation, move or rotate the follow object.

Unbind   Unbinds the object.

Precedence	Sets the precedence value for this joint. The IK solution for each joint is calculated in order of the highest **Precedence** first.
Child-> Parent	Automatically sets joint precedence in decreasing order from child to parent. Precedence values are assigned to selected objects only. To automatically assign Precedence values to the entire chain, select all objects in the chain and click **Child-> Parent** or **Parent->Child**.
Parent-> Child	Automatically sets joint precedence in decreasing order from parent to child.
Copy	Copies sliding or rotational joint settings so they can be pasted to another object. Sliding and rotational joint settings are copied to separate areas, so there can be one of each copied at the same time.
Paste	Pastes the last copied sliding or rotational joint settings. If the last copied joint settings cannot be pasted to the currently selected object, the **Paste** button is disabled.
Mirror Paste	When **X**, **Y** or **Z** is selected, the last copied settings are pasted and mirrored at the same time. When **None** is selected, no mirroring takes place.

Auto Termination

The options on this rollout work only with Interactive IK (when the **Inverse Kinematics on/off toggle** is on).

Auto Termination	When checked, the chain is terminated (made immobile) at the number of links up the chain specified by the **# of Links Up** value. For example, if **# of Links Up** is 3, the terminator will always be 3 links up from the current selection, no matter which object on the chain is selected.

Rotational and Sliding Joints

The settings in the Rotational Joints rollout set the limits for rotation of the currently selected object. The rollout has three sections, one for each of the X, Y and Z axes. Rotational limits are set according to the parent object's local axes.

The Sliding Joints rollout has the same settings as the Rotational Joints rollout, but the Sliding Joints parameters limit the motion of the object along the parent object's local axes, rather than the rotation.

Active	When checked, the object can be rotated or moved on the specified axis or axes.
Limited	When checked, the rotation or movement of the object is limited to the range set under **From** and **To**.
Ease	When checked, a joint resists rotation or movement when it gets close to its limit.
From **To**	Sets each end of the limited range of movement in degrees or units. These limits are relative to the object's orientation and position on frame 0. If these values are the same, no rotation can take place. These values can be negative numbers.
Spring Back	When checked, the joint has a tendency to spring back to its original position when it resists motion. The **Spring Back** value determines how strongly the joint springs back to its original position.
Spring Tension	Sets the tension in an imaginary spring in the joint. As this value is increased, the spring effect pulls harder on the joint as it moves away from its original position.
Damping	**Damping** is used to approximate friction or resistance in a joint. A value of 0.0 allows the object to move freely, while the maximum value of 1.0 creates extreme resistance which prevents the object from rotating or moving at all, creating the same effect as turning off the **Active** checkbox for the object. Values between 0.0 and 1.0 cause more resistance to rotation or movement as the value gets closer to 1.0.
	If **Damping** is set to 0.0, the object will rotate or move to its full limit before the parent link begins to move or rotate. If **Damping** is set to a value between 0.0 and 0.1, the object will rotate or move to a portion of its limit before the parent object begins to rotate or move. If more motion or rotation is applied after the parent object begins to move or rotate, the child object will continue to move or rotate, but at a reduced rate.
	In real life, only the most highly greased mechanical joints rotate and move without resistance. **Damping** is useful for approximating the behavior of real-life joints such as body limbs or mechanical cranks.

References another VRML world within the current VRML world.

General Usage

On the **Create** panel ![icon], click **Helpers** ![icon]. From the pulldown menu, choose **VRML 2.0**. Click **Inline**. Click and drag to create the icon. Enter a URL for the *.wrl* files with the **Insert URL** parameter.

VRML Inline

Insert URL   Specifies the URL of the *.wrl* files to display in place of the icon.

Bookmarks   Selects a URL from a list of bookmarks. You can enter new bookmarks, or click **Import List** to import the list of bookmarks from your browser.

Icon Size   Sets the diameter of the icon. If the icon is rotated or scaled, the referenced VRML scene is also rotated or scaled.

Panels

Lathe

Lathes a 2D shape to make a 3D object.

General Usage

Create a 2D shape. Select the shape. On the **Modify** panel , click **Lathe**. Change the settings under **Direction** and **Align** as desired.

Sub-Object Level

The **Lathe** modifier has one **Sub-Object** level, **Axis**. Use this sub-object level to move or rotate the axis around which the shape is lathed.

Parameters

Degrees	Sets the number of degrees around the axis that the shape will be lathed.
Weld Core	When checked, vertices at the center of the lathed object that occupy the same space are removed.
Flip Normals	Lathed objects are sometimes created with face normals pointing the wrong way. The most obvious symptom of this problem is that the object appears to be inside out. Click **Flip Normals** to flip all normals on the object in the opposite direction.
Segments	The number of segments around the lathed object.
Capping	**Cap Start** and **Cap End** cap the two ends of the lathed object. **Grid** creates caps with faces arranged in a grid. **Morph** creates cap faces in a pattern that can be repeated from one lathed object to another. Use **Morph** if you plan to morph the lathed object to another lathed object.
Direction	Determines the direction of the axis in the world coordinate system.
Align	Aligns the axis to the shape. **Min** and **Max** align the axis to either end of the shape. **Center** centers the axis on the shape.

Output	**Patch** makes the extruded object into a patch type object that can be modified with the **Edit Patch** modifier. **Mesh** makes the object a mesh type object that can be edited with the **Edit Mesh** modifier. **NURBS** makes the extruded object a NURBS type object. If you choose NURBS and collapse the object's stack, the **Modify** panel display will change to allow the editing of a NURBS surface. For information on NURBS surfaces, see the **Point Surface** entry in this section of the book.
Generate Mapping Coords	Generates cylindrical mapping coordinates along the length of the extruded object.
Generate Material IDs	Assigns different material IDs to the extruded object's sides and caps.

Usage Notes

The **Lathe** modifier is available only when a 2D shape is selected.

If you are unfamiliar with Lathe, the easiest way to work with it is to try different **Direction** axes and **Align** methods until the object looks right.

If the lathed object appears to be pinched at the top or bottom, checking **Weld Core** can sometimes eliminate this problem.

If the shape is open and **Degrees** is less than 360, the inside of the lathed object will appear invisible because face normals are pointing outward. For information on normals, see the **Normal** entry.

Lattice

Converts object segments into cylinders, making it look like a wireframe.

General Usage

Select an object. On the **Modify** panel 🥁, click **Lattice**. The object is converted to struts (cylinders replacing each segment) and junctions (hedra objects placed at each vertex).

Parameters

Geometry	Choose to display **Struts Only**, **Junctions Only** or **Both**.
Radius	The radius of the struts (cylinders replacing segments).
Segments	The number of segments along each strut.
Sides	The number of sides on each strut.
Material ID	Sets the material ID for struts.
Visible Edges	Generates struts only on visible face edges.
All Edges	Generates struts on all face edges. A polygon is sometimes made up of two or more faces, so turning on this option increases the number of struts.
End Cap	Applies end caps to struts.
Smooth	Applies smoothing to struts.
Geodesic Base Type	Choose the type of hedra object to appear at each junction.
Radius Segments Material ID Smooth	These parameters work in the same way as those under the **Struts** section, but are applied to **Junction** objects.
Mapping Coordinates	**None** applies no mapping coordinates to the lattice. **Reuse Existing** applies the mapping coordinates currently on the object to the lattice. **New** applies cylindrical mapping coordinates to the struts, and spherical to hedras at junctions.

Usage Notes

The **Lattice** modifier has rare and specialized uses. Most often, it is used to create a wireframe representation of an object.

Creates an extruded L-shaped object.

General Usage

On the **Create** panel , click **Geometry** . Select **Extended Primitives** from the pulldown menu. Click **L-Ext**.

Click and drag to set the overall size of the object. Move the cursor and click to set the height, then move the cursor and click again to set the width of both sections of the object.

The orientation of the object's local X, Y and Z axes is determined by the viewport in which you begin drawing. The object's local X and Y axes are set on the viewport's drawing plane.

The top of the L-extension, as viewed from the drawing viewport, is called the *front*. When you create a L-extension by drawing on the screen, each section is drawn with the same width. If you want varying widths on your L-extension, you must type in new values under the Parameters rollout after creating the object. You can also give each section a different number of segments.

Creation Method

Corners	Draws the L-extension starting at a front corner.
Center	Draws the L-extension from the center of the object.

Keyboard Entry

X, Y, Z	Sets the location of the L-extension base corner at the X, Y and Z location in the world coordinate system.
Side Length **Front Length** **Side Width** **Front Width** **Height**	See these listings under Parameters below.
Create	Creates the L-extension using the parameters entered.

Parameters

Side Length	Length of the side of the L-extension as viewed from the current viewport.
Front Length	Length of the top of the L-extension as viewed from the current viewport.

Panels

Side Width	Width of the L-extension side.
Front Width	Width of the L-extension front, which is the top of the L-extension as viewed from the drawing viewport.
Height	Height of the L-extension. Each section of the L-extension shares the same height.
Side Segs	Number of segments along the side of the L-extension.
Front Segs	Number of segments along the top of the L-extension.
Width Segs	Number of segments along the width of the L-extension.
Height Segs	Number of segments along the height of the L-extension.
Generate Mapping Coords	Generates mapping coordinates on the L-extension.

Level of Detail

Substitutes different versions of the same object with varying detail based on the object's size in the rendered image.

General Usage

Create two or more objects that are identical except for their level of detail. Align the objects to a common center with **Align** ▨. Choose *Group/Group* from the menu to group the objects. Select the group.

On the **Utilities** panel ↑, choose **Level of Detail**. Click **Create New Set** to use the selected group. The grouped objects are listed on the panel. Highlight each object and set **Thresholds** for each one.

Level of Detail

Create New Set	Selects the objects in the currently selected group for use with the utility. The objects are listed on the panel below along with an automatically set **Max Size**. The **Min Size** and **Max Size** for each object can be changed by highlighting the object on the list and changing **Min Size** and **Max Size** below.
Add To Set	Adds an object from the group to the selection list. In order to add an object to the list, it must be part of the group that was selected when **Create New Set** was clicked. To add the object to the list, click **Add To Set** and click on the object in the group.
Remove From Set	Removes the object highlighted on the list from the selection set.
Width	Sets the width of the intended rendering resolution. This value is used to calculate the **Min Size** and **Max Size** values.
Height	Sets the height of the intended rendering resolution. This value is used to calculate the **Thresholds** values.
Reset to Current	Resets the **Width** and **Height** to the current rendering resolution set on the Render Scene dialog. This dialog can be accessed by choosing *Rendering/Render* from the menu.
Display in Viewports	When checked, the highlighted object is displayed in viewports while all others in the group are hidden. This checkbox can be on for only one object in the group at a time. By default, this checkbox is on for the least detailed object in the group.

Threshold Units	Sets the type of measurement to be used when determining which object in the set should be substituted at which time. **Pixels** sets the **Threshold** values in pixels, while **% of Target Image** uses the size of the object as a percentage of the total rendered image size.
Thresholds	Sets the minimum and maximum size for the highlighted object. Units are based on the type of units selected under **Threshold Units**. When the object reaches the **Min Size** in the rendered image, the highlighted object is substituted. When the **Max Size** is reached, the highlighted object is replaced with the next object on the list. Changing **Min Size** also changes the **Max Size** of the previous object on the list, while changing **Max Size** changes the **Min Size** of the next object.

Usage Notes

The **Level of Detail** utility can save rendering time. When an object with high detail (many faces and vertices) is close to the camera, the detail is often necessary for a smooth rendering. However, when the same object is far from the camera, high detail on the object is not needed. Objects with high detail take longer to render than those with little detail. This utility substitutes objects with low detail when the object is far from the camera, which decreases rendering time.

All objects must be in a group before they can be used with this utility, and the group must be closed. Although material and mapping properties can be set after objects are selected for the **Level of Detail** utility, it is easier and quicker to assign materials and mapping before using this utility.

To apply a modifier to an object after it has been chosen for the **Level of Detail** utility, turn on **Display in Viewport** for the object, then open the group with the *Group/Open* menu option. Select the object, go to the **Modify** panel and modify the object as desired.

To assign a material to an object after it has been chosen for the **Level of Detail** utility, turn on **Display in Viewport** for the object. Drag the material from the Material Editor to the object. When the Assign Material dialog appears, choose **Assign to Object**. If you choose **Assign to Selection**, the material will be assigned to the entire group.

The **Level of Detail** utility works by setting up Visibility tracks for each object. To dismantle an entire Level of Detail setup, open Track View, expand each object's tracks. Highlight each Visibility track and remove it with **Delete Visibility Track** .

Panels

Creates an open or closed line shape, either curved or straight.

General Usage

On the **Create** panel, click **Shapes**. Click **Line**. Click on the screen to set the first point, then move the cursor and click as many times as desired to create more points. Right-click to end creation of the line.

General

See the General rollout under the **Arc** entry.

Creation Method

The Creation Method rollout is available only on the **Create** panel.

Initial Type   Sets the type of vertex when you click to create a vertex on the line. Corner creates a Corner type vertex, while Smooth creates a Smooth type vertex.

Drag Type   Sets the type of vertex when you click and drag to create a vertex. Corner creates a Corner type vertex, Smooth creates a Smooth type vertex, and Bezier creates Bezier type vertices.

Keyboard Entry

Under the Keyboard Entry rollout, a line is created one point at a time with the **Add Point** button. The **Close** and **Finish** buttons end creation of the line.

The Keyboard Entry rollout is available only on the **Create** panel.

X, Y, Z   Specifies the location of the point created when **Add Point** is clicked. The point is placed along the X, Y and Z axes in the world coordinate system.

Add Point   Adds a point to the line at the location specified by X, Y and Z.

Close   Closes the line by creating a segment that connects the last point with the first point.

Finish   Completes the line without closing it.

Linked Xform

A modifier that links one object to another.

General Usage

Select the object to be linked. On the **Modify** panel ⚙, click **More** if necessary and choose **Linked Xform**. Click **Pick Control Object** and pick the object to which you wish to link. Transform the control object to see the linked object transform along with it.

Parameters

Pick Control   To pick the object to control the currently selected ob-
Object   ject, click **Pick Control Object** and pick the object.

Usage Notes

When one object is linked to another with the **Linked Xform** modifier, the objects behave exactly as if they were linked with **Select and Link** 🔗. However, with the **Linked Xform** modifier, you can link sub-objects such as faces and vertices to another object. To do this, select the faces or vertices with the **Mesh Select** modifier, then apply a **Linked Xform** modifier. When the control object is moved, rotated or scaled, the selected faces or vertices are transformed in the same manner.

Sets limits on how a child object transforms when a parent is transformed.

General Usage

Select a child object. On the **Hierarchy** panel ⬚, choose **Link Info**. Set limits on how the child object can be transformed, and on which transforms it inherits from the parent.

Locks

The settings under the Locks rollout limit the movement, rotation and scaling of the selected object on its local axis. These limits do not apply to transforms inherited from a parent object. To set limits to inherited transforms, use the settings under the Inherit rollout.

Inherit

The settings under the Inherit rollout limit the transforms passed on from ancestors to the selected child object. The axes displayed refer to transforms along world coordinate axes for parent objects.

Usage Notes

The settings under the Inherit rollout work exactly the same as the settings on the **Link Inheritance (Selected)** utility, except that the settings here can be used for only one selected object while **Link Inheritance (Selected)** can be used for two or more selected objects.

Link Inheritance (Selected)

Limits the transforms passed on from parent objects to linked children.

General Usage

Select one or more child objects. On the **Utilities** panel , choose **Link Inheritance (Selected)**. Turn checkboxes on or off for any of the transforms and axes.

Usage Notes

The axes displayed refer to transforms along world coordinate axes for parent objects. For example, this utility is where you would disable a child object's rotation on a particular axis. When the object's parent or any ancestor is rotated around the world Y axis, the child will not rotate. However, the child might move to retain its relationship to the parent object as it rotates.

This utility works exactly the same as the Inherit rollout under **Link Info** option on the **Hierarchy** panel , except that this utility can be used for two or more selected objects while **Link Info** can only be used for one selected object.

Substitutes objects of varying levels of detail in a VRML scene based on their distances from the view.

General Usage

Create two or more objects that are identical except for the level of detail in each object.

On the **Create** panel [icon], click **Helpers** [icon]. From the pulldown menu, choose **VRML 2.0**. Click **LOD**. Click and drag to create the icon.

Use **Align** [icon] to align all the objects and the **LOD** icon to a common center.

Click **Pick Objects** and pick the objects. Set the **Distance** for each object.

Level of Detail

Pick Objects To pick objects to be used with **LOD**, click **Pick Objects** and select the objects. As each object is selected, it appears on the list. When finished, click **Pick Objects** again to turn it off.

Distance Sets the distance from the view, in units, at which the highlighted object is substituted.

Delete To delete an object from the list, highlight the object and click **Delete**.

Icon Size Sets the diameter of the icon. The icon size does not affect the operation of the **LOD** helper.

Usage Notes

Each object referenced by an **LOD** helper is saved as part of the *.wrl* file, so each **LOD** object increases the size of the file. A *.wrl* file with many **LOD** objects can become quite large. Your users' computers may not have as much memory or as much processing power as the computer you use, and a large *.wrl* file can take them a very long time to load. For a large VRML world with many objects, consider replacing **LOD** objects with one object of medium detail.

Loft

Lofts a shape along a path.

General Usage

Create two shapes, one for the path and one for the shape to be lofted.

Select the path. On the **Create** panel 🖌️, click **Geometry** 🔵. Choose **Loft Object** from the pulldown menu. Click **Loft**. Click **Get Shape**, and pick the shape to be lofted. The shape jumps to the end of the path, creating the loft object. To see the loft object in wireframe viewports, check Skin on the Skin Parameters rollout.

To add more shapes to the path, change the **Path** value and click **Get Shape** to get another shape for the path. To deform the loft object, to the **Modify** panel and use one of the options on the Deformations rollout.

Creation Method

Get Path	If the shape to be lofted was selected before Loft was clicked, click Get Path and pick the path. The beginning of the path moves to the shape's pivot point.
Get Shape	If the path was selected before Loft was clicked, click Get Shape and pick the shape to be lofted. The shape moves to the path, aligned so its pivot point sits at the beginning of the path.
Move **Copy** **Instance**	Determines how to use the picked shape or path. **Move** uses the shape or path itself. **Copy** uses a copy of the shape or path, and **Instance** uses an instance. If **Instance** is selected, the path or shape can later be modified, and the loft object will change accordingly.

Surface Parameters

Smooth Length	When checked, the loft object is smoothed along its path.
Smooth Width	When checked, the loft object is smoothed around the path.
Apply Mapping	Special mapping coordinates can be applied to the loft object. These mapping coordinates can be thought of as cylindrical mapping coordinates that twist and turn, following the contours of the path. Check **Apply Mapping** to apply the mapping coordinates to the loft object.

Length Repeat	The number of mapping tiles along the path.
Width Repeat	The number of mapping tiles going around the path.
Normalize	Affects the distribution of mapping tiles along the path. Checking **Normalize** distributes mapping tiles evenly along the path. When **Normalize** is unchecked, tiles are distributed according to the vertices on the path, which can result in tiles of different sizes along the path.

Skin Parameters

Cap Start	Places a cap at the beginning of the path.
Cap End	Places a cap at the end of the path.
Morph	Creates a cap with a repeatable number of vertices. Use this option on loft objects that will be used as morph targets.
Grid	Creates a cap with vertices and faces arranged in a grid. This type of cap works better with modifiers than the **Morph** type does.
Shape Steps	The number of steps between shape vertices. A curved or round shape is not actually curved, but is made to look curved with a series of straight steps between vertices. The more steps there are, the more curved the shape appears. The number of **Shape Steps** determines the number of segments around the loft object when **Optimize** is unchecked, or when the shape is curved. When a curved shape has been lofted, increasing the number of **Shape Steps** makes the loft object appear smoother.
Path Steps	The number of steps between vertices on the path. The number of **Path Steps** determines the number of segments along the length of the path. When a curved path has been used for the loft, increasing **Path Steps** makes the loft object appear smoother.
Optimize Shapes	Removes steps from straight segments of shapes.
Adaptive Path Steps	When unchecked, the vertices on the original path spline are used as path vertices. When checked, vertices are also placed at shapes on the path and at deformation

vertices. If there is only one shape on the path and deformation has not been used to alter the loft object, this checkbox has no effect.

Contour   When checked, the shape turns and twists along the path so it is always perpendicular to the path.

Banking   If the path is curved and has vertices on more than one plane, checking **Banking** will cause the shape to rotate as it banks around 3D curves. The degree of banking is automatically set according to the curvature of the path.

Constant Cross-Section   When checked, shapes are scaled up or down at sharp turns in the path to maintain the same overall width in the loft object. For example, if you create a rectangular path and loft a shape along it to make a picture frame, the corners are smaller than the rest of the frame when **Constant Cross-Section** is unchecked. This is because the path makes a 90-degree turn at corners, forcing the shape to a 45-degree angle to the path. Checking **Constant Cross-Section** ensures that the shape size at sharp corners in the path matches the width of the rest of the object.

Linear Interpolation   Determines the type of transition between shapes on the path. When **Linear Interpolation** is checked, the transition is straight. When unchecked, the transition creates a smooth curve from one shape to another. If there is only one shape on the path, this option has no effect.

Skin   When checked, the loft object's "skin" is displayed in all viewports.

Skin in Shaded   Displays the loft object's "skin" in shaded viewports, regardless of whether **Skin** is checked.

Path   Sets the current path level. Any shapes picked with **Get Shape** are placed at this level. In viewports, the current path level is indicated by a small yellow **X** on the path.

If **Percentage** is selected below, **Path** is the percentage of the total path length. If **Distance** is selected, **Path** is the distance from the path start in units. If **Path Steps** is selected, **Path** is the number of the path step, and the total number of path steps is displayed next to the **Path** value.

	Snap	Sets the snap increment. When **On** is checked, clicking on the **Path** spinner increments the **Path** level by the **Snap** amount.
	Pick Shape	Click **Pick Shape** to pick a shape from the loft object on the screen. The **Path** value changes to the level of the picked shape.
	Previous Shape	Changes the **Path** level to the level of the previous shape on the path.
	Next Shape	Changes the **Path** level to the level of the previous shape on the path.

Deformations

The Deformations rollout is available only on the **Modify** panel.

Scale	Scales the loft object.
Twists	Rotates shapes on the path.
Teeter	Tilts shapes so they are no longer perpendicular to the path.
Bevel	Creates beveled edges on the loft object. **Bevel** works similar to **Scale**, but also bevels nested shapes.
Fit	Fits the loft object to two profile shapes.

Each button on this rollout accesses a window specific to the deformation type. In each window (except **Fit**) is a red or green line called the *deformation curve*. Changing this line deforms the loft object.

By default, the deformation curve is straight with two end points. The values for points are displayed down the left side of the window. The default deformation curve does not deform the loft object in any way. For example, the points on the **Scale** deformation curve default to 100 at each end, meaning the object is at 100% of its original size.

The numbers across the window top refer to a percentage of the path. The left end of the deformation curve affects the loft object at the beginning of the path, while the right end affects the end of the path.

Deformation curves work with X, Y and Z directions. The Z direction always points up the path. The X and Y directions are the local X and Y axes of the first shape on the path. These X and Y axes are used when working with deformations.

To change a deformation curve, add points to the curve by clicking

Insert Corner Point ⊞ and clicking on the deformation curve. Click

Move Control Point ⊞ and click on a point to select it. When a point is selected, it turns white. Points can also be selected by holding down the <Ctrl> while clicking each point, or by drawing a bounding area around points. Click and drag on the point to move it.

When a point is selected, its percentage and value are displayed in entry boxes at the bottom center of the deformation window. You can type in a new percentage or value in these entry boxes.

Several of the buttons on deformation windows are common to all windows.

⊟	**Make Symmetrical**	When this button is on, the same curve or shape is used for both the X and Y axes.
⟋	**Display X Axis**	When this button is on, the deformation curve for the X axis is displayed, and can be changed. The X axis deformation curve is displayed as a red line.
⟍	**Display Y Axis**	When this button is on, the deformation curve for the Y axis is displayed, and can be changed. The Y axis deformation curve is displayed as a green line.
⤬	**Display X/Y Axes**	When this button is on, the deformation curve for both the X and Y axes are displayed, and can be changed.
⤲	**Swap Deform Curves**	Swaps the X and Y deformation curves.
⊞	**Move Control Point**	Moves selected control points. The buttons on this flyout can be used to constrain movement of control points to horizontal ↔ or vertical ↕.

Panels

⊥	**Scale Control Point**	Scales one or more selected control points toward or away from each other.
✳	**Insert Corner Point**	Inserts a linear control point on a deformation curve. To insert the point, click **Insert Corner Point** and click on the deformation curve. A linear control point is similar to a corner vertex on a spline. You can change the control point type by right-clicking on a selected control point and choosing a new type from the pop-up menu that appears.
✳	**Insert Bezier Point**	Inserts a Bezier control point on the deformation curve. Click **Insert Bezier Point** and click on the curve to set the point. A Bezier control point has handles that can be moved to change the shape of the curve around the point. This button is available from the **Insert Corner Point** flyout ✳.
⊖	**Delete Control Point**	Deletes selected control points.
✕	**Reset Curve**	Resets the deformation curve to the default curve, deleting any inserted control points.

The buttons at the lower right of each window can be used to control the display of the deformation curves in the window.

🖑	**Pan**	Click **Pan** and click and drag on the window to pan the display of deformation curves.
⧄	**Zoom Extents**	Zooms the window to the extents of displayed curves. **Zoom Horizontal Extents** ⧄ zooms to the horizontal extents, while **Zoom Vertical Extents** ⧄ zooms to the vertical extents.
⇥	**Zoom Horizontal**	Click **Zoom Horizontal** and click and drag in the window to zoom horizontally.
⇕	**Zoom Vertical**	Click **Zoom Vertical** and click and drag in the window to zoom vertically.
🔍	**Zoom**	Click **Zoom** and click and drag on the window to zoom in or out.

Zoom Region To zoom into a specified region, click **Zoom Region**. Click and drag on the window to set the region.

Bevel Deformation

The **Bevel** deformation window has one additional flyout with three buttons.

 Normal Bevel

Adaptive (Linear)

Adaptive (Cubic)

These buttons select the method used to calculated the bevel angle. **Normal Bevel** is selected by default. When the shape or path has sharp curves or corners, the bevel might not look right. If this happens, choose **Adaptive (Linear)** or **Adaptive (Cubic)** to adjust the bevel angle.

Fit Deformation

Fit deformation works differently from other deformation types. **Fit** deformation has no default deformation curve. The deformation curves are shapes created with ordinary MAX spline creation tools such as **Circle** and **Line**. The shapes are used to determine the profile of the loft object. To pick a shape for **Fit** deformation, click **Get Shape** and click on the shape in a viewport.

Mirror Horizontally Mirrors the deformation shape from left to right.

Mirror Vertically Mirrors the deformation shape up and down.

Rotate 90 CCW

Rotate 90CW

Sometimes, a deformation shape comes into the window with the wrong orientation. Use **Rotate 90 CCW** to rotate the shape 90 degrees counterclockwise, or **Rotate 90 CW** to rotate it 90 degrees clockwise.

Delete Curve Deletes the entire deformation shape.

Get Shape To get a shape from the scene for use as a deformation curve, click **Get Shape** and click on the shape.

Generate Path Replaces the path with a straight line path with one vertex on each end.

Sub-Object Levels

A loft object has two **Sub-Object** levels, **Shape** and **Path**. The **Shape** sub-object level has options for working with individual shapes on the path.

Panels

MapScaler

A modifier that retains the scale of a map even when the object is scaled.

General Usage

Apply a mapped material to an object. Select the object and go to the Modify panel ⚙️. Click **More** if necessary. Choose **MapScaler** from the World Space Modifiers list.

Scale the object. The map on the object stays the same size regardless of the size of the object.

Parameters

Scale
: The height of one tile area of the map, in units. To figure out the scale, find out how large an area you would like the map to cover in the final scene, and set Scale to the height of this area.

Wrap Texture
: When checked, MAX attempts to wrap the map evenly around the object.

Up Direction
: Sets the axis to be considered the V direction for mapping coordinates. For information on the V direction in mapping coordinates, see the **Mapping Coordinates** entry in the *Mat Editor* section.

Usage Notes

When an object is scaled, any maps applied to the object are scaled as well. The **MapScaler** modifier prevents maps on objects from being scaled when the object is scaled. This modifier is primarily for use in mapping walls for architectural renderings.

Material

A modifier that applies and animates material IDs.

General Usage

Select an object and go to the **Modify** panel ![icon]. Click **More** if necessary. Choose **Material**. Change the **Material ID** as desired.

Parameters

Material ID   Sets the material ID for the object or selected faces.

Usage Notes

You can use the **Material** modifier to change the **Material ID** for selected faces only by first using the **Mesh Select** modifier to select faces, then choosing the **Material** modifier.

The **Material ID** can be animated with the Material modifier.

For information on material IDs and how they work, see the **Multi/Sub-Object** entry in the *Mat Editor* section.

Creates, opens and runs MAXScript files.

General Usage

On the **Utilities** panel [icon], choose **MAXScript**. From the **Utilities** pulldown, choose the MAXScript you wish to run. Click **Run Script** to run the program.

You can also start a new script by clicking **New Script**, or open a script for editing with **Open Script**.

MAXScript

Open Listener	Opens the MAXScript Listener window. In this window, you can type MAXScript commands and have them executed immediately. Output from a currently running script also appears in this window.
New Script	Opens the MAXScript editor for a new script. MAXScript files are saved with the extension *.ms*.
Open Script	Displays a file selector for opening an existing script.
Run Script	Displays a file selector for selecting a script to run.
Utilities	When you run a script, it sometimes defines utilities. After a script is run, utilities defined by the script appear on this pulldown. To run a utility, choose it from the pulldown.
Close	Closes the utility.

Usage Notes

MAXScript is a utility for writing programs that control various aspects of MAX. In many ways, using **MAXScript** is like writing your own 3D Studio MAX plug-ins.

MAXScript includes its own programming language. The many intricacies of **MAXScript** are beyond the scope of this book. Consult the 3D Studio MAX online help for information on programming with **MAXScript**.

Measure

Displays dimension and volume information about selected objects.

General Usage

On the **Utilities** panel $\boxed{\mathsf{T}}$, choose **Measure**. Select one or more objects with any of the MAX selection functions. Information about the object appears on the panel.

Measure

The name of the selected object appears just below the Measure rollout title. If multiple objects are selected, then **Multiple Selected** is displayed.

Lock Selection
When this checkbox is checked, information about the current selection is retained even if another object is selected. Check this checkbox when you want to view the current object's information while creating or editing another object.

Surface Area
Displays the surface area of the object. If more than one object is selected, the total surface area is displayed. If only shapes are selected, no information appears here.

Volume
Displays the volume of the object. If more than one object is selected, the total volume is displayed. If only shapes are selected, no information appears here. If geometry with missing or deleted faces is selected, an asterisk (*) appears to the right of the volume value to indicate that the value shown may not be accurate.

Center of Mass
Displays the center of mass of the object. If more than one object is selected, the total center of mass of the selected objects is displayed. Center of mass is calculated with the assumption that each object has uniform mass throughout its volume. If only shapes are selected, no information appears here.

Create Center Point
Creates a helper point object at the center of the object or selection set. For information on helper point objects, see the **Point** entry in the *Panels* section of this book.

Length
Displays the length of the selected shape. If more than one shape is selected, the total length appears. If no shapes are selected, no information appears here.

Dimensions　Displays the dimensions of the objects along the world coordinate system. Dimensions are calculated according to the amount of space the object currently takes up in world space, not according to the object's original dimensions. Objects that have been rotated or scaled will show dimensions different from their original creation dimensions.

New Floater　Displays a floater with all information from the Measure rollout. The floater can be extended to the left or right to display long numbers. The **Lock Selection** checkbox on the floater performs the same function as the **Lock Selection** checkbox on the panel.

Close　Closes the utility.

Usage Notes

When the floater is displayed, you can view the length and dimensions of geometry and shapes while they are being created.

If any of the values displayed are too long to fit in the panel, click **New Floater** and extend the floater to the right to display the full length of the numbers.

MeshSmooth

A modifier that adds faces to an object and smooths sharp edges.

General Usage

Select an object. On the **Modify** panel ![icon], click **More** if necessary, and choose **MeshSmooth**. If the selected object is complex, wait a few moments while the **MeshSmooth** solution is calculated. Check **Smooth Result** to smooth the resulting object.

Parameters

Strength
A value from 0 to 1 that sets the size of new faces relative to existing faces. Smaller values create smaller faces, while larger values create larger faces. The default value of 0.5 makes the new faces about the same size as the existing faces. Changing the **Strength** does not change the number of new faces, just the size.

Relax Value
A value from -1 to 1 that affects each vertex's position relative to its neighboring vertices. The **Relax Value** looks at the vertices around each vertex and finds the average of all distances. When the **Relax Value** is 1, the vertex moves so it is the average distance from all neighbors, making the object smoother and rounder. At -1, it moves away from its neighbors by this average distance, making sharp edges. At 0, there is no change in vertex position.

Sharpness
A value from 0 to 1 that determines how sharp an angle at a vertex must be before new faces are added at that point. For each vertex, the average angle of all its edges is calculated. If the average angle is less than the **Sharpness** value times 180 degrees, faces are added at the vertex.

Operate On
Choosing **Faces** ![icon] smooths according to triangular faces, while **Polygons** ![icon] uses the object's polygons, which are made up of one or more faces. Each option creates a different pattern of faces.

Quad Output
When checked, the **MeshSmooth** modifier attempts to create only four-sided polygons on the mesh. This option creates yet another pattern of faces for each of the **Face** and **Polygon** methods.

Panels

Apply To Whole Mesh	When checked, any face, vertex or edge selection passed up the modifier stack is ignored, and **MeshSmooth** is applied to the entire mesh.
Iterations	The number of times the **MeshSmooth** is applied to the object. A higher number of Iterations takes longer to calculate. With a complex object, 3 or 4 Iterations can take several minutes to calculate.
Smooth Result	When checked, all faces are assigned the same smoothing group.
Separate by	When **Materials** is checked, new faces are never created between two adjacent faces with different material IDs. When **Smoothing Groups** is checked, new faces are never created between two edges with different smoothing groups.
Update Options	These options determine when the object is updated on the screen. **Always**, the default, immediately updates the object whenever a change is made. **When Rendering** updates the object when the scene is rendered. **Manually** updates the object only when the **Update** button is checked.

Panels

Mirror

Mirrors an object along any axis.

General Usage

Select an object. On the **Modify** panel 🔧, click **More** if necessary to access the list of modifiers not displayed. Choose **Mirror**. Select an axis or axes over which to mirror the object. Check **Copy** to copy the object.

Sub-Object Level

The **Mirror** modifier has one **Sub-Object** level, **Mirror Center**. Move the mirror center to change the base point from which the object is mirrored.

Parameters

Mirror Axis   Changes the axis along which mirroring takes place. The object's local axis is used for alignment.

Offset   Moves the mirrored object or copy in the direction of the mirror axis by the specified number of units.

Copy   Check **Copy** to make the mirrored object a copy of the original object.

Usage Notes

The **Mirror** modifier works similarly to the **Mirror Selected Objects** button ▶◀ on the Toolbar.

The advantage of using the **Mirror** modifier is that you can later change the modifier's parameters or delete the modifier from the stack altogether if necessary. You can also animate the **Mirror Center** sub-object to animate the mirroring of the object.

The disadvantage is that a mirrored copy created with the **Mirror** modifier is an instance by nature, and can only be moved by moving the original object, or by changing the **Offset** value and/or moving the **Mirror Center** sub-object on the **Mirror** modifier panel.

Modify

Modifies the selected object or objects.

General Usage

There are several operations that can be performed on the **Modify** panel 🐟.

- Apply a modifier to an object.
- Access a previously applied modifier so its parameters can be changed.
- Click **Edit Stack** 🗐 to change the object type or collapse the object.
- Access additional parameters for the object that aren't available on the **Create** panel 🛬.

Applying a Modifier

Select one or more objects and go to the **Modify** panel. Some modifiers appear on the Modifiers rollout. Click on a modifier to select it, or click the **More** button to see a list of all other modifiers, and choose from the list.

The list of modifiers changes depending on the type of object selected. Some modifiers can only be applied to shapes, while others can only be applied to 3D objects.

As each modifier is applied, it goes on to a list called the *modifier stack*. The last modifier applied is placed at the top of the stack. At the bottom of the stack is the creation level, showing the type of object created. At any time, you can access a previously applied modifier or the creation level from the stack by choosing it from the modifier stack pulldown on the Modifier Stack rollout. The panel changes to reflect the parameters of the modifier or creation level. You can then change the parameters as you like.

When sub-objects are selected with modifiers such as **Mesh Select**, **Edit Mesh**, **SplineSelect** or **Edit Spline**, and the **Sub-Object** button is not turned off before the next modifier is applied, the next modifier is applied to the sub-object selection. This is called *passing a selection up the stack*. The modifier used to make the selection appears on the stack with an asterisk (*) next to it. A modifier applied after the selection modifier also has an asterisk next to it to indicate that it is applied to a selection only. To disable the selection and apply a modifier to the entire object, apply another selection modifier such as **Mesh Select** or **SplineSelect**, and turn off the **Sub-Object** button. The next modifier added to the stack will be applied to the entire object.

Several options are available on the **Modify** panel for working with modifiers and the modifier stack.

Use Pivot Point	If multiple objects are selected, checking **Use Pivot Point** causes the modifier to use each object's pivot point as the center of the modifier for that object. If **Use Pivot Point** is unchecked, the center of the selection is used to modify each object. You must check or uncheck this option before applying the modifier in order for it to take effect.

 Pin Stack   When **Pin Stack** is clicked and you select another object, the information that was on the panel when you clicked **Pin Stack** will remain on the panel.

 Active/ inactive modifier toggle   Turns the highlighted modifier on or off in both viewports and the rendered image.

 Active/ inactive in viewport   This button turns the effect of the current modifier on or off in viewports. The modifier is still active in rendered images. This button is useful for times when you want to see how the object would look without a particular modifier. This button is available from the flyout under the **Active/inactive modifier toggle** button.

Show end end result on/off toggle   When the toggle is off, the effect of modifiers at or below the current level are displayed, and the effect of modifiers higher on the stack are not.

Make Unique   When a modifier is applied to a selection of objects, the modifier is an *instanced* modifier. If you later return to that modifier for one or more the selected objects, any changes made to the modifier will also affect the other objects in the original selection. There may be times when you want to make an instanced modifier unique to a selected object, meaning it will only affect the selected object and not the entire selection. To do this, select the object and click **Make Unique**.

Remove modifier from the stack   Removes the current modifier from the stack.

Using Edit Stack

The **Edit Stack** button 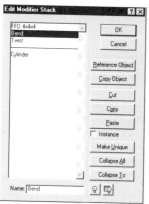 can be used to change the object type or edit the modifier stack. When you click **Edit Stack**, one of two things happens depending on the type of object selected. Either a pop-up menu appears, or the Edit Modifier Stack dialog appears.

If the pop-up menu appears, you can choose to change the object type to one of the types listed. You can also choose **Edit Stack**, which will access the Edit Modifier Stack dialog.

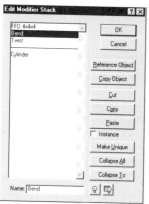

Here you can change the names of modifiers on the stack, and copy and paste modifiers. You can also collapse the stack to conserve memory and improve redraw time.

Reference Object	Turns an Instance object into a **Reference** object. A line is placed above the highest modifier on the stack. This line indicates that the next modifier applied to this object will be applied to this object only. For information on the difference between **Copy**, **Instance** and **Reference**, see the **Clone** entry in the *Menus* section.
Copy Object	Turns an **Instance** or **Reference** object into a **Copy**. All modifiers on the stack become unique to the object.
Cut **Copy** **Paste**	These options can be used to cut or copy one or more modifiers from the stack, and paste them elsewhere on the stack or onto another object's stack. To **Cut** or **Copy** modifiers, highlight them on the stack and click **Cut** or **Copy**. To paste modifiers, highlight the modifier or line just below where you want to paste the modifiers, and click **Paste**. If **Instance** is checked when **Paste** is clicked, instances of the modifiers are added to the stack.

Make Unique   Highlight a modifier and click **Make Unique** to make the modifier unique. See **Make Unique** under the **Applying a Modifier** section above.

Collapse All   Collapses the entire stack to an **Editable Mesh**, **Editable Patch** or **Editable Spline** entry. The stack then consists of one entry with no modifiers. Collapsing the stack causes the object and its modifications to take up less memory, which means it redraws faster. However, once the stack is collapsed, the object's modifiers can no longer be accessed and changed. Collapsing the stack also deletes any animation keys for the object's parameters such as **Height** or **Radius**. Animation keys created by turning on the **Animate** button and transforming the object are not affected.

Collapse To   Collapses the stack up to the currently selected modifier.

Name   Allows you to enter a new name for the modifier. A name can be a meaningful description such as "Taper 45 degrees" or "Top faces extruded".

Active/ inactive modifier toggle   Turns the highlighted modifier on or off in both viewports and the rendered image.

Active/ inactive in viewport   This button turns the effect of the current modifier on or off in viewports. The modifier is still active in rendered images.

Modify Parameters

For some objects, there are parameters on the **Modify** panel that don't appear on the **Create** panel. For example, the **Modify** panel for a **Loft** object has a Deformation rollout that does not appear on the **Create** panel. For information on these parameters, see the entry concerned.

Morphs from one object to another.

General Usage

Create two or more objects with the same number of vertices. Select one object. On the **Create** panel ![icon], click **Geometry** ![icon]. From the pulldown list, choose **Compound Objects**. Click **Morph**. Click **Pick Target**, and click on each morph target.

Move to the first frame where you want the object to morph to one of the morph targets. Highlight the morph target on the list, and click **Create Morph Key**. Continue to move to each key frame, select the morph target and click **Create Morph Key**. Click **Play Animation** ![icon] to see the object morph between targets.

Pick Targets

Pick Target	To pick objects to be used as morph targets, click **Pick Target** and pick each target object. Only objects with the same number of vertices as the original object can be picked.
Reference **Move** **Copy** **Instance**	Determines how morph targets are used for the morph operation. Choosing **Move** uses the object itself. Choosing **Reference**, **Copy** or **Instance** uses one of these clone types as the morph target, leaving the original object in the scene. For information on these clone types, see the **Clone** entry in the *Menus* section.

Current Targets

Morph Targets	Lists the currently active morph targets.
Morph Target Name	Displays the name of the highlighted morph target. You can enter a new name for the morph target here. Entering a new morph target name does not change the name of the original object.
Create Morph Key	Creates a key for the highlighted morph target at the current frame.
Delete Morph Target	Removes the highlighted morph target from the list.

Usage Notes

Morphing works by moving the vertices of one object to the positions of vertices in a another object. A *morph target* is an object to which another object will morph. Morph targets must have the same number of vertices as the original object.

The most stable method for making morph targets is to copy the original object and alter the copies without changing the number of vertices. You can, if you like, create morph targets with completely different creation methods. As long as they have the same number of vertices, the objects can be morphed. However, the vertex configuration on the objects may be so different that vertices will not change positions smoothly, making the object appear mangled while the morph is taking place.

There are many alternatives to morphing for changing the shape of an object. If you want an object to simply bend over a period of time, for example, you can apply a **Bend** modifier to the object and animate the bend's **Angle**, eliminating the need for morph targets. If you want a space warp to be applied gradually over time, you can often animate the space warp's parameters to get the effect you want.

You can use **Edit Mesh**, **Edit Spline** or **Edit Patch** to change a morph target's shape, as long as you don't delete vertices from the object. For example, you can create an open spline for creating a vase with the **Lathe** modifier. Before lathing, copy the spline, and use **Edit Spline** to move the vertices in the copy. Apply the **Lathe** modifier to both splines, using the same **Lathe** settings for both. The two lathed objects have the same number and configuration of vertices, and can be used as morph targets for each other.

Another useful tool for creating morph targets is the **Conform** compound object, which conforms an object to another object. Use **Conform** to create morph targets, then use **Morph** to morph between them. See the **Conform (Geometry)** entry for information on this tool.

The influence of each morph target can be adjusted in the Track View window. Click the **Track View** button to access the window. In the **Track View** Hierarchy, display the **Morph** track for the original object, and right-click on a key on the **Morph** track. The Key Info dialog appears. Here you can adjust each morph target's influence percentage.

Panels

Captures motion from a device and applies to selected tracks.

General Usage

Access **Track View** 🔲. Select a track for which you want to apply motion capture. Click **Assign Controller** 🔲 and select a motion capture controller. Click **Properties** 🔲 and choose a device for each component of the track for which you want to use motion capture.

Go to the **Utilities** panel 🔨. Click **Motion Capture**. When you are ready to record motion, click **Start** and use the motion capture device to input data. When recording is finished, the system will take a few moments to create keys.

Motion Capture

Start	Starts recording motion from the motion capture device.
Stop	Stops recording motion from the motion capture device.
Test	Allows you to test motion capture. To test, click **Test** and input data from the motion capture device. The selected tracks behave as if the motion were being captured. Using **Test** does not record any motion capture data.
Play During Test	Plays existing animation during a test. Any motion assigned to selected tracks is not played as the tracks are awaiting motion capture data from the device.
Start/Stop	If you are using a midi control device with stop, play and record controls, you can use this device to control the recording of motion capture data. To use this type of device, click **Start/Stop**. Select the device from the pulldown list. Specify the **Channel** to which the midi device is assigned, and specify the note event which will trigger the **Stop**, **Play** and **Record** functions.
Enable	Enables the use of the midi device specified under **Start/Stop** to record motion capture, in addition to the **Start**, **Stop** and **Test** buttons on the **Utilities** panel.

Motion Capture

Tracks	Lists all tracks with motion capture controllers assigned. To select a track, click on the track. When a track is selected, the box to the left of the track name turns red. You can also use the **All, Invert** and **None** buttons to control the selection.

The entry box just under **Tracks** is for named selection sets of tracks. To name a selection set, select one or more tracks. A track is selected when the box at the left of its name is red. When the desired tracks are selected, enter a selection set name in the entry box and press **<Enter>**. To choose a selection set, access the pulldown menu of selection set names and choose the selection set. To delete a selection set, choose the selection set and click **Delete Selection Set** ✕. A track may be in only one selection set at a time. |
All	Selects all listed tracks to receive motion capture data.
Invert	Inverts the selection of tracks to receive motion capture data.
None	Unselects all tracks.
Preroll	Sets the number of frames that will play before the **In** frame. Setting **Preroll** to a value less than 0 allows you a period of time to prepare for making motion.
In	Sets the first frame for which selected tracks will receive motion capture data.
Out	Sets the last frame for which selected tracks will receive motion capture data.
Live During Preroll	When checked, incoming motion capture data is displayed on objects during preroll, but is not recorded. This can cause sudden shifts of motion when the **In** frame is reached as the captured data may not match the incoming data at that point. When unchecked, the effects of incoming motion capture data are not displayed until the **In** frame is reached.
Per Frame	Sets the number of keys that will be generated per frame. Checking **1** sets one key per frame, while checking **2** creates 2 keys per frame.

Reduce Keys   When checked, the number of keys is reduced. The **Threshold** value is used to determine how much keys can be reduced.

Threshold   The number of increments of tolerance for key reduction. When **Reduce Keys** is on, MAX captures the motion then analyzes it to see where keys can be removed while still retaining motion similar to what was captured. Keys are reduced to as few as possible while keeping all motion within the **Threshold** increment. For example, if motion capture is used to set keys for a **Rotation** track while **Threshold** is 2.0, keys will be reduced while keeping all rotation to within 2 degrees of the original motion capture data. If motion capture is used for a **Position** track, **Threshold** refers to the number of units of tolerance.

Usage Notes

In order to use motion capture, you must first assign a motion capture controller to one or more tracks.

Whether or not the **Animate** button is on does not affect the recording of motion capture data.

Motor

A space warp that applies a specific amount of rotational force (torque) to an object or particles.

General Usage

Create a particle system object, or any other object.

Under the **Create** panel , click **Space Warps** . Select **Particles & Dynamics** from the pulldown menu. Click **Motor**. Click and drag on the screen to create the space warp object.

To use the space warp with a particle system, select the particle system and click **Bind to Space Warp** on the Toolbar. Click and drag from the object to the space warp to bind it to the space warp.

To use the space warp on an object, the space warp must be assigned as part of a dynamics simulation with the **Dynamics** utility. See **Dynamics** for more information.

Parameters

On Time
Off Time
The frame numbers where the rotational force begins and ends. For particles, when the **On Time** frame is reached, particles already in the scene react to the force as a group. Particles born between **On Time** and **Off Time** are affected as they are born. Particles that have already been affected continue in their current directions after the **Off Time** frame is reached. Particles born after the **Off Time** are not affected by the space warp.

Basic Torque
Sets the amount of rotational torque on the object or particle system. The amount of force is expressed as Newton-meters (**N-m**), Pounds-feet (**Lb-ft**) or Pounds-inches (**Lb-in**). With dynamics, the amount entered for **Basic Torque** is accurately represented in the scene. With particles, the amount is subjectively interpreted.

Feedback On
When checked, the rotational torque is decreased as objects approach the **Target Revs**. When unchecked, the rotational force is constant regardless of object speed.

Reversible
When checked, the rotational force is reversed when an object is above the **Target Revs**. This option is available only when **Feedback On** is checked.

Panels

Target Revs Specifies a target number of revolutions for use with the **Feedback On** and **Reversible** options. The units for **Target Revs** can be set to revolutions per hour (**RPH**), minute (**RPM**) or second (**RPS**). This option is available only when **Feedback On** is checked.

Gain Specifies how quickly the rotational force is affected when **Target Rev** is approached. When **Gain** is 100, the object's revolution rate is corrected immediately. When **Gain** is lower than 100, the correction time is slower, resulting in a looser interpretation of the **Target Revs. Gain** can also be over 100. This option is available only when **Feedback On** is checked.

The settings in the **Periodic Variation** section create noise to randomly affect the **Basic Force** value. Two sets of noise can be used.

Enable Enables periodic variation with the settings below.

Period 1 Sets the total time for the variation. For example, a **Period 1** of 50 cycles the effect every 50 frames.

Amplitude 1 The size of the effect. This value uses the same units as **Basic Torque**. If the amplitude is the same as the **Basic Torque**, the torque will range from zero to twice the **Basic Torque**.

Phase 1 The noise effect is created with a sine wave. **Phase 1** sets which phase the wave is in, such as whether it is just starting, is peaking or is just ending at any point in time. A phase of 1.0 shifts the wave by an entire wavelength, a phase of 2.0 shifts it by two entire wavelengths, etc. To make the noise look different, the phase can be changed to a fractional number such as 0.5. To make the noise change over time, animate the phase value.

Period 2
Amplitude 2 Sets the period, amplitude and phase for the second noise effect.
Phase 2

The **Particle Effect** section specifies a range for the **Motor** effect. The range is set in a spherical shape around the **Motor** space warp icon.

Enable Enables the range and displays it on the screen.

Range Sets the radius of the range.

Icon Size Sets the size of the icon in square units. The size of the icon has no relationship to the motor effect.

Panels

NavInfo

Sets navigational parameters for a VRML scene.

General Usage

On the **Create** panel , click **Helpers** . From the pulldown menu, choose **VRML 2.0**. Click **NavInfo**. Click and drag to create the icon. Change parameters as desired.

NavigationInfo

Type	Specifies the type of movement for the VRML world. **Walk** is used for navigating a VRML world on foot or with a vehicle that sits on or just above the ground plane. **Fly** is similar to walk, except that gravity and terrain following may be disabled or ignored in the browser. When **Walk** or **Fly** is selected, the browser will strictly support collision detection. **Examine** is used for examining individual objects, and can include, but does not require, the ability to move closer or farther away from objects, or to orbit around them. **None** disables all browser-specific navigation, requiring the user to navigate with mechanisms in the scene such as **Anchor** or **Inline** helpers.
Headlight	When checked, a direct light always points into the scene from the view. Check this option only if there are no lights in the scene.
Visibility Limit	Sets the distance from the view that can be seen, in units. Any objects beyond this distance are not visible.
Speed	Sets the speed at which the viewer can move through the scene, in units per second.
Collision	Sets the distance from the view and objects at which collision is detected. The **Collision** value effectively sets the width of the viewer.
Terrain	Sets the height of the view over the ground or surface. In effect, the **Terrain** value sets the height of the viewer.
Step Height	Sets the maximum height of objects that can be stepped on to elevate the view. All walls should be above this height, while staircase steps and other objects that the viewer can climb onto should be below this height.
Icon Size	Sets the diameter of the icon. The icon size does not affect the operation of the **NavInfo** helper.

Creates a many-sided shape, with either circular or straight segments.

General Usage

On the **Create** panel ![icon], click **Shapes** ![icon]. Click **NGon**. Under the Parameters rollout, set the number of sides with the **Sides** value. Click and drag on the screen to create the shape.

General

See the General rollout under the **Arc** entry.

Creation Method

Edge   When **Edge** is turned on, the shape is created starting from the shape edge when you click and drag on the screen. To keep a vertex at the first location clicked, drag up, down, left or right rather than diagonally.

Center   When **Center** is on, the shape is created from the center outward when you click and drag on the screen.

Keyboard Entry

X, Y, Z   Specifies the location of the center of the shape created when **Create** is clicked. The center is placed at the X, Y and Z location in the world coordinate system.

Radius   See these parameters under Parameters below.
Corner Radius

Create   Creates the shape with the parameters entered. The shape is placed on the current viewport's plane.

Parameters

Radius   Specifies the radius of the shape. When **Inscribed** is selected, the shape fits inside an imaginary circle with **Radius** as its radius. When **Circumscribed** is selected, the shape fits just outside the circle.

Sides   Sets the number of sides on the shape.

Corner   Sets the radius of fillets at corners of the shape.
Radius

Circular　　When checked, the shape is created with circular segments, and each vertex is Bezier type. When unchecked, the shape is created with straight segments, and each vertex is Bezier corner type.

Usage Notes

An **NGon** is useful for creating a circle with more than four vertices. It can also be used to create circular objects such as gears.

A modifier or space warp that changes the shape of an object with random bumpy patterns.

General Usage

To apply **Noise** as a modifier, select the object. Click on the **Modify** panel ⚙ and click **More** if necessary to access the full list of modifiers. Select **Noise**.

To apply **Noise** as a space warp, go to the **Create** panel 🔧. Click **Space Warps** and select **Modifier-Based** from the pulldown menu. Click **Noise**. Click and drag on the screen to create the **Noise** space warp's length and width, then move the cursor and click again to specify the space warp's height. Select the object and use **Bind to Space Warp** to bind it to the space warp. To modify the noise, select the space warp and go to the **Modify** panel.

Adjust the **Strength** along the desired axes, and reduce **Scale** if necessary.

Sub-Object Levels

The **Noise** modifier has two **Sub-Object** levels.

Gizmo   The gizmo controller for the noise effect. This controller can be transformed and animated to animate the noise effect.

Center   The center from which the noise effect emanates.

Space Warp Parameters

Length   Sets the dimensions of the space warp object.
Width
Height

Decay   Sets the amount by which the noise strength will diminish over distance. When **Decay** is 0, the noise strength is the same over the entirety of the object. When **Decay** is higher than 0, the noise decays (becomes less) farther from the space warp center.

Parameters

Seed   Sets the base number for calculating noise.

Scale   Sets the scale of the noise pattern. Often the default value of 100 produces very large noise that cannot be seen on the object. Change this value to a lower number such as 10 or 20 to see the effect.

Fractal
Uses a fractal equation to generate the noise pattern. Fractal patterns are often smoother and more realistic looking than non-fractal patterns.

Roughness
A value from 0 to 1 that determines the sharpness of the fractal noise. Higher values generate rougher noise than lower values.

Iterations
The number of times the noise is calculated when **Fractal** is checked. Setting this value to 1 is the same as unchecking **Fractal**.

Strength
Sets the strength of the noise along each of the object's local axes. Try strength values roughly equal to the radius or height of the object to see the noise effect. You must enter a value for at least one of the axes in order to generate noise.

Animate Noise
Enables animation of noise. You must animate the **Phase** value in order to animate noise.

Frequency
Noise is created with a sine wave. **Frequency** sets the amount of time taken to complete one sine wave. Lower values make animated noise quiver more slowly, while higher values make it quiver faster.

Phase
Noise is created with a sine wave, and animation of noise is achieved by shifting the sine wave over time. Phase refers to the starting point of the sine wave. Animate the **Phase** value to shift the starting point of the wave, and thus animate the noise effect. In order to animate the noise, the **Phase** value must be animated and the **Animate Noise** option must be checked.

Usage Notes

In order to see how **Noise** affects an object, the object must have sufficient detail.

To produce noise, you must enter a value for at least one of the axes listed under **Strength**. Try values of 20, 30 and 40 for X, Y and Z to start. If you still don't see any noise, reduce **Scale** to 10. Experiment with different **Strength** and **Scale** settings to see the effect.

To animate the noise effect, you can animate the **Gizmo** and **Center** rather than the **Phase** value.

Panels

Unifies or flips normals.

General Usage

On the **Modify** panel 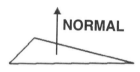, click **More** if necessary. Choose **Normal**. Choose to **Unify Normals** and/or **Flip Normals**.

Parameters

Unify Normals   Models imported from other 3D programs sometimes have normals pointing in seemingly random directions. **Unify Normals** attempts to turn normals in the right direction with a complex calculation. This process sometimes doesn't yield the expected results.

Flip Normals   Flips normals in the opposite direction.

Usage Notes

A normal is an imaginary arrow pointing out from a face.

NORMAL

Normals are used to determine the "inside" and "outside" of an object. Faces with normals pointing toward the view or camera are considered to be outside, while normals pointing away from the view or camera are considered to be inside.

When creating and rendering simple objects such as primitives, the concept of face normals doesn't really come into play. Only when you do something a little unusual, like placing a camera inside a sphere, will you see how the direction of face normals affects a rendering.

When a camera is placed inside a sphere (with appropriate lighting, of course), it essentially sees nothing. The face normals on the sphere all point outward from the sphere, which means the face normals are pointing away from the camera no matter where it looks. In this case, nothing appears to render from the camera view.

This phenomenon also occurs when an object is cut in half. The inside of the object is exposed, but doesn't seem to be there in a rendering. This is because the face normals of the cutaway part of the object are pointing away from the view or camera.

Panels

Normal

Normals pointing away from the camera are often seen in models imported from other programs, particularly CAD programs. If you render an imported model and parts of it seem to be missing, you most likely have some face normals pointing the wrong way.

There are several ways to deal with faces that are pointed away from the camera:

- Turn on the **Force 2-Sided** checkbox on the Render Scene dialog when rendering
- Turn on the **2-Sided** checkbox under the Basic Parameters rollout of the object's material
- Use a **Double Sided** material
- Apply a **Normal** modifier to the object and turn on the **Unify Normals** or **Flip Normals** checkboxes

You can apply the **Normal** modifier to specific faces only by using the **Mesh Select** modifier to select faces before applying the **Normal** modifier.

Panels

Selects a NURBS curve to which to apply another modifier.

General Usage

Create a **Point Curve** or **CV Curve**, as a sub-object of either a NURBS curve or NURBS surface. On the **Modify** panel ⚡, click **More** if necessary and choose **NCurve Sel**. Click **Sub-Object** and select the sub-object level you wish to select.

Click on the NURBS sub-object on the screen to select it. The sub-object turns red when selected. Apply another modifier such as **Taper** or **Bend**. The new modifier is applied only to the selected NURBS sub-object.

Usage Notes

For information on creating NURBS surfaces and sub-objects, see the **Point Surf** entry.

NSurf Sel

Selects a NURBS surface to which to apply another modifier.

General Usage

Create a **Point Surf** or **CV Surf** as a sub-object of a NURBS surface. On the **Modify** panel ⚡, click **More** if necessary and choose **NSurf Sel**. Click **Sub-Object** and select the sub-object level you wish to select.

Click on the NURBS sub-object on the screen to select it. The sub-object turns red when selected. Apply another modifier such as **Taper** or **Bend**. The new modifier is applied only to the selected NURBS sub-object.

Usage Notes

NSurf Sel can be likened to the **Mesh Select** modifier, which selects faces, vertices and edges at the sub-object level of a mesh object.

For information on creating NURBS surfaces and sub-objects, see the **Point Surf** entry.

Creates a cylinder with convex caps.

General Usage

On the **Create** panel ![icon], click **Geometry** ![icon]. Select **Extended Primitives** from the pulldown menu. Click **OilTank**. Click and drag to set the radius of the cylinder, then move the cursor and click to set the height. Move the cursor and click again to set the cap height.

Creation Method

Edge	Draws the oiltank starting from the outer edge of the cylinder.
Center	Draws the oiltank starting from the center of the cylinder.

Keyboard Entry

X, Y, Z	Sets the location of the oiltank base center along the X, Y and Z axes of the coordinate system.
Radius Height Cap Height Overall Centers Blend	See these listings under Parameters below.
Create	Creates the oiltank using the parameters entered.

Parameters

Radius	Radius of the cylindrical center portion of the oiltank.
Height	When the **Overall** option is on, this value refers to the height of the entire oiltank. When the **Centers** option is on, this value refers the height the cylindrical midsection of the oiltank.
Cap Height	Sets the height of the oiltank caps. **Cap Height** cannot be larger than the **Radius** value.
Overall	When this option is on, the **Height** value refers to the height of the entire oiltank, including the caps.
Centers	When this option is on, the **Height** value sets just the height of the straight part of the oiltank.

Blend	A **Blend** value over 0 creates a bevel between each cap end and the midsection. The bevel extends into the midsection and caps by half the number of units specified by the **Blend** value. The **Blend** value cannot be greater than half the midsection length.
Sides	Sets the number of sides on the oiltank.
Height Segs	Sets the number of segments along the length of the cylindrical portion of the oiltank.
Smooth	When the **Smooth** checkbox is on, the surface renders as a smooth, rounded object. When **Smooth** is off, the surface of the oiltank renders as faceted.
Slice On	Slices the oiltank around its local Z axis according to the **Slice From** and **Slice To** angles. The slice function starts at the **Slice From** angle and moves counterclockwise to the **Slice To** angle, cutting out the portion of the oiltank in this area. The sliced area is capped with two flat surfaces. Slice angles are measured from the object's local X axis, which is considered to be zero degrees.
Generate Mapping Coords	Generates mapping coordinates on the oiltank. Cylindrical mapping coordinates are applied to the cylindrical midsection of the oiltank while planar mapping coordinates are applied to the caps. Sliced areas receive planar mapping.

Creates an omnidirectional light.

General Usage

On the **Create** panel , click **Lights** . Click **Omni**. Click and drag in a viewport to place the light. Move the light or change parameters as desired.

General Parameters

See the General Parameters rollout under the **Target Spot** entry.

Projector Parameters

The options on the Projector Parameters rollout set up a map to be projected from the light source.

Projector	Turn on this checkbox to enable the projection map.
Map	Before clicking **Map**, set up a map in the Material Editor to be projected from the light. Click and drag the map to the **Map** button, or click **Map** and choose the projector map from the Material/Map Browser.

Shadow Parameters

See the Shadow Parameters rollout under the **Target Spot** entry.

Usage Notes

Omni lights shine in all directions. Omni lights are useful for general illumination of scenes, or to soften shadows cast by other lights in the scene.

To create a white light, set **R**, **G** and **B** to the same values. Use the **V** value to increase or decrease the intensity of the light.

A negative **Multiplier** with a gray or white light creates a dark spot where the light shines in the scene, simulating a shadow. A light with a negative **Multiplier** takes less time to render than a shadow, and can be used instead of a shadow when shorter rendering times are needed.

Optimize

Reduces the number of faces on an object.

General Usage

Select an object. On the **Modify** panel ![icon], click **More** if necessary. Choose **Optimize**. Adjust the **Face Thresh** value if desired.

Parameters

Level of Detail	Two sets of settings can be active for this modifier at the same time. To work with these levels of detail, choose the **L1** option for **Viewports**. Change parameters as desired. Choose **L2** for **Viewports** and try different parameter settings. Determine the settings you prefer, and choose the appropriate option for **Renderer**. The object will render with the settings for the selected option.
Face Thresh	Sets the threshold angle for faces. Adjacent faces that meet at an angle lower than the **Face Thresh** angle will be collapsed into one face. Higher values remove more faces. For very complex objects with lots of curves, this value can be set as high as 50 or 60 and still retain sufficient detail.
Edge Thresh	Sets the threshold angle for edges that bound only one face. For example, an edge at the top edge of an uncapped cylinder bounds only one face. In general, the **Edge Thresh** angle should be lower than the **Face Thresh** angle.
Bias	The **Optimize** modifier sometimes produces long, thin faces that make it impossible to bend or twist the object afterward. **Bias** is a value from 0 to 1 that helps eliminate this type of face on the object. A **Bias** of 0 has no effect on these faces, while higher values eliminate these faces by making many smaller faces from each long, thin face. To start with, work with the default value of 0.1. If there are still too many thin faces on the object, increase the **Bias**.
Max Edge Len	Specifies a maximum length for edges of optimized faces. When this value is 0, this parameter does not affect the optimization process. Higher values limit the face edge length. Setting this parameter assists in the elimination of long, thin faces addressed by the **Bias** parameter.

Panels (side tab)

Auto Edge When checked, edges of faces with angles within the **Face Thresh** are not displayed. When checked, these face edges are displayed.

Material Boundaries When checked, only faces with the same material ID are collapsed.

Smooth Boundaries When checked, only faces sharing at least one smoothing group number are collapsed.

Update On complex objects, the time to calculate optimization can take some time. When **Manual Update** is unchecked, optimization is calculated after each parameter is changed. When **Manual Update** is checked, optimization is calculated only when the **Update** button is clicked.

Last Optimize Status Displays the original and current face and vertex count for the object.

Usage Notes

The **Optimize** modifier can be applied only to specific faces by selecting faces with the **Mesh Select** modifier prior to applying the **Optimize** modifier.

Parameters

Assigns transform controllers to an object.

General Usage

Select an object. Go to the **Motion** panel ⌾. Click **Parameters**. High-light an animation track on the list and click **Assign Controller** ▷. Choose a controller from the list.

When a controller is selected for an animation track, this panel changes to reflect the controller options. Some controllers require information to be entered or objects to be selected.

For information on the rollouts that appear when you choose a controller, look up the controller name in the 3D Studio MAX online help. For information on the default controllers, look up **PRS Controller** in the online help.

A particle system that emits particles from an object.

General Usage

Create an object to be used as the particle emitter. Go to the **Create** panel ⬃ and click **Geometry** ⬁. Select **Particle Systems** from the pulldown menu. Click **PArray**. Click and drag on the screen to create the **PArray** icon. Click the **Pick Object** button under the Basic Parameters rollout, and click on the emitter object.

Click **Play Animation** ▶ to see the default particle animation. Change parameters as desired.

Basic Parameters

Pick Object To pick the emitter object, click **Pick Object** and click on the object to be used as an emitter. The emitter name appears below the **Pick Object** button.

Particle Formation Determines how particle emission points are distributed over the emitter object. These options are available for all particle types except **Object Fragments**.

Over Entire Surface distributes the particles at random across the emitter object surface. **Along Visible Edges** distributes particles at random on the edges of object faces. **At All Vertices** emits particles from object vertices. **At Distinct Points** chooses a set number of points scattered at random across the object, and emits particles only from those points. The number of points is set with the **Total** value. **At Face Centers** emits points from the center of the object's faces.

Icon Size Sets the dimensions of the **PArray** icon cube. The size of the icon does not affect the **PArray** effect.

Icon Hidden When checked, the icon is hidden. **PArray** parameters can be accessed by clicking on a particle.

Viewport Display Determines how particles will display in viewports. **Dots** displays small specks. **Ticks** displays small crosses. **Mesh** displays mesh objects depending on the particle type chosen. This option is not available when either the **Constant** or **Facing** particle type is selected. **Bbox** displays instanced objects as bounding boxes. This option is only available when **Instanced Object** or **Object Fragments** is chosen as the particle type.

Percentage of Particles sets the percentage of particles that appear in viewports. When using a large number of particles, set this value to 10 or 20 to keep viewports uncluttered and reduce redraw time.

Particle Generation

Particle Quantity	Determines the number of particles. When **Use Rate** is selected, the value below **Use Rate** sets the number of particles born on each frame. When **Use Total** is selected, the value below **Use Total** sets the total number of particles born between the **Emit Start** and **Emit Stop** frames. These options are not available when **Object Fragments** is chosen as the particle type.
Speed	The speed of particles in units per frame. **Variation** varies the speed by a percentage of the **Speed** value. For example, if **Speed** is 20 and **Variation** is 10, particle speeds will vary by 10% of 20 = 2 units per frame, resulting in speeds varying from 18 to 22 units per frame.
Divergence	The number of degrees of variation in a particle's direction. If **Divergence** is 0, particles travel straight out from the emitter. If **Divergence** is greater than 0, a particle's direction of travel can diverge from its original direction by an angle ranging from 0 to the **Divergence** angle.
Emit Start Emit Stop	The frames at which particles start and stop emitting.
Display Until	The frame at which all particles will disappear, regardless of whether the **Emit Stop** frame has been reached.
Life	Sets the number of frames that each particle will live. **Variation** is the number of frames by which **Life** can vary.
Subframe	When any of these checkboxes is checked, particles can be born between frames, and will already be in motion by the time the next frame is generated. To enable this feature for a still emitter, check **Creation Time**. If the emitter moves, check **Emitter Translation**. If the emitter rotates, check **Emitter Rotation**.
Size	The dimensions of the particle. This parameter has no effect when the **Object Fragments** particle type is selected. If **Instanced Geometry** is the particle type, the **Size** value is multiplied by the instanced object size. **Variation** is the percentage by which **Size** can vary.

Grow For   The number of frames from birth over which each particle will grow from nearly invisible to its full size.

Fade For   The number of frames before death over which a particle will shrink from its full size to nearly invisible.

Uniqueness   **Seed** sets the base number for calculation of the particle pattern. Click **New** to randomly generate a new **Seed**.

Particle Type

Standard Particles   Uses one of the particle types listed in the **Standard Particles** section of this rollout.

Triangle renders each particle as a triangular face. **Cube** renders each as a cube.

Special uses three intersecting rectangular faces. This type of particle is effective when used with a face-mapped material.

Facing renders each particle as a rectangular face that always faces the view. **Constant** renders each particle as an octagon that always faces the view. Both these types always face the view regardless of the options set on the Particle Rotation rollout.

Tetrahedron creates pyramids, **SixPoint** creates 2D six-pointed stars, and **Sphere** uses spheres.

MetaParticles   Uses spheres as metaball particles. The options for these particles are on the **MetaParticles Parameters** section of this rollout. When metaball particles get close to or hit each other, a joining skin is created between them, similar to the skin created when water droplets meet.

Tension is a value from 0.1 to 10 that sets how likely particles are to join when they are close to one another. When **Tension** is between 0 and 1, a skin is formed between particles as soon as they are near each other. With higher values, the skin is not created until particles are touching or intersecting. When **Tension** is above 1, **Viewport** and **Render** coarseness below may have to be increased in order to see particles.

The settings under **Evaluation Coarseness** determine how detailed particles will appear. Lower values make objects appear more detailed, but take longer to display and render. **Viewport** sets the detail in viewports, and

Render sets the detail when rendering. **Automatic Coarseness** sets **Viewport** coarseness to roughly the particle size, and **Render** coarseness to roughly half the **Viewport** coarseness.

Object Fragments	Uses emitter faces as particles. The options for these particles appear in the **Object Fragment Controls** section of this rollout. **Thickness** specifies the thickness of faces, in units, as they are emitted.

There are three methods for breaking the object into particles. **All Faces** uses all emitter faces as particles. **Number of Chunks** uses groups of faces as particles using the **Minimum** value to determine the minimum number of face groupings, although more may be used depending on how the object's faces are arranged. **Smoothing** breaks apart all faces whose angles between one another are greater than the **Angle** value.

Instanced Geometry	Uses an instance of an object in the scene as individual particles. The selected object can be animated, can have linked children, and/or can be a group. The options for these particles are set on the **Instancing Parameters** section of this rollout.

Click **Pick Object** to pick the object to be used as a particle. When **Use Subtree Also** is checked, all linked children of the selected object are used, or if the object is part of a group, the entire group is used.

The settings under **Animation Offset Keying** determine how the instanced object's animation will be used. **None** uses no animation. When **Birth** is selected, as each particle is born it is animated starting at frame 0 of the instanced object's animation. **Random** uses the **Birth** method but adds a random number to each starting frame number. The random number can be any number from 0 to the **Frame Offset** value.

The settings in the **Mat'l Mapping and Source** section choose the material source for all particle types.

Time Distance	These options are available only when **Icon** is chosen as the material source. When **Icon** is chosen, maps are applied to particles starting at the bottom edge of the map and work upward as the particle moves away from the emitter. The bottom of the map is applied to particles when they are born. Choosing **Time** causes higher areas

of the map to be applied to particles over time, with the upper edge of the map applied to particles when the number of frames specified by **Time** has passed. Choosing **Distance** uses higher areas of the map as particles move away from the emitter, reaching the top edge of the map when the particle has moved away from the emitter by the number of units specified by **Distance**.

Get Material From
Click this button each time you change the material source (**Icon**, **Picked Emitter**, **Instanced Geometry**). The previous material/map assignment remains assigned until it is overwritten with this button.

Icon
Uses the material assigned to the **PArray** icon. Mapping is applied according to the **Time** and **Distance** settings above.

Picked Emitter
Uses the material and mapping assigned to the emitter. Each particle takes its material and mapping from the portion of the emitter from which it is born.

Instanced Geometry
Uses the material and mapping assigned to instanced geometry. This option is available only when **Instanced Geometry** is selected as the particle type.

Fragment Materials
Assigns different material IDs to different parts of fragments. A **Multi/Sub-Object** material can then be assigned to the emitter, which will in turn apply to fragments. **Outside ID** sets the material ID for the outside faces of fragments. **Edge ID** sets the material ID for the portions of fragments extruded with the **Thickness** parameter in the **Object Fragments** section. **Backside ID** sets the material ID for the back side of fragments. These options are available only when **Object Fragments** is selected as the particle type. For information on **Multi/Sub-Object** materials, see the **Multi/Sub-Object** entry in the *Mat Editor* section of this book.

Particle Rotation

The options on the Particle Rotation rollout specify how each particle will spin. Each particle spins around only one axis throughout its life, but each particle may have a different spin axis.

Spin Time
Sets the number of frames over which each particle makes a full spin. **Variation** sets the percentage of possible variation in each particle's **Spin Time**.

Panels

Phase   Sets the initial particle spin, in degrees. **Variation** sets the percentage of possible variation in the **Phase**.

Random   Causes each particle to spin about a different random axis.

Direction of   Causes particles to spin about the axis pointing in the
Travel/MBlur   direction of travel. When **Stretch** is above 0, each particle is stretched by a percentage of its length. The percentage is equal to the particle speed times the **Stretch** value. When a material with a **Particle MBlur** type **Opacity** map is assigned to particles, the stretched portion of the particle appears more transparent, creating motion blur. For information on **Particle MBlur**, see the **Particle MBlur** entry in the *Mat Editor* section of this book.

User Defined   Creates a vector around which particles spin. The direction of the vector is set by the **X Axis**, **Y Axis** and **Z Axis** values. **Variation** is the number of degrees by which a particle's spin axis can vary from the vector.

Object Motion Inheritance

The options on the Object Motion Inheritance rollout determine how particles will react when the emitter's position is animated.

Inheritance   The percentage of particles that will move along with the emitter. When **Inheritance** is below 100, the motion of some particles will not be affected by the emitter's motion, creating a trail of particles behind the emitter.

Multiplier   The fraction of the emitter motion that will be transferred to the particles that move along with the emitter. **Variation** provides a percentage of variation for the **Multiplier** value.

Bubble Motion

The options on the Bubble Motion rollout provide controls for making particles wobble as they move, similar to the motion of bubbles as they rise to the top of water. Bubble motion is created with a sine wave.

Amplitude   The amplitude of the wave, or the number of units by which the particle moves off its course. Each particle moves to one side of its course by the number of units set by **Amplitude**, then to the other side, completing one bubble motion cycle. **Variation** is the percentage of variation in the **Amplitude**.

Period	The number of frames over which each particle moves through one bubble motion cycle. **Variation** is the percentage of variation in the **Period**.
Phase	A value from -360 to 360 that sets where the bubble motion cycle starts. A **Phase** of 0 starts the bubble motion at the beginning of the cycle, while a **Phase** of -180 or 180 starts the motion in the middle of the cycle. **Variation** sets the percentage of possible variation in the **Phase** from one particle to another.

Particle Spawn

The options on the Particle Spawn rollout can be used to cause particles to spawn (give birth to) more particles when they die or collide with a deflector. Spawned particles also spawn further particles when they die or collide with a deflector.

None	Spawns no particles.
Die on Collision	Causes particles to die when they collide with a deflector.
Spawn on Collision	Causes each particle to spawn when it collides with a deflector.
Spawn on Death	Causes each particle to spawn when it dies.
Spawn Trails	Causes each particle to spawn at each frame of its life.
Spawns	The maximum number of spawns beyond the original generation of particles.
Affects	The percentage of particles that will spawn new particles.
Multiplier	Multiplies the number of particles spawned. Paths of multiplied particles vary from the paths of other particles only when **Chaos** is above 0. **Variation** varies the **Multiplier**.
Chaos	A percentage of 180 degrees for varying spawned particle direction. The direction of a spawned particle can vary from its parent particle's direction by a random percentage, ranging from 0 to the **Chaos** percentage.

The options under the **Speed Chaos** section affect the speed of spawned particles.

Factor	Affects particle speed by a percentage of the parent particle's speed. How this factor is used depends on whether **Slow, Fast, Both, Inherit Parent Velocity** or **Use Fixed Value** are selected below.
Slow **Fast** **Both**	**Slow** slows down particles, **Fast** speeds them up, and **Both** slows down some particles and speeds others up. The parent particle's speed is multiplied by a random percentage up to the **Factor** percentage to determine how the spawned speed is changed. These settings are only in effect when **Inherit Parent Velocity** is unchecked.
Inherit **Parent** **Velocity**	Uses the parent particle's velocity as a base speed, and speeds up each spawned particle by a random percentage up to the **Factor** percentage. **Inherit Parent Velocity** always speeds up spawned particles regardless of whether **Slow, Fast** or **Both** is selected above.
Use Fixed **Value**	Whichever method is chosen above for setting spawned particle speed, checking **Use Fixed Value** uses the **Factor** percentage itself on all speed calculations, not a random percentage. This causes all spawned particles to have the same speed.

The options under **Scale Chaos** affect the scale of spawned particles.

Factor	A random percentage range used to scale particles.
Up **Down** **Both**	The **Up** and Down options scale spawned particles, making them larger or smaller than their parents by a random percentage up to the **Factor** percentage. **Both** scales some particles up and others down.
Use Fixed **Value**	Scales all particles up or down by the **Factor** percentage, with no random calculations. This causes all spawned particles to be the same size.

The **Lifespan Value Queue** allows you to specify alternative life spans for spawned particles, rather than using the **Life** value on the Particle Generation rollout. The first number in the queue (list) specifies the life span of the first generation of spawned particles, the second value specifies the life span of the second generation, and so on. If there are more generations than values specified, spawned particles take on the life span of the last value entered.

Panels

Add   Adds the **Lifespan** value to the queue.

Delete   To delete a value from the queue, highlight the value and click **Delete**.

Replace   Replaces the highlighted value with the **Lifespan** value.

The **Object Mutation Queue** creates a list of objects that replace the current object when a new generation is spawned. These options are available only when **Instanced Geometry** is selected as the particle type. The first generation of spawned particles turns into the first object on the list, the second generation turns into the second object, and so on.

Pick   To pick an object for the queue, click **Pick** and pick the object.

Delete   To delete an object from the list, highlight the object and click **Delete**.

Replace   To replace an object on the list, highlight the object, click **Replace** and pick another object.

Load/Save Presets

PArray parameters can be saved in a named preset that can be loaded later on.

Preset Name   Enter a name for the current preset.

Saved Presets   Lists previously saved presets.

Load   Loads the highlighted preset.

Save   Saves the current **PArray** settings with the name entered for **Preset Name**.

Delete   Deletes the highlighted preset.

Usage Notes

Particles work with the concept of birth and death. When particles leave the emitter, they are born. When they have existed for the length of time set by the **Life**, they die (disappear).

How long particles remain on the screen is determined by the **Life Span** and **Display Until** values. If the particle stream doesn't extend far enough, you can increase the **Life** value to make particles last longer, or you can increase the **Speed** to make the particles move faster.

Panels

PatchDeform

A modifier that deforms an object according to a patch object's shape.

General Usage

Create a patch object. Create an object with several segments. Select the object. On the **Modify** panel ![icon], click **More** if necessary, and choose **PatchDeform** under World Space Modifiers or Object Space Modifiers. Click **Pick Patch** and click on the patch object. The object deforms according to the patch object.

Sub-Object Level

The **PatchDeform** Object Space Modifier has one **Sub-Object** level, **Gizmo**. The gizmo is an instance of the patch object oriented so its local XY plane aligns with the world XY plane. Click **Sub-Object** to access the patch object.

Parameters

Pick Patch	To pick the patch, click **Pick Patch** and click on the patch. The gizmo is an instance of the patch object. With the **PatchDeform** World Space Modifier, the gizmo's XY plane is aligned with the object's local XY plane. With the **PatchDeform** Object Space Modifier, the gizmo's XY plane is aligned with the world coordinate system's XY plane.
U Percent	A percentage of the total width of the patch along its local X axis, where the object's pivot point sits. **Percent** can be a negative or positive number.
U Stretch	Stretches the deformed object along the patch's local X axis.
V Percent V Stretch	The same as **U Percent** and **U Stretch**, but along the patch's local Y axis.
Rotation	Rotates the object around the patch gizmo by the specified number of degrees.
Move to Patch	When clicked, the object is moved to the patch. The object is reoriented so its local axis matches the patch's local axis. This option is available only for the **PatchDeform** World Space Modifier.
Patch Deform Plane	Reorients the object's local XY, YZ, or ZX plane to the patch gizmo. **Flip** flips the object in the opposite direction.

PathDeform

A modifier that deforms an object to follow a path.

General Usage

Create an object with several segments. Create a shape. Select the object. On the **Modify** panel , click **More** if necessary, and choose **PathDeform** under World Space Modifiers or Object Space Modifiers. Click **Pick Path** and click on the shape. The object deforms to follow the shape.

Sub-Object Level

The **PathDeform** Object Space Modifier has one **Sub-Object** level, **Gizmo**. The gizmo is an instance of the path. Click **Sub-Object** to access the path instance.

Parameters

Pick Path
: To pick the path, click **Pick Path** and click on the path.

Percent
: The percentage of the length of the path where the object's pivot point sits. Increase or decrease the **Percent** to position the object along the path. **Percent** can be a negative or positive number.

Stretch
: Stretches the deformed object to a multiple of its original length.

Rotation
: Rotates the object around the path by the specified number of degrees.

Twist
: Twists the object around the path by the specified number of degrees.

Move to Path
: When clicked, the object is moved to the path. The object is reoriented so its local axis matches the path's local axis. This option is available only for the **PathDeform** World Space Modifier.

Path Deform
: With the **PathDeform** World Space Modifier, the object is reoriented so its local **X, Y** or **Z** axis points in the direction of the beginning of the path. With the **PathDeform** Object Space Modifier, the path is reoriented so its beginning points toward the object's local **X, Y** or **Z** axis. **Flip** flips the axis in the opposite direction.

Path Follow

A space warp that makes particles follow a spline path.

General Usage

Create a particle system object, and create a shape to be used as a path for the particles to follow.

Under the **Create** panel ![icon], click **Space Warps** ![icon]. Select **Particles & Dynamics** from the pulldown menu. Click **Path Follow**. Click and drag on the screen to create the space warp object.

Click **Pick Shape Object** and pick the path for the particles to follow.

Select the particle system object and click **Bind to Space Warp** ![icon] on the Toolbar. Click and drag from the particle system object to the space warp to bind it to the space warp. The particles follow the path.

Parameters

Pick Shape Object	Selects the shape for the particles to follow.
Unlimited Range	When checked, all particles follow the path throughout the animation. When particles reach the end of the path, they continue along a repeated version of the path. When this checkbox is unchecked, the **Range** below is used to determine when particles will follow the path.
Range	Determines when particles will follow the path when **Unlimited Range** is unchecked. If **Unlimited Range** is unchecked and the particle system object is positioned away from the start of the path, particles will move at first as if there is no space warp. When particles come within the **Range** distance of the start of the path, they will then follow the path. When particles reach the end of the path, they continue to follow a repeated version of the path until they reach the **Range** distance from the end of the path. After this, particles continue along in the same direction to the end of their lives.
Along Offset Splines	When this option is on and the particle system object is moved away from the path, the particles follow an offset path.

Along Parallel Splines	When this option is on, particles follow the exact path chosen. If the particle system object is moved away from the path, particles follow a duplicate, parallel version of the path starting at the particle system object.
Constant Speed	When checked, particles move at a constant speed along the path. When unchecked, particles move at varying speeds depending on the distance between vertices.
Stream Taper	When **Stream Taper** is greater than 0, particles converge or diverge on the path over time. When **Converge** is selected, particles move closer to the path over time. When **Diverge** is selected, particles move away from the path over time. When **Both** is selected, some particles converge while others diverge. **Variation** sets the percentage of random variation for the taper.
Stream Swirl	Sets the number of turns particles make around the path. **Clockwise** and **Counterclockwise** set the direction of the turns. **Bidirectional** causes particles to move in both directions. **Variation** sets the percentage of random variation for the swirl.
Start Frame	The frame at which the **Path Follow** space warp begins to affect particles.
Travel Time	The number of frames it takes each particle to traverse the path.
Variation	The percentage of variation for the **Travel Time**.
Last Frame	The frame at which the **Path Follow** space warp ceases to influence the particles.
Seed	Sets the base number to be used in calculating the **Path Follow** effect.
Icon Size	Specifies the size of the **Path Follow** icon. The size of the icon does not affect the operation of the space warp.

Usage Notes

To make particles follow the path exactly, move the particle system object to the start of the path.

PBomb

A space warp that provides an additional push to particles.

General Usage

Create a particle system object.

On the **Create** panel ![icon], click **Space Warps** ![icon]. Select **Particles & Dynamics** from the pulldown menu. Click **PBomb**. Click and drag on the screen to create the space warp.

Select the particle system object and click **Bind to Space Warp** ![icon] on the Toolbar. Click and drag from the particle system object to the space warp to bind it to the space warp.

Select the space warp object. Click **Modify** ![icon] to access the parameters. Change the parameters as desired.

PBomb stands for *particle bomb*. **PBomb** provides an extra surge or push to moving particles, or moves stationary particles.

Parameters

Blast Symmetry	Specifies the shape of the surge. The **PBomb** icon changes to reflect the selection. **Spherical** pushes particles outward from the center of a spherical icon. **Cylindrical** pushes the particles outward along the body of a cylindrical icon. **Planar** pushes particles perpendicular to a flat rectangular icon in both directions away from the icon.
Chaos	Varies the force on particles. This value has an effect only when **Duration** is greater than 0.
Start Time	The frame at which the impulse starts.
Duration	The number of frames over which the impulse is applied, after the first impulse. Setting **Duration** to 0 will create one impulse at the **Start Time** frame.
Strength	The increase in speed of particles, in units per frame.
Unlimited Range	When this option is selected, **PBomb** affects particles no matter how far they are from the **PBomb** icon.
Linear	When this option is selected, the force against particles is limited to the area set by the **Range**. The force at the center of the space warp icon is set by the **Strength** value, and decays in a linear fashion to zero force at the outer edge of the **Range**.

Exponential	When this option is selected, the force against particles is limited to the area set by the **Range**, and decays exponentially from full **Strength** at the center of the space warp icon to zero force at the outer edge of the **Range**.
Range	Sets a range for **Linear** and **Exponential** decay. **Range** is set in units from the center of the space warp icon, and is indicated by a spherical icon when **Linear** or **Exponential** is selected.
Icon Size	Sets the size of the icon in square units. The size of the icon has no effect on the **PBomb** space warp.

Usage Notes

The **PBomb** space warp works only on particle systems. **PBomb** works best with a **PArray** particle system. See the **PArray** entry for information on setting up a **PArray** particle system.

To explode a 3D object rather than a particle system, use the **Bomb** space warp. See the **Bomb** entry for more information.

PBomb can be used to explode a 3D object indirectly by exploding a **PArray** particle system set up to work with an **Object-based Emitter**. Be sure to bind the **PArray** icon, not the object instanced as the emitter, to the **PBomb** space warp.

PCloud

A particle system that fills an object with particles, then emits the particles.

General Usage

On the **Create** panel ![icon], click **Geometry** ![icon]. Select **Particle Systems** from the pulldown menu. Click **PCloud**. Click and drag on the screen to create one dimension of the PCloud icon, then move the cursor and click again to set the other dimension. Choose an emitter type under the Basic Parameters rollout, and set the **Speed** on the Particle Generation rollout.

Click **Play Animation** ![icon] to see the default particle animation. Change parameters as desired.

Basic Parameters

Pick Object
: To pick the emitter object, click **Pick Object** and click on the object to be used as an emitter. This object is in effect only when **Object-based Emitter** is selected below.

Particle Formation
: Determines the type of object used as an emitter. **Box Emitter**, **Sphere Emitter** and **Cylinder Emitter** each create a **PCloud** emitter in the specified shape. **Object-based Emitter** allows you to choose an object from the scene as the emitter with the **Pick Object** button above.

Display Icon
: Sets the dimensions of the **PCloud** icon, which is also the emitter if **Box Emitter**, **Sphere Emitter** and **Cylinder Emitter** are chosen above. **Rad/Len** sets the height of a **Box Emitter,** and the radius of a **Sphere Emitter** or **Cylinder Emitter**. **Width** sets the width of a **Box Emitter,** and **Height** sets the height of a **Box Emitter** or **Cylinder Emitter**.

Icon Hidden
: When checked, the icon is hidden. When the icon is hidden, you can access the **PCloud** parameters by clicking on any one of the particles.

Viewport Display
: Determines how particles will display in viewports. **Dots** display particles as small specks. **Ticks** displays particles as small crosses. **Mesh** displays particles as mesh objects. The mesh object displayed depends on the particle type chosen. This option is not available when either the **Constant** or **Facing** particle type is selected.

Percentage of Particles sets the percentage of total particles that appear in viewports. When using a large number of particles, set this value to 10 or 20 to keep viewports uncluttered and reduce redraw time.

Particle Generation

Particle Quantity	Determines the particle quantity. When **Use Rate** is selected, the value below **Use Rate** sets the number of particles born on each frame. When **Use Total** is selected, the value below **Use Total** sets the total number of particles born between the **Emit Start** and **Emit Stop** frames.
Speed	The speed of particles in units per frame. **Variation** varies the speed by a percentage of the **Speed** value. For example, if **Speed** is 20 and **Variation** is 10, particle speeds will vary by 10% of 20 = 2 units per frame, resulting in speeds varying from 18 to 22 units per frame.
Random Direction	Causes particles to move in random directions.
Enter Vector	Causes particles to move in the direction of the vector specified by the **X**, **Y** and **Z** axis settings.
Reference Object	Causes objects to move in the direction specified by the local Z axis of a picked object. Click **Pick Object** to pick the object. **Variation** is the percentage of possible variation from this course.

See the **PArray** entry for the remainder of the **PCloud** parameters. All remaining parameters are the same, except that **Object Fragments** are not available as **PCloud** particles.

Pivot (Doors)

Creates one or two doors hinged at the frame.

General Usage

On the **Create** panel ⚞, click **Geometry** ⚟. From the pulldown menu, choose **Doors**. Click **Pivot**. Click and drag to create the width of the door, move the cursor and click to set the depth or height, then move the cursor and click again to set the remaining dimension.

Creation Method

See the Creation Method rollout under the **BiFold** entry.

Parameters

See the Parameters rollout under the **BiFold** entry.

Leaf Parameters

See the Leaf Parameters rollout under the **BiFold** entry.

Usage Notes

Pivot doors are set up in exactly the same manner as **BiFold** doors. The difference is that **Pivot** doors are hinged at the frame, while **BiFold** doors are hinged at the center of two doors.

Adjusts pivot points.

General Usage

Select one or more objects. On the **Hierarchy** panel ⚏, click **Pivot**. Click **Affect Pivot Only** or **Affect Object Only**, and transform the object or pivot point.

Adjust Pivot Only

Affect Pivot Only	When this option is selected, pivot points can be transformed with transforms from the Toolbar. The pivot point appears in viewports as a red, green and blue axis.
Affect Object Only	When this option is selected, objects can be transformed while the pivot point stays still.
Affect Hierarchy Only	When this option is selected, objects linked to selected objects are transformed, while the selected objects stay still. This option works only with rotate and scale.
Center to Object/Pivot	When **Affect Pivot Only** is selected, this button is labeled **Center to Object**, and centers the pivot point on the object. When **Affect Object Only** is clicked, this button is labeled **Center to Pivot**, and centers the object on the pivot.
Align to Object/Pivot	When **Affect Pivot Only** is selected, this button is labeled **Align to Object**, and reorients the pivot point to match the orientation of the object. When **Affect Object Only** is clicked, this button is labeled **Align to Pivot**, and reorients the object to match the orientation of the pivot point.
Align to World	Aligns the pivot point or object's orientation to the world coordinate system.
Reset Pivot	Resets each pivot point to its default position and orientation in relationship to its object.
Don't Affect Children	When this button is clicked, a parent object can be transformed without transforming its linked children.
Transform	Orients the pivot point to match the orientation of the world coordinate system. Clicking this button has the same effect as clicking **Align to World** when **Affect Pivot Only** is selected.

Panels

Scale

If an object is scaled using the **Select and Non-uniform Scale** button then linked to a parent object, it will sometimes behave unpredictably when the parent object is transformed. Clicking **Scale** when the child object is selected eliminates this problem. **Scale** resets the scaling of the object so it becomes part of the object's definition. There is no apparent change in the object when this button is clicked.

Panels

Pivoted

Creates a window with one sash hinged at the center.

General Usage

On the **Create** panel [icon], click **Geometry** [icon]. From the pulldown menu, choose **Windows**. Click **Pivoted**. Click and drag to create the width of the window, move the cursor and click to set the depth or height, then move the cursor and click again to set the remaining dimension.

Creation Method

See the Creation Method rollout under the **Awning** entry.

Parameters

Height **Width** **Depth**	Sets the dimensions of the window.
Frame	Sets the dimensions of the window frame. **Horiz. Width** and **Vert. Width** affect the size of the glazing in addition to the frame size. **Thickness** also affects the thickness of the rails in the window sashes.
Glazing	Sets the **Thickness** of the glass.
Width	Sets the width of the rails in the sash.
Vertical	When checked, the window rotates vertically. The effect of turning on this checkbox is visible only when **Open** is above 0.
Open	Opens window panels by the specified percentage. 100 percent is equal to 180 degrees.
Generate Mapping Coords	Generates mapping coordinates on the window.

Point

Creates a point helper object.

General Usage

On the **Create** panel [icon], click **Helpers** [icon]. Click **Point**. Click and drag on the screen to create and place the point.

Parameters

Show Axis Tripod   When checked, the tripod axis is displayed.

Axis Length   Sets the size of each axis on the tripod axis that appears when the point object is unselected.

Usage Notes

A point helper object has many uses. It can be used as the center of the **Reference Coordinate System** so objects can be rotated about the point object. It can also be used to mark reference points in your scene.

Creates a NURBS point curve.

General Usage

On the **Create** panel 🐾, click **Shapes** 🖾. Select **NURBS Curves** from the pulldown menu. Click **Point Curve**. Click on the screen to set the first point on the curve, then move the cursor and click as many times as desired to create more points. Right-click to end creation of the curve.

Keyboard Entry

On the Keyboard Entry rollout, a point curve is created one point at a time with the **Add Point** button. The Keyboard Entry rollout is available only on the **Create** panel.

X, Y, Z Specifies the location of the point created when **Add Point** is clicked. The point is placed along the X, Y and Z axes in the world coordinate system.

Add Point Adds a point at the location specified by **X**, **Y** and **Z**.

Close Closes the curve by creating a segment that connects the last point with the first point.

Finish Completes the curve without closing it.

Curve Approximation

Point curves are approximated with steps between control points, just as standard splines are. For information on the parameters on this rollout, see the General rollout under the **Arc** entry.

Modify Panel

On the **Modify** panel, a NURBS curve can be used as a parent NURBS object for more NURBS curves. A NURBS curve cannot be used as a parent NURBS object to create NURBS surfaces.

Attach Attaches standard or NURBS shapes to the parent curve object. Click **Attach** and click on the shape. The shape becomes a CV curve and the object's stack can no longer be accessed. To edit an attached curve, click **Sub-Object**. Choose **Curve** to select or transform the curve, or choose **Curve CV** move the control points.

Attach Multiple	Attaches several curves at once. Click **Attach Multiple** to access a dialog similar to the one that appears when **Select by Name** ⧉ is clicked. All available shapes appear on the list. Select multiple shapes and click **Attach**.
Reorient	When checked, attached shapes are reoriented so their local axes align with the parent NURBS curve's local axes.
Import	Imports a shape while maintaining its modifier stack. To import a curve, click **Import** and click on the curve. To edit an imported object, see **Extract Import** in the **Sub-Object Imports** section below.
Import Multiple	Imports several shapes at once.
Display	**Lattices** displays lattices on CV curves. **Curves** displays curves. **Dependents** displays dependent objects created with the Create Points and Create Curves rollouts.
NURBS Creation Toolbox	Displays a toolbox that can be used to create points and curves. All the options on the toolbox appear on the Create Points and Create Curves rollouts below.

For the remaining entries on this rollout, see the General rollout under the **Arc** entry.

For information on the parameters on the Curve Approximation rollout, see the Curve Approximation rollout earlier in this entry.

For information on the parameters on the Create Points and Create Curves rollouts, see the **Point Surf** entry.

Usage Notes

A **CV Curve** is a spline surrounded by *control vertices* (CVs). Control vertices may or may not lie on the object's surface. The control vertices are used to change the shape of the curve. NURBS curves are convenient for making smooth curves quickly.

NURBS curves created on the **Create** panel cannot be used to make NURBS surfaces. NURBS curves can be used with a **Lathe** or **Extrude** modifier, or with **Loft**, to make a 3D object.

To make a NURBS curve that will be used to make a 3D object, create a NURBS surface with **Point Surf** or **CV Surf**. Go to the **Modify** panel and create the curve with the options on the Create Curves rollout.

Creates a NURBS point surface.

General Usage

On the **Create** panel ![icon], click **Geometry** ![icon]. Select **NURBS Surfaces** from the pulldown menu. Click **Point Surf**. Click and drag on the screen to create the point surface.

Go to the **Modify** panel ![icon] to edit the surface or create more NURBS surfaces.

Keyboard Entry

X, Y, Z   Sets the location of the point surface center at the X, Y and Z location in the world coordinate system.

Length   See these listings under Create Parameters below.
Width
Length Points
Width Points

Create   Creates the point surface using the parameters entered.

Create Parameters

Length   Sets the size of the point surface along its local X axis.

Width   Sets the size of the point surface along its local Y axis.

Length Points   Sets the number of control points along the surface's local X axis.

Width Points   Sets the number of control points along the surface's local Y axis.

Generate Mapping Coords   Generates planar mapping coordinates on the point surface.

Modify Panel

Once a NURBS surface has been created, you can use it to create more NURBS surfaces and curves. To do this, select the surface and access the **Modify** panel. Several options appear on the panel that were not available on the **Create** panel.

Sub-Object Levels

At each sub-object level, sub-objects can be selected, moved, hidden and deleted.

Curve Affects point curves and CV curves. Here, the curves themselves can be selected and transformed.

Curve CV Affects control points on a CV curve. The weight of control points can also be set.

Imports Affects imported objects. To make an imported object editable, select **Instance** and click **Extract Import**. The instance can then be selected as a separate object and modified, and changes will be passed to the object imported to NURBS.

Point Affects points on a point surface or point curve.

Surface CV Affects control points on a CV surface. The weight of control points can also be set here. All attached primitives and splines are CV surfaces, as are surfaces created with the **CV Surf** option.

Surface Affects point surfaces and CV surfaces. Here, surfaces can be selected and transformed.

General

Attach Attaches objects to the parent surface. Click **Attach** and click on the object. The object becomes a CV surface and the object's stack can no longer be accessed. To edit an attached object, click **Sub-Object**. Choose **Surface** to select or transform the surface, or choose **Surface CV** to move the control points.

Attach Multiple Attaches several objects at once. Click **Attach Multiple** to access a dialog similar to the one that appears when **Select by Name** is clicked. All available objects appear on the list. Select multiple objects and click **Attach**.

Reorient When checked, attached objects are reoriented so their local axes align with the parent NURBS curve's local axes.

Import	Imports an object while maintaining its modifier stack. To import an object, click **Import** and click on the object. To edit an imported object, access the **Imports** sub-object level, select the imported object, select **Instance** and click **Extract Import**. When the extracted object is edited, the instanced import will change as well.
Import Multiple	Imports several objects at once.
Display	**Lattices** displays lattices on any CV surfaces and curves. **Curves** displays curves, and **Surfaces** displays surfaces. **Dependents** displays dependent objects created with the options on the Create Points, Create Curves and Create Surfaces rollouts.
NURBS Creation Toolbox	Displays a toolbox that can be used to create points, curves and surfaces. All the options on the toolbox appear on the Create Points, Create Curves and Create Surfaces rollouts below.

Surface Approximation

This rollout determines how NURBS surfaces and curves will be approximated by viewports and the renderer.

Viewports Renderer	Two sets of parameters can be set with this rollout. Choose **Viewports** to set the parameters for viewports, or **Renderer** to set them for the renderer.
Iso Parametric Lines	Sets the number of lines used to approximate surfaces in viewports. **U Lines** sets the number of lines going around the object's local X axis, while **V Lines** sets the number of lines around the Z axis.
	Iso Only displays only the lines, even in shaded viewports. **Iso And Mesh** displays the lines and the mesh approximation, while **Mesh Only** displays the mesh approximation only.

The settings in the **Mesh Parameters** section determine how the mesh will be generated. When **Viewports** is selected, you must select **Mesh Only** as the display method in order to see the effects of these options.

Parametric	Generates faces based on an equation using the **U Steps** and **V Steps** values.
Spatial	Uses the **Edge** value to specify the maximum edge length for faces, in units. Lower values make more faces.

Panels

Curvature	Uses the curvature of the surface to determine how faces will be generated. **Distance** is the percentage by which the generated surface is allowed to deviate from the actual NURBS surface. The percentage is calculated from the diagonal measurement of the surface's bounding box. **Angle** is the maximum angle between faces on the surface. Lower values increase the number of faces.
View-Dependent	When checked, the renderer takes the object's distance into account when calculating the detail on the object. Objects far from the view receive fewer faces than those close to the view.
Merge	Surface sub-objects are often not perfectly joined. **Merge** sets the tolerance for nearly-joined objects. When **Merge** is 0, nearly-joined objects render with a gap between them. The **Merge** value multiplied by 0.1 and the bounding box diagonal give you the number of allowable units between two surfaces in order for the renderer to consider them to be joined.

Curve Approximation

Point curves are approximated with steps between control points, just as standard splines are. For information on the parameters on this rollout, see the General rollout under the **Arc** entry.

Create Points

The options on the Create Points rollout create points that can be used to create curves with the Create Curves rollout options. After a point is created, it can be edited at the **Point Sub-Object** level.

Point	Creates a point. Click **Point**, and click and drag on the screen to place the point.
Point Point	Creates a point at the location of another point, or relative to a point's location. Click **Point Point** and click on an existing point. The Offset Point rollout appears at the bottom of the panel. Choose **At Point** to leave the point at the same location as the picked point, or choose **Offset** to offset the point. Change the **X Offset, Y Off-set** and **Z Offset** values to move the point along the NURBS parent object's local X, Y and Z axes.

Curve Point Creates a point relative to a curve. The point can also be used to trim the curve. Click **Curve Point** and click on the curve. The Curve Point dialog appears at the bottom of the rollout.

The **U Position** sets the point's location along the length of the curve. **On Curve** leaves the point on the curve. **Offset** moves the point along the NURBS parent surface's local X, Y and Z axes. **Normal** moves the point a specified **Distance** away from the curve in the direction perpendicular to the curve. **Tangent** moves the point along the curve's tangent by the distance specified by **U Tangent**.

Trim Curve trims the curve to the point. **Flip Trim** trims the curve in the opposite direction.

Curve-Curve Creates a point at the intersection of two curves. The point can be used to trim the curves. Click **Curve-Curve** and click at the intersection of two curves. The Curve-Curve Intersection rollout appears at the bottom of the dialog. Check **Trim Curve** to trim either or both curves, and check **Flip Trim** to flip the trim.

Surf Point Creates a point relative to a surface. Click **Surf Point** and click on the surface. The Surface Point dialog appears at the bottom of the rollout.

U Position and **V Position** set the point's location along the surface's local X and Y axes. **On Surface** leaves the point on the surface. **Offset** moves the point along the NURBS parent surface's local X, Y and Z axes. **Normal** moves the point a specified **Distance** away from the surface in the direction perpendicular to the surface. **Tangent** moves the point along the surface's tangent plane by the distance specified by **U Tangent** and **V Tangent**.

Create Curves

The options on the Create Curves rollout create new curves from scratch, or create curves from existing curves or points. To select or transform a curve, use the **Curve** sub-object level. Control points for CV curves are edited at the **Curve CV** sub-object level. Points on a point curve are edited at the **Point** sub-object level.

Point Curve Creates a new point curve.

CV Curve Creates a new CV curve.

Curve Fit	Creates a point curve through a series of existing points. Points can be created with the options on the Create Points rollout. To create the curve, click **Curve Fit** and click on each point. Right-click when finished creating the curve.
Transform	Creates a copy of a curve. Click **Transform** and click and drag on a curve to copy it.
Blend	Connects two curves with a smooth curve. Click **Blend**. Move the cursor to the end of one curve until a blue rectangle appears. Click and drag to the end of the second curve until the blue rectangle appears, and release the mouse. The Blend Curve dialog appears at the bottom of the panel. **Tension 1** affects how far the new curve continues to move in the direction of the first curve's end, while **Tension 2** affects how soon the new curve takes on the curvature of the second curve's start.
Offset	Creates an outline copy of a curve. Click **Offset** and click and drag on a curve. The Offset Curve dialog appears at the bottom of the panel. **Distance** sets the new curve's distance from the original curve.
Mirror	Mirrors a curve. Click **Mirror** and click on the curve to mirror. The Mirror Curve rollout appears at the bottom of the panel. Choose the **Mirror Axis** and set the **Offset** distance for the new curve. These options work similarly to those for the **Mirror** modifier.
Chamfer	Creates a straight curve between two intersecting curves. Click **Chamfer**. Move the cursor to the end of one curve until the cursor changes to a blue rectangle. Click and drag to the end of the second curve until the blue rectangle appears, and release the mouse.
	The Chamfer Curve rollout appears at the bottom of the panel. **Length 1** and **Length 2** refer to the distance from the beginning of each line where the chamfer takes place. **Trim Curve** and **Flip Trim** trim each curve and flip the trim.
	If the bevel between the two curves appears as an orange line, the **Chamfer** operation didn't work. Try changing **Length 1** and **Length 2** until the bevel turns green, indicating that it worked.

Fillet

Creates a rounded corner between two curves. Click **Fillet**. Move the cursor to the end of one curve until a blue rectangle appears. Click and drag to the end of the second curve until the blue rectangle appears, and release the mouse.

The Fillet Curve rollout appears at the bottom of the panel. Radius sets the radius of the fillet arc. **Trim Curve** trims either or both curves where the fillet meets the curve. **Flip Trim** flips the trim. If the fillet between the two curves appears as a straight orange line, try increasing the **Radius** of the fillet.

U Iso Curve
V Iso Curve

Creates definition curves in a surface. Click **U Iso Curve** or **V Iso Curve** and move the cursor to a surface. If **U Iso Curve** is selected, curves along the NURBS parent surface's local X axis turn blue to indicate that they can be picked. Click to create the curve on the surface. The **V Iso Curve** option creates curves along the parent surface's local Y axis.

Create Surfaces

The options on the Create Surfaces rollout allow you to make a variety of surfaces. **Point Surf** and **CV Surf** make new surfaces from scratch, while the remaining options create new surfaces using those existing in the NURBS object already.

To select or transform a surface, use the **Surface** sub-object level. Control points on a CV surface are edited at the **Surface CV** sub-object level, while points on a point surface are edited at the **Point** sub-object level.

Point Surf

Creates a new point surface.

CV Surf

Creates a new CV surface.

Transform

Creates a copy of a surface. Click **Transform**, and click and drag on a surface to make a copy.

Blend

Creates another surface that blends two surfaces together. Click **Blend** and move the cursor over the edge of a surface. When the edge turns blue, click and drag the cursor to another until the edge turns blue. Release the mouse. A new **Blend** surface appears between the two edges.

Panels

When **Blend** is clicked, the Blend Surf rollout appears at the bottom of the panel where you can adjust parameters associated with the new surface. **Tension 1** and **Tension 2** control the tension at the first and second edges selected, respectively. **Flip End 1** and **Flip End 2** flip the **Blend** surface at either end.

Offset

Creates a copy of a surface and exaggerates the curvature of the surface as it moves further from the original object. Click **Offset** and click and drag on a surface. When an **Offset** is clicked, the Offset Surface rollout appears at the bottom of the panel. **Distance** sets the new surface's distance from the original object.

Mirror

Mirrors a surface. Click **Mirror** and click on the surface to mirror. The Mirror Surface rollout appears at the bottom of the panel. Choose the **Mirror Axis** and set the **Offset** distance for the new surface. These options work similarly to those for the **Mirror** modifier.

Extrude

Extrudes a curve. **Extrude** can only be applied to a NURBS curve, which can be created with the **Point Curve** and **CV Curve** options on the Create Curves rollout. To extrude a curve, click **Extrude** and click on the curve. The Extrude Surface rollout appears at the end of the panel, where you can change the extrude **Amount** and **Direction**.

Lathe

Lathes a curve. **Lathe** can only be applied to a NURBS curve. To lathe a curve, click **Lathe** and click on the curve. The Lathe Surface rollout appears at the end of the panel, where you can change the number of **Degrees** and the axis **Direction**, and **Align** the axis. These options work similarly to those for the Lathe modifier.

Ruled

Creates a surface between two curves. **Ruled** can only be used on two NURBS curves. To make a **Ruled** surface, click and drag on a curve, move the cursor to another curve, and release the mouse. The Ruled Surf rollout appears at the bottom of the panel. **Flip Beginning** flips the first curve selected, while **Flip End** flips the second curve selected.

U Loft   Creates a loft type surface from two or more curves. To create the surface, click **U Loft** and click on each curve to be included in the surface. Right-click when finished selecting curves. When **U Loft** is clicked, the U Loft Surface rollout appears at the bottom of the panel. The **U Curves** listing displays the name of each curve selected. When **Display While Creating** is checked, each portion of the **U Loft** surface displays as it is generated.

Cap   Caps an open surface edge. Cap is particularly useful for capping the top of an extruded or lathed surface. Click **Cap** and move the cursor over the open edge until it turns blue. Click to set the cap.

Usage Notes

Panels

NURBS surfaces are smooth, rounded surfaces that can be created and edited in a variety of ways. NURBS surfaces are altered by moving points, which are similar to vertices. But on a NURBS surface, when points are pulled in any direction, the rest of the object deforms accordingly, similar to clay being pushed or pulled into a shape.

There are two types of NURBS objects, *point* and *CV*. With a point object, the deformation points are right on the object. With a CV surface, the points can be on or around the object. Even though a point is not right on the object, it still deforms the object when it is pulled.

Like ordinary MAX objects, NURBS surfaces can be created with the use of splines. NURBS splines are called *curves*. A NURBS curve looks and acts like an ordinary MAX spline, except that it is deformed by control points. NURBS curves can also be of the point or CV type.

Once you have constructed a few NURBS surfaces, you can use them to create more surfaces. For example, use **Blend** on the Create Surfaces rollout to make a smooth transition between NURBS surfaces.

NURBS objects work differently from ordinary MAX objects. If you want to create a NURBS surface from several other surfaces, you might think that you should create all the surfaces on the **Create** panel, then work with them all under the **Modify** panel. This is not the way it works.

To use NURBS, you must first create one, and only one, NURBS surface. You then go to the **Modify** panel and use the options under the Create Surfaces rollout to create other NURBS surfaces as sub-objects of the first NURBS surface. Only in this way can you use the surfaces together to make one object. The original NURBS surface can be part of the final object, or it can be deleted later. Either way, you need that original NURBS object to get going.

You can also create an original NURBS object by creating a primitive, then going to the **Modify** panel and clicking **Edit Stack** 🔲. Choose **NURBS Surface** from the pop-up menu. The primitive is converted to a NURBS surface and can be used to create more NURBS surfaces on the **Modify** panel.

Sample Usage

This sample usage will show you how to create a simple NURBS surface.

1. On the **Create** panel, click **Geometry**. Select **NURBS Surfaces** from the pulldown menu. Click **Point Surf**. Click and drag on the screen to create a point surface. The size or placement of the surface is not important.

2. Go to the **Modify** panel. Several options for NURBS surfaces appear. Under the Create Surfaces rollout, click **Point Surf**. Create another point surface next to the first one.

You have just created a NURBS surface sub-object. This might seem confusing as you haven't yet accessed the sub-object level, but this is the way NURBS works. Sub-objects are created at the object level. To edit the new sub-object, however, you have to go to the sub-object level.

3. Click **Sub-Object**. Make sure the **Surface** sub-object is selected. Select the new point surface. Move the surface so it sits next to and below the original surface.

Now you're ready to use the two surfaces to make another surface.

4. Turn off the **Sub-Object** level. Under the Create Surfaces rollout, click **Blend**. Move the cursor to the edge of the first surface. Put the cursor on the edge nearest the second surface. When the edge turns blue, click and drag from the edge to the second surface. Move the cursor until the second surface edge nearest the first surface turns blue. Release the mouse.

A new surface is created between the first and second surfaces. Other surfaces can be created and joined in the same way.

From this sample usage, you can see that creating a NURBS surface consists of the following steps:

1. Create a NURBS surface.
2. Go to the **Modify** panel.
3. Create more NURBS surfaces and/or curves.
4. Use the NURBS surfaces or curves with the tools on the **Modify** panel to create more surfaces.

Polygon Counter

A utility that displays the polygon count for selected objects and the entire scene.

General Usage

On the **Utilities** panel ⊤, choose **Polygon Counter**. The floating Polygon Count dialog appears. Under Selected Objects, enter the maximum number of polygons you would have to have per object as the **Budget** value. Under **All Objects**, enter a **Budget** value equal to the maximum number of polygons for the scene. As you create each object, watch the Polygon Count dialog to ensure the object is within your maximum bounds.

Polygon Count

Polygon counts for both selected objects and the entire scene are provided on the dialog. For both sets of information, a band of colored dashes provides a visual checking method. When dashes are all green, the polygon count is well under the **Budget** value. When yellow dashes appear, the polygon count is approaching or has just reached the **Budget** value. Red dashes mean the polygon count is well over the **Budget** value.

Selected Object	Displays information about selected objects. For **Budget**, enter the desired maximum polygon count. The **Current** value is the polygon count for currently selected objects.
All Objects	Displays information about all objects in the scene. For **Budget**, enter the desired maximum polygon count for all objects. The **Current** value is the polygon count for all objects in the scene.

Usage Notes

The **Polygon Counter** utility does not reduce polygons or prevent you from creating an object with a polygon count over the **Budget** value. The **Polygon Count** dialog merely provides information for you to use in any way you see fit.

Preserve

A modifier that preserves the original look of an object.

General Usage

Create an object and make a copy of the object. Apply an **Edit Mesh** modifier to one of the objects, and move some of its vertices.

Select the modified object. On the **Modify** panel ![icon], click **More** if necessary and choose **Preserve**. Click **Pick Original** and pick the copy of the selected object. The modified object changes to look more like the original object.

Parameters

Pick Original — To pick the original object to which to compare the selected object, click **Pick Original** and click on the object.

Iterations — The number of times the **Preserve** result is calculated. Higher numbers match the original object more closely, but take longer to calculate.

Edge Lengths — A value that determines how important it is to preserve edge lengths. This value is compared to the **Face Angles** and **Volume** values to determine which is the most important to preserve.

Face Angles — Determimes how important it is to preserve face angles.

Volume — Determines how important it is to preserve the original object volume.

Selection — Determines the selection used in calculating the **Preserve** result. If none of these options is checked, the last selection on the stack is used. **Apply to Whole Mesh** applies **Preserve** to the entire mesh. **Selected Verts Only** uses the last vertex selection on the stack. **Invert Selection** inverts the last vertex selection on the stack.

Usage Notes

Objects modified with **Edit Mesh** sometimes have stretched faces or sharp edges where vertices have been moved. The **Preserve** modifier can sometimes smooth out these problems.

Creates a three-sided prism.

General Usage

On the **Create** panel 🖎, click **Geometry** 🗔. If necessary, select **Standard Primitives** from the pulldown menu. Click **Prism**. Click and drag to set the size of one side of the triangle base. Move the cursor and click to set the second and third sides, then move the cursor again and click to set the height.

The orientation of the object's local X, Y and Z axes is determined by the viewport in which you begin drawing. The object's local X and Y axes are set on the viewport's drawing plane.

Creation Method

Isosceles   Creates a prism with an icosceles triangle (triangle with two equal sides) as its base.

Base/Apex   Creates a prism with any triangle as its base. To use an equilateral triangle as the prism base, turn on this option, and hold down the **<Ctrl>** key while drawing the base triangle.

Keyboard Entry

X, Y, Z   Sets the location of the prism base center at the X, Y and Z location in the world coordinate system.

Side 1 Length See these listings under Parameters below.
Side 2 Length
Side 3 Length
Height

Create   Creates the object using the parameters entered.

Parameters

Side 1 Length Sets the length of the triangle along the prism's local X axis. Side 1 is the first side set when the base triangle is drawn.

Side 2 Length Sets the length of the right side of the base triangle as viewed from the drawing viewport.

Side 3 Length Sets the length of the left side of the base triangle as viewed from the drawing viewport.

Height	Sets the height of the prism.
Side 1 Segs	Sets the number of segments along side 1 of the base triangle. Side 1 is the first side set when the base triangle is drawn.
Side 2 Segs	Sets the number of segments along side 2 of the base triangle. Side 2 is the right side of the base triangle as viewed from the drawing viewport.
Side 3 Segs	Sets the number of segments along side 3 of the base triangle. Side 3 is the left side of the base triangle as viewed from the drawing viewport.
Height Segs	Sets the number of segments along the height of the prism.
Generate Mapping Coords	Generates planar mapping coordinates for each surface of the prism.

Creates a window with three sashes, two of which are hinged.

General Usage

On the **Create** panel ![icon], click **Geometry** ![icon]. From the pulldown menu, choose **Windows**. Click **Projected**. Click and drag to create the width of the window, move the cursor and click to set the depth or height, then move the cursor and click again to set the remaining dimension.

Creation Method

See the Creation Method rollout under the **Awning** entry.

Parameters

Height **Width** **Depth**	Sets the dimensions of the window.
Frame	Sets the dimensions of the window frame. **Horiz. Width** and **Vert. Width** affect the size of the glazing in addition to the frame size. **Thickness** also affects the thickness of the rails in the window sashes.
Glazing	Sets the **Thickness** of the glass.
Width	Sets the width of the rails in the sashes.
Middle Height	The height of the middle sash in relationship to the window frame.
Bottom Height	Sets the height of the bottom sash in relationship to the window frame.
Open	Opens the two bottom window panels by the specified percentage. 100 percent is equal to 90 degrees.
Generate Mapping Coords	Generates mapping coordinates on the window.

Protractor

Creates a protractor helper object that measures the angle between two objects.

General Usage

On the **Create** panel ![icon], click **Helpers** ![icon]. Click **Protractor**. Click and drag on the screen to create and place the protractor. Pick two objects. The angle between the protractor and the two objects appears next to **Angle**.

Parameters

Pick Object 1   Click **Pick Object 1** and pick the first object. Click **Pick**
Pick Object 2   **Object 2** and pick the second object.

Angle   Displays the angle between the two picked objects and the protractor. The **Angle** value cannot be changed directly.

Usage Notes

To view the angle while moving the picked objects, select the protractor and go to the **Modify** panel ![icon]. Click **Pin Stack** ![icon]. Select and move either of the other objects. The protractor angle remains on the panel and is updated as you move the objects.

Creates a sensor that triggers an action or animation when the view enters the sensor.

General Usage

On the **Create** panel [icon], click **Helpers** [icon]. From the pulldown menu, choose **VRML 2.0**. Click **ProxSensor**. Click and drag to create the sensor box. Adjust the dimensions of the box and click **Pick Action Objects** to choose the objects that will animate when the view enters the sensor.

Level of Detail

Length
Width
Height   Sets the dimensions of the **ProxSensor**.

Pick Action Objects   To cause objects to animate when the view enters the sensor, click **Pick Action Objects** and select the objects. As each object is selected, it appears on the list. Animated objects, cameras, lights or **AudioClip** helpers can be selected. When finished, click **Pick Action Objects** again to turn it off.

Enable   Enables the sensor.

Delete   To delete an object from the list, highlight the object and click **Delete**.

Usage Notes

The **ProxSensor** icon does not appear to viewers in the VRML scene.

When the view exits the **ProxSensor** area, objects specified with **Pick Action Objects** stop animating.

Push

A space warp that applies a specific amount of force to an object or particles.

General Usage

Create a particle system object, or any other object.

Under the **Create** panel ![icon], click **Space Warps** ![icon]. Select **Particles & Dynamics** from the pulldown menu. Click **Push**. Click and drag on the screen to create the space warp object.

To use the space warp with a particle system, select the particle system and click **Bind to Space Warp** ![icon] on the Toolbar. Click and drag from the object to the space warp to bind it to the space warp.

To use the space warp on an object, the space warp must be assigned as part of a dynamics simulation with the **Dynamics** utility. See **Dynamics** for more information.

Parameters

On Time   The number of the frame where the force begins. For particle systems, when the **On Time** frame is reached, particles that are already in the scene will react to the force as a group, while particles that are born between **On Time** and **Off Time** will be affected as they are born.

Off Time   The number of the frame where the force ends. For particle systems, particles that are born after this frame will not respond to the space warp. Particles that have already been affected will continue in their current directions after the **Off Time** frame is reached.

Basic Force   Sets the amount of force on the object or particle system. The amount of force is expressed as **Newtons** or **Pounds**, depending on the button selected below. With dynamics, the amount entered for **Basic Force** is accurately represented in the scene. With particles, the amount is subjectively interpreted, and must be adjusted by visually checking the effect in the scene.

Feedback On   When checked, the force is decreased as objects approach the **Target Speed**. When unchecked, the force is constant regardless of object speed.

Reversible   When checked, the force is reversed when an object is above the **Target Speed**. This option is available only when **Feedback On** is checked.

Target Speed Specifies a speed with which to compare object speed so the **Feedback On** and **Reversible** options can be used. **Target Speed** is expressed in units per frame.

Gain Specifies how quickly the force is affected when **Target Speed** is approached. When **Gain** is 100, the object's speed is corrected immediately. When **Gain** is lower than 100, the correction time is slower, resulting in a looser interpretation of the **Target Speed**. **Gain** can also be over 100. This option is available only when **Feedback On** is checked.

The settings in the **Periodic Variation** section create noise to randomly affect the **Basic Force** value. Two sets of noise can be used.

Enable Enables periodic variation with the settings below.

Period 1 Sets the time over which the first variation takes place. For example, setting **Period 1** to 50 causes the effect to cycle every 50 frames.

Amplitude 1 The size of the effect. This value uses the same units as **Basic Force**. If the amplitude is the same as the **Basic Force**, the force will range from zero to twice the **Basic Force**.

Phase 1 The noise effect is created with a sine wave. **Phase 1** sets which phase the wave is in, such as whether it is just starting, is peaking or is just ending at any point in time. A phase of 1.0 shifts the wave by an entire wavelength, a phase of 2.0 shifts it by two entire wavelengths, etc. To make the noise look different, the phase can be changed to a fractional number such as 0.5. To make the noise change over time, animate the phases values.

Period 2
Amplitude 2 Sets the period, amplitude and phase for the second noise effect.
Phase 2

The **Particle Effect** section specifies a range for the **Push** effect. The range is set in a spherical shape around the **Push** space warp icon.

Enable Enables the range and displays it on the screen.

Range Sets the radius for the **Push** effect. **Range** is specified in units from the space warp.

Icon Size Sets the size of the icon in square units. The size of the icon has no relationship to the push effect.

Pyramid

Creates a pyramid.

General Usage

On the **Create** panel ![icon], click **Geometry** ![icon]. If necessary, select **Standard Primitives** from the pulldown menu. Click **Pyramid**. Click and drag to set the size of the pyramid base, then move the cursor again and click to set the height.

The orientation of the object's local X, Y and Z axes is determined by the viewport in which you begin drawing. The object's local X and Y axes are set on the viewport's drawing plane.

Creation Method

Base/Apex	Draws the pyramid starting at a corner of the pyramid base.
Center	Draws the pyramid from the center of the pyramid base.

Keyboard Entry

X, Y, Z	Sets the location of the pyramid base center at the X, Y and Z location in the world coordinate system.
Width **Depth** **Height**	See these listings under Parameters below.
Create	Creates the object using the parameters entered.

Parameters

Width	The size of the pyramid base on the object's local X axis.
Depth	The size of the pyramid base on the object's local Y axis.
Height	Sets the height of the pyramid.
Width Segs	Sets the number of segments along the width.
Depth Segs	Sets the number of segments along the depth.
Height	Sets the number of segments along the height of the pyramid.
Generate Mapping Coords	Generates planar mapping coordinates for each of the five pyramid surfaces.

Creates a patch object with rectangular polygons.

General Usage

On the **Create** panel 🖎, click **Geometry** 🗇. Choose **Patch Grids** from the pulldown menu. Click **Quad Patch**. Click and drag on the screen to create the patch object.

Keyboard Entry

X, Y, Z	Sets the location of the patch grid center at the X, Y and Z location in the world coordinate system.
Length **Width**	Sets the dimensions of the patch grid.
Create	Creates the object using the parameters entered.

Parameters

Length **Width**	Sets the dimensions of the patch grid.
Length Segs **Width Segs**	Sets the number of segments within each grid area along the **Length** and **Width**. The number of patch vertices is determined by the number of segments plus 1. For example, when **Length Segs** is 3, there are 4 patch vertices along the **Length**.
Generate Mapping Coords	Generates planar mapping coordinates on the patch grid.

Usage Notes

Patch grids are flat, rectangular objects that can be smoothly deformed. A patch grid consists of a *surface* and a *lattice*. The surface is the grid itself, made up of vertices, edges and deformable patches. The lattice is a grid-shaped collection of vertices, edges and patches. When any of the lattice sub-objects are transformed or altered, the change is passed on to the surface to deform it accordingly. Patch vertices have movable Bezier handles for fine control over the deformation.

Compare the patch object type with an **Editable Mesh**, which is deformed by moving the vertices, faces and edges, and does not deform smoothly when these sub-objects are transformed. Patch objects behave more like **NURBS** surfaces than **Editable Mesh** objects.

You can turn any object into a patch object by applying an **Edit Patch** modifier to it. However, for most objects, the **Edit Patch** modifier determines the number and configuration of vertices and patches on the lattice. When you create a **Quad Patch**, you can control the number of vertices and patches on the lattice exactly with the **Length Segs** and **Width Segs** values. The number of patch vertices along each dimension is set by the number of segments plus 1.

A **Quad Patch** is made with rectangular polygons, while a **Tri Patch** is made with triangular polygons.

Rectangle

Creates a rectangle shape.

General Usage

On the **Create** panel , click **Shapes** . Click **Rectangle**. Click and drag on the screen to create the rectangle. A rectangle has four Bezier Corner vertices.

General

See the General rollout under the **Arc** entry.

Creation Method

Edge	When **Edge** is turned on, the rectangle is created starting from the edge when you click and drag on the screen.
Center	When **Center** is on, the rectangle is created from the center outward when you click and drag on the screen.

Keyboard Entry

X, Y, Z	Specifies the location of the center of the shape created when **Create** is clicked. The center is placed at the X, Y and Z location in the world coordinate system.
Length **Width**	See these listings under Parameters below.
Create	Creates the shape with the parameters entered. The shape is placed on the current viewport's plane.

Parameters

Length	Sets the length of the shape along its local Y axis.
Width	Sets the width of the shape along its local X axis.

Relax

A modifier that relaxes sharp edges in an object.

General Usage

Select an object. On the **Modify** panel ![icon], click **More** if necessary, and choose **Relax**.

Parameters

Relax Value   The **Relax** modifier works by looking at the vertices around each vertex and finding the average of all distances. When the **Relax Value** is 1, the vertex moves so it is the average distance from all neighbors, making the object smoother and rounder. At -1, it moves away from its neighbors by this average distance, making sharp edges. The **Relax Value** can be any number from -1 to 1.

Iterations   The number of times the average distance between vertices is calculated to come up with the **Relax** result.

Keep Boundary Pts Fixed   When checked, open edges on the object are not relaxed, and retain their original position and size while the rest of the object is relaxed.

Rescale World Units

Rescales all parameters for selected objects or an entire scene.

General Usage

Select one or more objects if desired. On the **Utilities** panel , choose **Rescale World Units**. Click **Rescale**. On the Rescale World Units dialog, enter a **Scale Factor,** and choose whether to affect the scene or the selection. Click **OK** to scale world units.

Rescale World Units

Rescale   Accesses the Rescale World Units dialog.

Scale Factor   A multiplier for the scale operation. For example, a **Scale Factor** of 2 will double all parameters for the scene or selected objects.

Scene   Causes the scale to affect all objects in the scene.

Selection   Scales parameters for selected objects only.

OK   Performs the scale operation.

Cancel   Cancels the scale operation. No objects or parameters are scaled.

Usage Notes

The **Rescale World Units** utility scales all parameters for objects, including creation parameters and map size parameters in the Material Editor. For example, after being scaled with a **Scale Factor** of 2, a sphere's **Radius** will double, as will the **Size** parameter for any 2D or 3D maps assigned to the object, such as **Noise** or **Dent**.

Reset XForm

Rotates the selected object's pivot point to match the world coordinate system without rotating the object.

General Usage

Select one or more objects. On the **Utilities** panel 🔧, choose **Reset XForm**. Click **Reset Selected** to reset the pivot point of the object to match the world coordinate system.

Any rotation or scaling previously applied to the object with **Select and Rotate** ↻ or any of the scale transform buttons will be placed in an **Xform** modifier at the top of the object's modifier stack.

Usage Notes

After using the **Reset Transform** utility, you can set an object's rotation to the world coordinate system or to reset its scale to the original values by going to the **Modify** panel 🔧 and clicking **Remove modifier from the stack** to remove the newly created **Xform** modifier from the object's modifier stack. By removing this modifier, you reset the object's rotation so its original local axes match the world coordinate system.

This orientation might not be the object's original orientation. For example, if you create a cylinder starting in the Front viewport, the cylinder's local axes do not match the world coordinate system. If you use **Reset Transform** and remove the **Xform** modifier created by this utility, the cylinder's axes will be aligned to the world coordinate system, which will make the cylinder look as if it were created in the Top viewport rather than the Front viewport.

This utility does not work on groups. To reset the transform of a group, go to the **Hierarchy** panel 品, click **Pivot** and click **Transform** under the Adjust Transform rollout.

Ring Array

Creates an array of linked boxes around a dummy object.

General Usage

On the **Create** panel [icon], click **Systems** [icon]. Click **Ring Array**. Click and drag on the screen to create the array. Change the **Number** value to set the number of linked boxes.

Parameters

Radius　Sets the distance from the center of the dummy object to the center of each linked box.

Amplitude　The positions of the linked boxes are set with a sine wave. **Amplitude** is the height of the sine wave, in units. When **Amplitude** is 0, the sine wave's amplitude is 0, and all boxes are aligned with the dummy object. When **Amplitude** is increased, the boxes move along the local Z axis according to the sine wave.

Cycles　The number of sine wave cycles across all linked objects. When **Cycles** is 0, there is no sine wave effect. When **Cycles** is 1, one sine wave is passed through all the objects to shift their positions, resulting in a tilted ring. Higher values make more cycles pass through the objects, resulting in more and more random motion.

Phase　The phase of the sine wave. When **Phase** is a whole number such as 0, 1 or 2, the sine wave is at the beginning of its cycle. Fractional values such as 0.5 shift the sine wave to other parts of its cycle. Animating the **Phase** value animates the linked objects as they move to follow the shifting sine wave.

Number　Sets the number of linked boxes around the dummy object.

Usage Notes

Ring Array is a quick method for creating a circular array of linked objects.

The dummy object at the center of the array is a 40x40x40 cube. You can change its dimensions by scaling it. The boxes linked to the dummy object are ordinary 20x20x20 boxes. You can change the dimensions of a box by selecting the box, going to the **Modify** panel [icon] and changing the box's creation parameters.

Ripple

A modifier or space warp that deforms an object with an animatable ripple.

General Usage

Create an object with several segments along each of its dimensions.

To apply **Ripple** as a modifier, first select the object. On the **Modify** panel ⚡|, click **More** if necessary to access the full list of modifiers. Select **Ripple**. The object deforms according to the wave parameters. Change or animate parameters as desired. You can also access the **Sub-object** level and transform the gizmo to change or animate the ripple.

To apply **Ripple** as a space warp, go to the **Create** panel 🗲|. Click **Space Warps** 〜 and select **Geometric/Deformable** from the pulldown menu. Click **Ripple**. Click and drag on the screen to create the ripple's radius, then release the mouse and drag to specify the ripple's amplitude. Click to set the amplitude. Select the object, and click **Bind to Space Warp** 🔧 on the Toolbar. Click and drag from the object to the ripple space warp to bind it to the ripple. The object deforms according to the ripple parameters. Select the ripple space warp. Go to the **Modify** panel to access the ripple parameters. Change or animate the parameters as desired. You can also move the ripple space warp to change or animate the effect.

Parameters

Amplitude 1 The amplitude (height) of the ripple along the ripple object's local X axis, in units. This value is set to the same value as **Amplitude 2** when you create the ripple, but can be changed afterward.

Amplitude 2 The amplitude (height) of the ripple along the ripple object's local Y axis, in units.

Ripple Length The length of one ripple cycle. By default, the ripple space warp displays two full ripple cycles to make the effect easy to see. This makes the actual length of the ripple space warp object on the screen twice the **Ripple Length**.

Phase A ripple is created with a sine wave. **Phase** sets which phase the wave is in, such as whether it is just starting, is peaking or is just ending at any point in time. A **Phase** of 1.0 shifts the wave by an entire wavelength, a **Phase** of

2.0 shifts it by two entire wavelengths, etc. To make the wave look different, **Phase** can be changed to a fractional number such as 0.5. To make the ripple undulate over time, animate the **Phase** value.

Decay   Sets the amount by which the amplitude will diminish over distance. When **Decay** is 0, ripple height (amplitude) is the same over the entirety of the ripple object and each bound object. When **Decay** is higher than 0, the amplitude decays as the ripple moves away from the space warp center. Try fractional values such as 0.005 or 0.2 for best results.

Space Warp Parameters

The **Ripple** space warp has three additional parameters. These parameters make the space warp object easier to see and work with, but do not influence the smoothness of the ripple on the bound object.

Circles   The number of circular segments on the space warp object. This value affects the radius of the space warp object on the screen, but does not change the effect.

Segments   The number of segments on the space warp object. Segments are arranged like spokes on a wheel. The number of segments can affect the amount of detail in the ripple.

Division   Scales the space warp without altering the ripple effect. Higher values make the space warp smaller, while lower values make it larger.

Modify Panel

When a **Ripple** space warp is bound to an object, it is placed on the object's stack as an entry called **Ripple Binding**. This stack entry can be accessed by selecting the bound object and going to the **Modify** panel. The **Ripple Binding** entry has one parameter, **Flexibility**. This value multiplies the amplitude of the ripple, and can be used to animate the amplitude.

Usage Notes

To animate the ripple, either the parameters on the ripple's Parameters rollout can be animated, or the ripple space warp or gizmo itself can be transformed (moved, rotated or scaled) over time. To make a ripple undulate, animate the **Phase** value on the Parameters rollout.

Scatter

Randomly scatters duplicates of an object in space or across the surface of another object.

General Usage

Select the object to be scattered. Under the **Create** panel ![icon], click

Geometry ![icon]. Choose **Compound Objects** from the pulldown menu. Click **Scatter**. Under the **Source Object Parameters** section of the Scatter Objects rollout, set the number of **Duplicates**. Under the Scatter Objects rollout in the **Distribution** section, choose whether to use a distribution object or transforms to scatter the source object.

If you choose the distribution object method, click **Pick Distribution Object** and click on the object. This scatters the source object over the faces of the distribution object. Each scatter object is associated with a particular face on the distribution object.

If you choose the transform method, you must change one or more of the settings under the Transforms rollout to see the objects scatter.

Pick Distribution Object

On the Pick Distribution Object rollout, the object to receive scattered objects is chosen.

Pick Distribution Object	To pick the object over which to distribute the scatter object, click this button and pick the object on the screen. You can also press the **<H>** key to pick from a list. After you pick the object, the name of the object appears next to **Object** above the button.
Reference	Uses a reference of the object as the distribution surface.
Move	Uses the object itself as the distribution surface.
Copy	Uses a copy of the object as the distribution surface.
Instance	Uses an instance of the picked object as the distribution surface.

Scatter Objects

Use Distribution Object	Positions the scatter objects along the surface of the picked distribution object.

Panels

Use Transforms Only	Scatters objects in 3D space according to the settings under the Transform rollout.
Objects	Lists the scatter and distribution objects. If you wish to access an object to alter it, highlight the object. The object can now be accessed in the modifier stack.
Source Name	Here you can change the name of the source (scatter) object for the purposes of the **Scatter** function.
Distribution Name	Here you can change the name of the distribution object for the purposes of the **Scatter** function.
Duplicates	Enter the number of source object duplicates to make and scatter. This value can be animated.
Base Scale	Scales the source object and duplicates to this percentage.
Vertex Chaos	Applies random noise to the vertices of the source object, making the duplicates rough in shape.

The remainder of the parameters determine how the source object is spread across the distribution object and how it is displayed. These options are in effect only if you have chosen **Use Distribution Object** above.

Perpendicular	When this checkbox is on, each duplicate is oriented perpendicular to the face it touches on the distribution object. When this checkbox is off, the duplicate maintains the same orientation as the source object.
Use Selected Faces Only	Distributes the duplicates to selected faces on the distribution object only. If you haven't selected faces, you can do so by highlighting the distribution object name under **Objects,** and accessing the distribution object on the modifier stack. Apply a **Mesh Select** modifier and select faces. Return to the **Scatter** level of the modifier stack when finished selecting faces.
Area	Distributes scatter objects evenly over the surface area of the distribution object.

Panels

Panels

Even	Distributes scatter objects evenly over the distribution object's faces. This option works with faces, as opposed to **Area** which works with surface area. For example, on a flat box with one segment along each dimension, faces on the top and bottom have a much larger surface area than faces on the sides. When this type of object is used as the distribution object, the **Area** and **Even** options scatter the objects quite differently.
Skip N	Skips a specified number of faces when placing the scatter object.
Random Faces	Distributes scatter objects randomly over the distribution object's surface.
Along Edges	Distributes scatter objects randomly over the distribution object's face edges.
All Vertices	Places a scatter object on each of the distribution object's vertices. When this option is turned on, the number of vertices, not the **Duplicates** value, determines the number of scatter objects.
All Edge Midpoints	Places a scatter object at the midpoint of each face edge. When this option is turned on, the number of face edges, not the **Duplicates** value, determines the number of scatter objects.
All Face Centers	Places a scatter object at the center of each triangular face on the distribution object. When this option is turned on, the number of triangular faces, not the **Duplicates** value, determines the number of scatter objects.
Result	Displays the scattered objects.
Operand	Displays only the original scatter source object, not the scattered objects.

Transforms

The settings under the Transforms rollout apply random transforms. If you have not chosen a distribution object, you must change some of the settings on the Transforms rollout in order to scatter objects. If you have chosen a distribution object, you can also use transforms to change the scattered objects.

Rotation	The settings under this section apply rotation randomly to objects. The X, Y and Z parameters specify the maximum rotation, in degrees, along each axis. When **Use Maximum Range** is checked, the maximum of the X, Y and Z values is used for all three values, and the remaining two values are disabled.
Local Translation	The settings under this section move objects along their local axes by a random distance. If a distribution object has been chosen, these settings move the scatter object by a random distance after they have been placed on the distribution object. The X, Y and Z parameters specify the maximum movement distance, in units, along each axis. When **Use Maximum Range** is checked, the maximum of the X, Y and Z values is used for all three values, and the remaining two values are disabled.
Translation on Face	The settings under this section move each object along its associated distribution object face. These settings are only in effect if a distribution object is used. The A and B parameters specify movement along the two edges of the face. The N value specifies the maximum distance along the normal, which moves scatter objects away from the distribution object faces. When **Use Maximum Range** is checked, the maximum of the A, B and N values is used for all three values, and the remaining two values are disabled.
Scaling	The settings under this section scale objects randomly. The X, Y and Z parameters specify the maximum scaling along each axis. When **Use Maximum Range** is checked, the maximum of the X, Y and Z values is used for all three values, and the remaining two values are disabled. **Lock Aspect Ratio** provides uniform scaling of duplicates based on the X scaling parameter.

Display

The settings on the Display rollout allow you to customize the display of the scatter and distribution objects.

Proxy	Displays the scatter objects as wedges, speeding up viewport redraw time. Rendered images will still show the original mesh objects.
Mesh	Displays the scatter objects as the original mesh objects.

Display	Specifies the percent of total duplicate objects that appear in viewports. If you have a large number of duplicates, reduce this percentage to speed up redraw time. Rendering will show all objects regardless of how many are displayed in viewports.
Hide Distribution Object	Hides the distribution object in both viewports and rendered images.
Seed	Sets the base value for the **Scatter** effect.

Load/Save Presets

On the Load/Save Presets rollout, you can save **Scatter** settings for later use.

Preset Name	Enter a name for the **Scatter** preset.
Saved Presets	Displays previously saved **Scatter** presets. To load a preset, highlight it from the list and click **Load**.
Load	Loads the highlighted preset.
Save	Saves the current settings to the preset name entered in **Preset Name**.
Delete	To delete a preset, highlight it from the **Saved Presets** list and click **Delete**.

A space warp that uses a spherical shape to deflect particles.

General Usage

Create a particle system object.

Under the **Create** panel , click **Space Warps** . Select **Particles & Dynamics** from the pulldown menu. Click **SDeflector**. Click and drag on the screen to create the space warp object. Move the space warp so it sits in the path of oncoming particles.

Select the particle system object and click **Bind to Space Warp** on the Toolbar. Click and drag from the particle system object to the space warp to bind it to the space warp.

Select the space warp object. Go to the **Modify** panel to access the parameters. Change the parameters as desired.

Basic Parameters

Bounce
: Sets the speed at which particles bounce off the deflector. When **Bounce** is 0, particles don't bounce off the deflector. When **Bounce** is 1.0, particles bounce at the same speed they have when they hit the deflector. Values between 0 and 1.0 make particles move at a fraction of the hit speed after bouncing. Values over 1.0 make particles move faster than the hit speed after bouncing.

Variations
: The percentage of variation for bounce speed. When **Variations** is 0%, no variation takes place. Higher values vary the bounce between 0 and 100 minus the specified percentage. For example, when **Variations** is 10%, particle speed can vary from 90% to 100% of the speed set by the **Bounce** value.

Chaos
: The percentage of variation for the bounce angle. The **Chaos** percentage uses a percentage of up to 90 degrees to vary the bounce angle. When **Chaos** is 0%, particles bounce off the space warp at exactly the same angle at which they hit it. When **Chaos** is 100%, particles bounce off the space warp at angles varying from 0 to twice the angle at which they hit. Intermediate **Chaos** values create variations of smaller degrees. For example, when **Chaos** is 20%, particles can bounce off the space warp at angles varying from the bounce angle to 18 degrees off the bounce angle (20% of 90 degrees).

Panels

Inherit Vel.   Sets how much of the particle velocity is due to the velocity of the space warp itself. A deflector can be passed through particles to affect their movements. The **Inherit Vel.** percentage determines how much of the deflector's velocity is passed on to the particles. When **Inherit Vel.** is 100, all of the deflector's velocity is passed on to particles. Smaller percentages pass on lesser amounts of velocity. This setting is particularly useful when passing a deflector through a series of stationary particles. In order to move stationary particles, the **Inherit Vel.** value must be set higher than 0%.

Diameter   Sets the diameter of the space warp object. Since particles bounce off the perimeter of the space warp, this value affects the bouncing of particles as well.

Section

Creates a shape based on a cross section of a 3D object.

General Usage

On the **Create** panel [icon], click **Shapes** [icon]. Click **Section**. Click and drag on the screen to create the section plane. Move or rotate the section plane to intersect a 3D object. A shape displays where the plane intersects the mesh, indicating the current cross section. When the desired cross section is displayed, click **Create Shape**.

Section Parameters

Create Shape Click to create the cross section shape. A dialog appears where you can enter a name for the shape. Each shape is created as an **Editable Spline**.

Update If **When Section Moves** is chosen, the cross section changes as the section plane is transformed. If **When Section Selected** is chosen, the cross section is not re-generated until the section plane is selected again, or **Update Section** is clicked. When **Manually** is chosen, the cross section is updated only when **Update Section** is clicked.

Section Extents Determines the extents used by the section plane to make cross sections. When **Infinite** is selected, the section plane extends on its plane to infinity, intersecting any objects on the same plane regardless of whether they are within the section plane boundary. When **Section Boundary** is selected, cross section shapes are generated only for mesh objects that touch the section plane. When **Off** is selected, no cross section shapes can be generated.

The color swatch sets the color of cross sections that appear when the section plane intersects an object.

Section Size Sets the **Length** and **Width** of the section plane.

Usage Notes

If the section plane intersects more than one object, the shape created with **Create Shape** will contain a cross section spline from each object intersected.

ShapeMerge

Uses a shape to cut an object or make a sub-object selection.

General Usage

Create an object. Create one or more shapes that partly or completely overlap the object.

Select the object. On the **Create** panel ![icon], click **Geometry** ![icon]. Choose **Compound Objects** from the pulldown menu. Click **ShapeMerge**. Click **Pick Shape** and pick a shape.

Select **Cookie Cutter** on the Parameters rollout to cut the object based on the shape. You can also select **Merge** and choose one of the sub-object types under the **Output Sub-Mesh Selection** section to make a sub-object selection.

Pick Operand

Pick Shape
: To pick a shape for the **ShapeMerge** operation, click **Pick Shape** and pick the shape from the scene. You can click **Pick Shape** repeatedly to pick as many shapes as desired.

  When a shape is picked, it is projected through the object along its local Z axis. New faces are created on the object according to the projection of the shape.

Reference
Move
Copy
Instance
: Determines how the picked shape is used with the **ShapeMerge** operation. Choosing **Move** uses the shape itself. Choosing **Reference**, **Copy** or **Instance** uses one of these clone types for the operation, leaving the original shape in the scene. For information on these clone types, see the **Clone** entry in the *Menus* section.

Parameters

Operands
: Displays the operands used in the **ShapeMerge** operation. To access the creation parameters for an operand, highlight the operand and access the modifier stack pulldown. Select the operand from the pulldown to access its creation parameters.

Delete Shape
: To delete a shape from the operands list, highlight the shape name and click **Delete Shape**.

Cookie Cutter
: Cuts the object based on the projection of the picked shapes through the object. Checking **Invert** inverts the cut.

Merge	Creates a selection set based on the projection of shapes. A sub-object type must be selected under the **Output Sub-Mesh Selection** in order to set the selection.
Output Sub-Mesh Selection	Choose a type of sub-object selection to pass on to the next modifier applied to the object. The sub-object selection does not appear red or change in any way to indicate that it is selected. When you apply another modifier to the object, however, the sub-object selection will be used. Selecting **None** does not make a sub-object selection, while **Face**, **Edge** and **Vertex** make face, edge and vertex selections respectively.
Display	An object resulting from a **ShapeMerge** operation is called the *result*. Choose to display the **Result** or **Operands**.
Update	Determines when the result is updated on the screen. **Always**, the default, immediately updates the result whenever a change is made. **When Rendering** updates the result when the scene is rendered. **Manually** updates the result when the **Update** button is checked.

Usage Notes

ShapeMerge is used primarily for making odd-shaped objects, or for making an odd-shaped selection set based on a shape.

The shapes used with **ShapeMerge** can be animated to animate the object's cut or selection.

Skew

A modifier or space warp that skews an object.

General Usage

To apply **Skew** as a modifier, select the object. Click on the **Modify** panel 🎱 and click **More** if necessary to access the full list of modifiers. Select **Skew**.

To apply **Skew** as a space warp, go to the **Create** panel 🎣. Click **Space Warps** 〰️ and select **Modifier-Based** from the pulldown menu. Click **Skew**. Click and drag on the screen to create the **Skew** space warp's length and width, then move the cursor and click again to specify the space warp's height. Select the object and use **Bind to Space Warp** 📐 to bind it to the space warp. To modify the skew, select the space warp and go to the **Modify** panel.

Increase the **Amount** value to skew the object. You can also access the **Sub-Object** level and transform the gizmo to change the effect. Animate the **Amount** value to animate the skew.

Sub-Object Levels

The **Skew** modifier has two **Sub-Object** levels.

Gizmo   The controller for the skew. This controller can be transformed and animated to animate the skew.

Center   The center or base point of the skew.

Space Warp Parameters

Length   Sets the dimensions of the space warp object.
Width
Height

Decay   Sets the amount by which the skew will diminish over distance. When **Decay** is 0, the skew amount is the same over the entirety of the object. When **Decay** is higher than 0, the skew decays (becomes less) as the distance between the bound object and space warp increases.

Parameters

Amount   Sets the amount of the skew in units.

Direction   Rotates the skew effect around the **Skew Axis** by the specified number of degrees. If you cannot get the skew effect to go in the desired direction, try setting **Direction** to 90.

Bend Axis   Sets the axis for the skew effect. The **Skew Axis** uses the gizmo's or space warp's local axis as a reference, not the local axis of the object being skewed.

Limit Effect   When checked, the skew is limited to parts of the object within the **Upper Limit** and **Lower Limit**. Other parts of the object will move in accordance with the skew effect, but will not themselves skew. When **Limit Effect** is checked, the upper and lower limits appear as two rectangles spanning the length and width of the space warp or gizmo.

Upper Limit   Sets the upper limit for the skew. The limit is set in units above the center of the gizmo or space warp. This value must be a positive number. Only vertices that lie between the **Upper Limit** and **Lower Limit** are bent.

Lower Limit   Sets the lower limit for the skew. The limit is set in units below the center of the gizmo or space warp. This value must be a negative number. Only vertices that lie between the **Upper Limit** and **Lower Limit** are bent.

Sliding

Creates two doors, one of which slides.

General Usage

On the **Create** panel ![icon], click **Geometry** ![icon]. From the pulldown menu, choose **Doors**. Click **Sliding**. Click and drag to create the width of the door, move the cursor and click to set the depth or height, then move the cursor and click again to set the remaining dimension.

Parameters

Height **Width** **Depth**	Sets the overall dimensions of the door.
Flip **Front Back**	Flips the doors between the front and back of the frame.
Flip Side	Moves the sliding door to the other side of the frame.
Open	Opens door panels by the specified percentage. At 100 percent the door is fully open.
Frame	Sets parameters for the frame. When **Create Frame** is unchecked, the frame is removed and the doors remain. **Width** and **Depth** set the width and depth of the frame. **Door Offset** moves the door or doors away from the center of the frame by the specified number of units.
Generate **Mapping** **Coords**	Generates mapping coordinates on the doors.

Leaf Parameters

See the Leaf Parameters rollout under the **BiFold** entry.

Creates a window with two sashes, one of which slides.

General Usage

On the **Create** panel [icon], click **Geometry** [icon]. From the pulldown menu, choose **Windows**. Click **Sliding**. Click and drag to create the width of the window, move the cursor and click to set the depth or height, then move the cursor and click again to set the remaining dimension.

Creation Method

See the Creation Method rollout under the **Awning** entry.

Parameters

Height **Width** **Depth**	Sets the overall dimensions of the window.
Frame	Sets the dimensions of the window frame. **Horiz. Width** and **Vert. Width** affect the size of the glazing in addition to the frame size. **Thickness** also affects the thickness of the rails in the window sashes.
Glazing	Sets the **Thickness** of the glass.
Rail Width	Sets the width of the rails in the sashes.
# Panels Horiz **# Panels Vert**	Sets the number of horizontal and vertical panels in the sashes.
Chamfered Profile	When checked, rails have chamfered edges similar to those of a wooden window.
Hung	When checked, the panels slide vertically. When unchecked, panels slide horizontally.
Open	Opens the sliding window by the specified percentage. At 100 percent the window is fully open.
Generate Mapping Coords	Generates mapping coordinates on the window.

Smooth

Assigns smoothing groups to selected faces.

General Usage

Select an object. Use the **Mesh Select** or **Edit Mesh** modifier to select faces on the object. On the **Modify** panel ![icon], click **More** if necessary, and choose **Smooth**.

Parameters

Auto Smooth	When checked, the object is automatically smoothed based on the **Threshold** angle. Faces that are at an angle that is less than the **Threshold** angle are assigned the same smoothing group.
Prevent Indirect Smoothing	Corrects auto smoothing. If **Auto Smooth** is checked and the smoothed object doesn't look right, work with the **Threshold** angle first. If smoothing still doesn't work properly, check **Prevent Indirect Smoothing** to apply a more advanced computation to automatic smoothing. This option takes much longer to calculate, and should be used only when absolutely necessary.
Smoothing Groups	Click on a smoothing group number to assign selected faces to the smoothing group.

Usage Notes

Smoothing groups are numbers assigned to faces that help MAX determine how an object should be rendered.

When two faces share the same smoothing group number, they are rendered as part of the same smooth surface. When two faces have different smoothing group numbers, they are rendered with a visible edge between them. Each face can belong to more than one smoothing group.

The **Smooth** and **Edit Mesh** modifiers have an **Auto Smooth** option that automatically assigns faces to smoothing groups based on a **Threshold** angle. If the angle between two faces is less than the **Threshold** angle, then the two faces are assigned the same smoothing group. Increasing the **Threshold** smooths out sharper areas of the mesh.

You can also assign specific smoothing group numbers to selected faces with the **Edit Mesh** modifier under the **Face Sub-Object** level.

Note that smoothing groups do not change an object's geometry in any way, only the way it is rendered.

A particle system that emits flakes.

General Usage

On the **Create** panel [icon], click **Geometry** [icon]. Select **Particle Systems** from the pulldown menu. Click **Snow**. Click and drag on the screen to create the **Snow** emitter.

Click **Play Animation** [icon] to see the default particle animation. Change parameters as desired.

Parameters

Viewport Count	The maximum number of particles at a time in viewports.
Render Count	The maximum number of particles at a time in rendered images.
Flake Size	The radius of rendered particles, in units.
Speed	Sets the initial speed of each particle in units per frame. Variation sets the possible variation in speed in units per frame. The actual speed of particles after they are born can be affected by a space warp.
Tumble	A value from 0 to 1 that determines how far off its original axis each particle can tumble. When **Tumble** is 0, particles do not tumble. At 0.1, particles wobble. When **Tumble** is 1, they spin wildly. Each particle tumbles around a different random axis.
Tumble Rate	A value that sets the speed at which particles tumble. Higher values make particles tumble faster.
Flakes Dots Ticks	Choose the type of particle to display in viewports.
Render	Choose the type of particle to show in renderings. **Six Point** displays a six-point star. **Triangle** displays a triangle. **Facing** creates rectangles that always face the camera or Perspective view, but don't necessarily face orthographic views.
Start	The frame at which particles begin to emit.

Panels

Life	The number of frames that each particle exists before dying (disappearing).
Birth Rate	Sets the number of particles born (emitted) on each frame. If this value is greater than the **Maximum Sustainable Rate**, particles are emitted in bursts. The **Birth Rate** can be animated. This value can only be entered directly when **Constant** is unchecked.
Constant	When checked, particles are born at the rate set by **Maximum Sustainable Rate**. When unchecked, the **Birth Rate** value is used.
Maximum Sustainable Rate	Displays the birth rate that can be sustained with the current settings. This value is automatically calculated from the **Render Count** divided by **Life**, and cannot be changed directly.
Width Length	Sets the dimensions of the emitter.
Hide	When checked, the emitter is hidden in viewports. The emitter does not render even if it is left unhidden.

Usage Notes

Snow and **Spray** are the original particle systems from 3D Studio MAX R1. Particle systems that are new in R2 are **Blizzard**, **PArray**, **PCloud** and **Super Spray**. These new particle systems have many more options for controlling particles.

Plays sound in a VRML scene.

General Usage

On the **Create** panel ![icon], click **Helpers** ![icon]. From the pulldown menu, choose **VRML 2.0**. Click **Sound**. Click and drag to create the icon. Click **Pick AudioClip** to pick the AudioClip helper set up with the desired sound file.

The actual sound file that plays from a **Sound** helper is set with the **AudioClip** helper. This **AudioClip** helper must be set up before you can set up a **Sound** helper.

A **Sound** helper can be rotated and animated to create sound effects.

Time Sensor

Intensity Sets the volume of the sound. A value of 1.0 plays the sound at normal volume.

Priority When more than one **Sound** helper is present in the scene and the browser cannot play them all, **Priority** helps the browser choose which sounds to play. A **Priority** of 0 is least important, while 1 is most important.

Spatialize When checked, the sound comes from the **Sound** helper. When unchecked, the sound is ambient.

Min Front
Min Back
Max Front
Max Back Sets the areas in which sound will be audible. The **Min Front** and **Min Back** values control a blue ellipsoid around the **Sound** helper. Inside the blue ellipsoid, the sound is at full volume. **Max Front** and **Max Back** control the size of a red ellipsoid around the **Sound** helper. When the view is outside the red ellipsoid, the sound cannot be heard. When the view is between the red and blue ellipsoid boundaries, the sound is at an intermediate volume depending on how close the view is to the boundaries. The **Min Front** and **Max Front** values control the ellipsoid sizes in the direction of the Sound helper arrow, while the **Max Front** and **Max Back** values work in the direction opposite the arrow.

Pick AudioClip To pick an AudioClip to play in the scene, click **Pick AudioClip** and pick the AudioClip helper. The AudioClip helper name appears above the button.

Icon Size Sets the diameter of the icon. The size of the icon does not affect the operation of the helper.

Sphere

Creates a sphere with four-sided polygons.

General Usage

On the **Create** panel ![icon], click **Geometry** ![icon]. If necessary, select **Standard Primitives** from the pulldown menu. Click **Sphere**. Click and drag to set the radius of the sphere.

The orientation of the object's local X, Y and Z axes is determined by the viewport in which you begin drawing. The object's local X and Y axes are placed on the viewport's drawing plane.

Creation Method

Edge	Draws the sphere starting at one corner of the sphere.
Center	Draws the sphere starting at the center of the sphere.

Keyboard Entry

X, Y, Z	Sets the location of the sphere center at the X, Y and Z location in the world coordinate system.
Radius	Sets the radius of the sphere.
Create	Creates the object using the parameters entered.

Parameters

Radius	Sets the radius of the sphere.
Segments	Sets the number of segments on the sphere.
Smooth	When the **Smooth** checkbox is on, the surface renders as a smooth, rounded object. When **Smooth** is off, the surface of the sphere renders as faceted. Turning **Smooth** on or off does not affect the geometry, only its appearance in a rendering.
Hemisphere	When this value is above 0.0, the sphere is cut to a fractional size. A **Hemisphere** value of 0.5 cuts the sphere in half. A value of 0.25 cuts off ¼ of the sphere, while 0.75 cuts off ¾ of the sphere.
Chop	When the **Hemisphere** value is above 0.0 and **Chop** is selected, segments from the cut portion of the sphere are lost, leaving a total number of segments less than the **Segments** value. Compare with **Squash**.

Panels

Squash	When the **Hemisphere** value is above 0.0, selecting **Squash** retains the specified number of segments, squeezing them into the remaining part of the sphere. Compare with **Chop**.
Base to Pivot	When this option is checked, the sphere's pivot point is placed at the base of the sphere. When this option is unchecked, the pivot point is placed at the sphere's center.
Generate Mapping Coords	Generates spherical mapping coordinates on the sphere.

Usage Notes

A sphere created with the **Sphere** option is built with longitude and latitude lines. The **GeoSphere** option creates a sphere with triangular polygons. Either type of sphere will work for most applications. The difference in the two types of spheres is sometimes apparent when creating a hemisphere or when modifying the sphere after creation.

Panels

SphereGizmo

Creates a sphere to contain an atmospheric effect.

General Usage

On the **Create** panel ![icon], click **Helpers** ![icon]. Select **Atmospheric Apparatus** from the pulldown menu. Click **SphereGizmo**. Click and drag on the screen to set the radius of the SphereGizmo.

Sphere Gizmo Parameters

Radius   Sets the radius of the **SphereGizmo**.

Hemisphere   Makes the **SphereGizmo** into a hemisphere by cutting off the lower part along the object's local Z axis. When you click and drag to create the **SphereGizmo**, the object's local X and Y axes are placed on the drawing plane, with the Z axis pointing up out of the viewport. When you check **Hemisphere**, the **SphereGizmo** appears the same in the drawing viewport, with the cutoff visible in other viewports.

Seed   Atmospheric effects rely on a seed number for calculating their effects. For example, a combustion effect with identical parameters in two equal-sized **SphereGizmos** will produce different versions of combustion if each has a different seed. Two **SphereGizmos** with the same seed will produce exactly the same combustion effect. If you have more than one **SphereGizmo** in your scene and you want to be sure they have different effects, make sure the **Seed** is set differently for each **SphereGizmo**.

New Seed   Randomly generates a new **Seed**.

Usage Notes

Atmospheric effects are created with the help of the **BozGizmo**, **SphereGizmo** and **CylGizmo** object. This type of gizmo is used to set the boundaries and the **Seed** for each of these effects. For information on atmospheric effects that use gizmos, see the **Environment** entry in the *Menus* section.

Deforms an object into a spherical shape.

General Usage

Select an object. On the **Modify** panel ⛏️, click **More** if necessary, and choose **Spherify**.

Parameters

Percent   Sets the percentage of distortion to be applied to the object. When **Percent** is 100 and the object has sufficient detail to be deformed to a sphere, the object becomes a sphere.

Usage Notes

Not all objects can be deformed to a perfect sphere with **Spherify**. An object must have sufficient detail in order to approximate a sphere. Boxy objects with few segments will be only partially deformed with **Spherify**.

Spherify can be used to make morph targets, where one copy of the original object can be morphed to a spherified copy. See the **Morph** entry in this section of the book for information on how to morph objects.

You can also use **Conform (Compound Objects)** to deform an object to another object. See the **Conform (Compound Objects)** entry in this section of the book for information on this process.

Spindle

Creates a cylinder with peaked caps.

General Usage

On the **Create** panel ![icon], click **Geometry** ![icon]. Select **Extended Primitives** from the pulldown menu. Click **Spindle**. Click and drag to set the radius of the cylinder. Move the cursor and click to set the spindle height, then move the cursor and click again to set the height of the peaked caps.

Creation Method

Edge	Draws the spindle starting from the outer edge of the cylinder.
Center	Draws the spindle starting from the center of the cylinder.

Keyboard Entry

X, Y, Z	Sets the location of the spindle base center along the X, Y and Z axes of the coordinate system.
Radius **Height** **Cap Height** **Overall** **Centers** **Blend**	See these listings under Parameters below.
Create	Creates the spindle using the parameters entered.

Parameters

Radius	Radius of the cylindrical center portion of the spindle.
Height	When the **Overall** option is on, **Height** refers to the height of the entire spindle. When the **Centers** option is on, this value refers the height the cylindrical midsection of the spindle.
Cap Height	Sets the height of the spindle caps. **Cap Height** cannot be greater than half the **Height** when **Overall** is selected.
Overall	When this button is selected, the **Height** value refers to the height of the entire spindle, including the caps.

Centers	When this button is on, the **Height** value sets just the height of the straight part of the spindle.
Blend	A **Blend** value over 0 creates a bevel between each cap end and the midsection. The bevel extends into the midsection and caps by half the number of units specified by the **Blend** value. The **Blend** value cannot be greater than half the midsection length.
Sides	Sets the number of sides on the spindle.
Height Segs	Sets the number of segments along the length of the cylindrical part of the spindle.
Smooth	When the **Smooth** checkbox is on, the surface renders as a smooth, rounded object. When **Smooth** is off, the surface of the spindle renders as faceted. Turning **Smooth** on or off does not affect the geometry, only its appearance in a rendering.
Slice On	Slices the spindle around its local Z axis according to the **Slice From** and **Slice To** angles. The slice function starts at the **Slice From** angle and moves counterclockwise to the **Slice To** angle, cutting out the portion of the spindle in this area. The sliced area is capped with two flat surfaces. Slice angles are measured from the object's local X axis, which is considered to be zero degrees.
Generate Mapping Coords	Generates mapping coordinates on the spindle. Cylindrical mapping coordinates are applied to the cylindrical midsection of the spindle while planar mapping coordinates are applied to the caps. Sliced areas receive planar mapping.

SplineSelect

A modifier that selects vertices, segments or splines on a shape.

General Usage

Create a shape. On the **Modify** panel 🐾, click **More** to access the list of modifiers. Choose **SplineSelect**. Choose the desired **Sub-Object** level and select vertices, segments or splines.

Sub-Object Levels

The SplineSelect modifier has three **Sub-Object** levels, **Vertex**, **Segment** and **Spline**. For each sub-object level, the parameters are similar.

Get Vertex/ Segment/ Spline Selection	Takes the last vertex, segment or spline selection and creates a selection based on the currently active sub-object.
Copy	Copies a named selection set of vertices, segments or splines so the selection can used with another **SplineSelect** modifier elsewhere on the stack. When you click **Copy**, the list of selection sets of vertices, segments or splines for this object at this level of stack appears. Choose a selection set.
Paste	Pastes a selection set of vertices, segments or splines from another **SplineSelect** modifier. **Paste** is enabled only when a **Copy** has been performed on a **SplineSelect** modifier elsewhere on the stack for the same object. Vertex selections can be pasted to segments or splines, and vice versa. You can also paste to the same modifier and sub-object for which a **Copy** was performed. In this case, you will be prompted for a new name for the selection set.

Usage Notes

The **SplineSelect** modifier is primarily for selecting splines to be deleted with the **DeleteSpline** modifier. Other modifiers such as **Extrude** and **Lathe** work on the entire shape regardless of whether a vertex, segment or spline selection is passed up the stack.

Panels

A particle system that emits drops.

General Usage

On the **Create** panel ![icon], click **Geometry** ![icon]. Select **Particle Systems** from the pulldown menu. Click **Spray**. Click and drag on the screen to create the **Spray** emitter.

Click **Play Animation** ![icon] to see the default particle animation. Change parameters as desired.

Parameters

The options on the **Spray** Parameters rollout are very similar to those for the **Snow** entry. See the **Snow** entry for all parameters except the following.

Drop Size   Sets the size of particles. If **Tetrahedron** is selected as the particle type for rendering, **Drop Size** is the length of the tetrahedron in units. If **Facing** is selected as the rendering particle type, **Drop Size** is one half the width of the square in units.

Drops   Choose the type of particle to display in viewports.
Dots
Ticks

Render   Choose the type of particle to show in renderings. **Tetrahedron** displays a tall pyramid. **Facing** creates rectangles that always face the camera or Perspective view, but don't necessarily face orthographic views.

Star

Creates a star shape.

General Usage

On the **Create** panel , click **Shapes** . Click **Star**. Click and drag on the screen to create one radius for points, then release the mouse. Move the cursor inside or outside the first radius and click again to create the second radius.

General

See the General rollout under the **Arc** entry.

Keyboard Entry

The Keyboard Entry rollout is available only on the **Create** panel.

X, Y, Z	Specifies the location of the center of the star created when **Create** is clicked. The center is placed at the X, Y and Z location in the world coordinate system.
Radius 1 **Radius 2**	See these listings under Parameters below.
Create	Creates the shape with the parameters entered. The shape is placed on the current viewport's plane.

Parameters

Radius 1	Specifies the radius of the star's inner or outer points.
Radius 2	Specifies a second radius for the star's inner or outer points. **Radius 2** may be either smaller or larger than **Radius 1**. The greater the difference between **Radius 1** and **Radius 2**, the sharper the star's points will be.
Points	Sets the number of points on the star.
Distortion	Sets the number of degrees over which to rotate the points set by **Radius 2**.
Fillet Radius 1	Rounds the angle at which the **Radius 1** points meet. This value is set in degrees.
Fillet Radius 2	Rounds the angle at which the **Radius 2** points meet. This value is set in degrees.

STL-Check

Checks an object for suitability for export to an *.stl* file.

General Usage

Select an object. On the **Modify** panel ⬚, click **More** if necessary, and choose **STL-Check**.

Parameters

Open Edge   Checks for open edges. An open edge is an edge that bounds only one face. Open edges can point out "holes" in an object.

Double Face   Checks for coincident faces.

Spike   Checks for faces that share only one edge with the rest of the object.

Multiple Edge   Checks for faces that share more than one edge.

Everything   Checks for all the errors above.

Don't Select   Detects errors, but does not select the offending edges or faces.

Select Edges   Selects edges causing an error.

Select Faces   Selects faces causing an error, or faces adjoining edges causing an error.

Change Mat-ID   When checked, faces that are in error are assigned the specified material ID. This can be useful for displaying erroneous faces in shaded viewports, where a **Multi/Sub-Object** material is assigned to the object before the **STL-Check** modifier is applied.

Check   When checked, the object is checked immediately for errors.

Status   Displays the number of errors found in the mesh.

Usage Notes

An *.stl* file is a file used for stereo lithography, where a machine creates a physical model from a 3D mesh. **STL-Check** checks for open edges, coincident faces and other situations that will cause problems when the *.stl* file is imported to the machine.

Stretch

A modifier or space warp that stretches an object.

General Usage

Create an object to stretch. This object should have several segments so it can stretch smoothly.

To apply **Stretch** as a modifier, select the object. Click on the **Modify** panel ![icon] and click **More** if necessary to access the full list of modifiers. Select **Stretch** .

To apply **Stretch** as a space warp, go to the **Create** panel ![icon]. Click **Space Warps** ![icon] and select **Modifier-Based** from the pulldown menu. Click **Stretch** . Click and drag on the screen to create the **Stretch** space warp's length and width, then move the cursor and click to specify the space warp's height. Select the object, and click **Bind to Space Warp** ![icon] on the Toolbar. Click and drag from the object to the space warp to bind it to the space warp. To modify the stretch, select the stretch space warp and go to the **Modify** panel.

Increase the **Stretch** value to stretch the object. You can also access the **Sub-Object** level and transform the gizmo to change the effect. Animate the **Stretch** value to animate the stretch.

You can also access the **Sub-Object** level and transform the gizmo to change the effect. Animate the **Stretch** value to animate the stretch.

Sub-Object Levels

The **Modify** panel for the **Stretch** modifier has two sub-object levels.

Gizmo	The controller for the stretch. This controller can be transformed and animated to animate the stretch.
Center	The center or base point of the stretch.

Space Warp Parameters

Length **Width** **Height**	Sets the dimensions of the space warp object.
Decay	Sets the amount by which the stretch will diminish over distance. When **Decay** is 0, the stretch is the same over the entirety of the object. When **Decay** is higher than 0, the stretch decays (becomes less) as the distance between the bound object and space warp increases.

Parameters

Stretch   Sets the amount of stretch to be applied to the object. Larger values stretch the object more. **Stretch** can be a negative value.

Amplify   Amplifies the stretch along the minor axes.

Stretch Axis   Sets the axis for the stretch. The **Stretch Axis** uses the gizmo's or space warp's local axis as a reference, not the local axis of the object being stretched.

Limit Effect   When checked, the stretch effect is limited to parts of the object within the **Upper Limit** and **Lower Limit**. Other parts of the object will move in accordance with the stretch effect, but will not stretched. When **Limit Effect** is checked, the upper and lower limits appear as two rectangles spanning the length and width of the space warp or gizmo.

Upper Limit   Sets the upper limit for the stretch. The limit is set in units above the center of the gizmo or space warp. This value must be a positive number. Only vertices that lie between the **Upper Limit** and **Lower Limit** are stretched.

Lower Limit   Sets the lower limit for the stretch. The limit is set in units below the center of the gizmo or space warp. This value must be a negative number. Only vertices that lie between the **Upper Limit** and **Lower Limit** are stretched.

Strokes

Launches commands with mouse strokes.

General Usage

On the **Utilities** panel ![icon], choose **Strokes**. Click the **Draw Strokes** button and click and drag the mouse. The command assigned to the stroke is launched. Close the Strokes dialog to end the use of strokes.

Usage Notes

Strokes are commands that can be enacted by clicking and dragging with the mouse. To use strokes, go to the **Utilities** panel and choose **Strokes**. Click **Draw Strokes** and click and drag in a viewport. The pattern of the cursor motion is interpreted and used to launch a command.

If a command has not been assigned to the stroke, or if the stroke is not recognized, the Stroke Not Found dialog appears. Click **Continue** to try the stroke again, or click **Define** to define the stroke.

Clicking **Define** accesses the Define Stroke dialog.

The pattern of the stroke is displayed at the upper left of the dialog. Stroke patterns are defined by a grid, and lines on the grid are labeled with the letters A through L. Each stroke is named by the lines that the cursor passes through during the stroke. For example, a stroke that passes through the lines labeled E, K, H and B is named EKHB.

Choose the command to be enacted by the currently displayed stroke pattern. If the command requires an object to be selected, the options at the lower left of the dialog become available. Choose a method for selecting objects to which the command will be applied.

Click **Review** to access the Review Strokes dialog. Here you can see all commands defined by strokes, and the stroke pattern that enacts each one.

If your mouse has a middle button, it can be set up to enact strokes at all times. To do this, choose *File/Preferences* from the menu to access the Preference Settings dialog. Under the **Viewports** tab, select **Stroke** in the **Middle Mouse Button** section. Any action with the middle mouse button is now treated as a stroke regardless of whether you are on the **Utilities** panel.

Sunlight

Creates a directional light controlled by a compass.

General Usage

On the **Create** panel [icon], click **Systems** [icon]. Click **Sunlight**. Click and drag on the screen to create the compass, then move the cursor and click again to set the light. Adjust the time, date and location settings to make the light shine in the direction of the sun. Move the compass so it sits on the ground plane.

Control Parameters

Azimuth	The angle of the sun in degrees off North. The **Azimuth** angle is calculated in a clockwise direction from North, which is 0. This angle is calculated from the date, time and location settings, and cannot be changed directly.
Altitude	The angle of the sun off the horizon. With **Sunlight**, the horizon is set at the compass plane. This angle is calculated from the date, time and location settings, and cannot be changed directly.
Hours Mins Secs	Sets the time of day. These settings can be animated to make the light move over time.
Month Day Year	Sets the date. These settings can be animated to make light move over time.
Time Zone	The time zone for the location. This number is used in conjunction with the **Hours**, **Mins** and **Secs** settings to determine the light direction. **Time Zone** is automatically set when you select a city with **Get Location**. Change the **Time Zone** if the location you are simulating is in a different time zone than the city selected with **Get Location**.
Daylight	When checked, daylight savings time is taken into account when the light direction is calculated.
Get Location	Click **Get Location** to access a dialog where a major city can be selected.
Latitude Longitude	Sets the latitude and longitude of the location being simulated. These values are automatically set when a city is selected with **Get Location**, but can be changed.

Orbital Scale   Sets the distance of the light from the compass rose. Use **Orbital Scale** to move the light outside all objects in the scene, but it is not necessary to move the light much farther than that. Direct lights shine parallel light rays, so accurate sunlight and shadows will appear in the rendering as long as the light is outside all objects.

North Direction   Rotates the compass rose in the clockwise direction by the specified angle. Use this setting to align the compass rose with the actual North direction in your scene.

Usage Notes

Sunlight uses a **Direct Light**. When the light is selected, the **Modify** panel contains parameters for an ordinary **Direct Light** only. For information on these parameters, see the **Direct Light** entry.

To change the date, time and other **Sunlight** parameters, select the light and go to the **Motion** panel.

When the compass rose is selected, the **Modify** panel contains parameters for the compass. For information on these parameters, see the **Compass** entry.

A particle system that emits particles from a single point.

General Usage

On the **Create** panel 🖑, click **Geometry** 📦. Select **Particle Systems** from the pulldown menu. Click **Super Spray**. Click and drag on the screen to create the icon.

Click **Play Animation** ▶ to see the default particle animation. Change parameters as desired.

Basic Parameters

Off Axis — Moves the particle stream around the emitter's local Y axis by the specified number of degrees. **Spread** fans out the particles on the local XZ plane.

Off Plane — Rotates the particle stream around the emitter's local Z axis. **Spread** fans out the particles on the XY plane.

Icon Size — Sets the dimensions of the **Super Spray** icon. The size of the icon does not affect the **Super Spray** effect.

Emitter Hidden — When checked, the icon/emitter is hidden. When the emitter is hidden, you can access the **Super Spray** parameters by clicking on any one of the particles.

See the **PArray** entry for the remainder of the **Super Spray** parameters. All remaining parameters are the same, except that **Object Fragments** are not available as **Super Spray** particles.

Surface Approximation

Sets surface approximation for selected NURBS objects.

General Usage

Select one or more NURBS surfaces. On the **Utilities** panel 🔧, click More if necessary, and choose **Surface Approximation**. Change parameters as desired, and click **Set Selected** to apply the parameters to selected NURBS surfaces.

Surface Approximation

See the Surface Approximation rollout under the **Point Surf** entry for information on the parameters on this rollout. This rollout is identical to the Surface Approximation rollout on the **Modify** panel for NURBS surfaces, except for the addition of the following parameters.

Set Selected   Click **Set Selected** to apply the current parameters to selected NURBS objects. This option has no effect on selected non-NURBS objects.

Reset   Resets all parameters on the rollout to their default values.

SurfDeform

A modifier that deforms an object according to a NURBS surface's shape.

General Usage

Create a NURBS point or CV surface. Deform the NURBS surface.

Create an object with several segments. Select the object.

On the **Modify** panel , click **More** if necessary, and choose **SurfDeform**. Click **Pick Surface** and click on the NURBS surface. The object deforms according to the NURBS surface.

Parameters

All parameters for the **SurfDeform** modifier are the same as those for the **PatchDeform** modifier. See the **PatchDeform** entry in this section of the book for information on these parameters.

Tape

Creates a measuring tape helper object.

General Usage

On the **Create** panel [icon], click **Helpers** [icon]. Click **Tape**. Click and drag on the screen to create the tape measure and its target.

Parameters

Length	Displays the length of the tape. When Specify Length is checked, a new length can be entered.
Specify Length	When checked, a new **Length** can be entered.
World Space	Displays the angle from the tape to its target in world space.

Usage Notes

A **Tape** object is useful for measuring the distance between two objects or points. **Tape** works best when used with **3D Snap** on [icon]. To set the snap points, right-click on the **3D Snap** button to access the Grid and Snap Settings dialog.

To view the tape **Length** while moving the tape target, select the tape (not its target) and go to the **Modify** panel [icon]. Click **Pin Stack** [icon]. Select and move the target. The **Length** remains on the panel and updates as you move the target.

Taper

A modifier or space warp that tapers an object.

General Usage

Create an object to taper. This object should have several segments in order to taper smoothly.

To apply **Taper** as a modifier, select the object. Click on the **Modify** panel ![icon] and click **More** if necessary to access the full list of modifiers. Select **Taper**.

To apply **Taper** as a space warp, go to the **Create** panel ![icon]. Click **Space Warps** ![icon] and select **Modifier-Based** from the pulldown menu. Click **Taper**. Click and drag on the screen to create the **Taper** space warp's length and width, then move the cursor and click to specify the space warp's height. Select the object, and click **Bind to Space Warp** ![icon] on the Toolbar. Click and drag from the object to the space warp to bind it to the space warp. To modify the taper, select the taper space warp and go to the **Modify** panel.

Increase the **Amount** value to taper the object.

Sub-Object Levels

The **Modify** panel for the **Taper** modifier has two sub-object levels.

Gizmo — The controller for the taper. This controller can be transformed and animated to animate the taper.

Center — The center or base point of the taper.

Space Warp Parameters

Length
Width
Height — Sets the dimensions of the space warp object.

Decay — Sets the amount by which the taper will diminish over distance. When **Decay** is 0, the taper angle is the same over the entirety of the object. When **Decay** is higher than 0, the taper decays (becomes less) as the distance between the bound object and space warp increases.

Parameters

Amount Sets the severity of the taper. **Amount** is multiplied by the original diameter of the gizmo or space warp to determine how much its size will increase or decrease between its center and top along the **Primary** axis. Although **Amount** can be any number, values between 0.2 and 2 usually work well.

Curve Curves the taper. **Curve** is multiplied by the original diameter of the gizmo or space warp to determine how much its size will change between its center and bottom along the **Primary** axis.

Primary Sets the main axis along which the gizmo or space warp changes in diameter. **Primary** uses the gizmo's or space warp's local axis as a reference, not the local axis of the object being tapered.

Effect The space warp or gizmo is also tapered along one or both of the remaining axes. Choose one or more axes for the secondary effect of the taper.

Symmetry When checked, the taper is symmetrical both above and below the center of the gizmo or space warp. Checking Symmetry has the same effect as setting **Curve** to the negative **Amount** value.

Limit Effect When checked, the taper is limited to parts of the object within the **Upper Limit** and **Lower Limit**. Other parts of the object will be repositioned in accordance with the taper, but will not taper. When **Limit Effect** is checked, the upper and lower limits appear as two rectangles spanning the length and width of the space warp or gizmo.

Upper Limit Sets the upper limit for the taper. The limit is set in units above the center of the gizmo or space warp. This value must be a positive number. Only vertices that lie between the **Upper Limit** and **Lower Limit** are tapered.

Lower Limit Sets the lower limit for the taper. The limit is set in units below the center of the gizmo or space warp. This value must be a negative number. Only vertices that lie between the **Upper Limit** and **Lower Limit** are tapered.

Creates a target camera.

General Usage

On the **Create** panel [icon], click **Cameras** . Click **Target**. Click and drag in a viewport to place the camera and target. Move the camera or target, or change parameters as desired.

Parameters

Lens	Sets the lens length for the camera.
FOV	Sets the field of view for the camera. For information on the relationship between the lens length and **FOV**, see the **Field-of-View** entry in the *Toolbar* section. The **Aperture Width** setting on the Render Scene dialog also affects the relationship between the lens length and the FOV. See the **Render** entry in the *Menus* section for information on the **Aperture Width** setting.
FOV Direction	Determines which direction is used to measure the **FOV**. Choose horizontal, vertical or diagonal measurement from the flyout.
Orthographic Projection	Removes perspective from the camera view.
Stock Lenses	Choose from nine lens lengths. The **FOV** changes accordingly when a lens length is chosen.
Show Cone	Shows the camera cone even when the camera is unselected.
Show Horizon	Displays a dark gray horizon line in the camera view.
Near Range Far Range	These values set the near and far ranges for atmospheric effects that work with a camera, such as fog.
Show	Displays the **Near Range** as a yellow line around the camera cone, and displays the **Far Range** as a brown line. If **Near Range** is 0 it is not visible on the screen.

Panels

Clipping Planes

Clipping planes determine how close or far away an object can be in relationship to the camera before it is considered to be invisible to the camera. When an object is very close to or far away from the camera, it is "clipped" from the scene and will no longer render in the camera view. **Near Clip** sets the clip distance in units for objects close the camera, while **Far Clip** sets the clipping distance in units for objects far away. When **Clip Manually** is unchecked, the default **Near Clip** and **Far Clip** values are used. When **Clip Manually** is checked, you can enter new **Near Clip** and **Far Clip** values.

Clipping planes rarely come into play in the normal course of creating scenes with 3D Studio MAX. At times, files imported from other programs or created with largely different unit setups will have extremely large or small objects. If objects in an imported scene don't render, check the size of the objects in the scene and make sure they are not so large or small that they are clipped from view.

Usage Notes

After a camera is set up, you can change a viewport to the camera view by activating the viewport and pressing the <C> key. If more than one camera is in the scene, you will be prompted to choose a camera for the view.

A **Target** camera is similar to a **Free** camera. The only difference is that **Target** cameras have a target that can be moved and animated as a separate object. The direction of a **Free** camera is controlled by rotating the camera itself, while the direction of a **Target** camera is controlled by the position of the target. Deciding which type of camera to use depends on how you plan to animate the camera, or which type of camera you find easier to work with.

Creates a direct light with a target.

General Usage

On the **Create** panel ![icon], click **Lights** ![icon]. Click **Target Direct**. Click and drag in a viewport to place the light and target. Move the light or target, or change parameters as desired.

General Parameters

See the General Parameters rollout under the **Target Spot** entry.

Directional Parameters

Hotspot	Sets the radius of the area where the light shines at full intensity. The **Hotspot** cannot be larger than the **Falloff**. The **Hotspot** area is indicated in viewports by light blue lines emanating from the light source.
Falloff	Sets the radius of the area where the light intensity fades from full intensity to zero. When the **Falloff** radius is nearly the same as the **Hotspot** radius, the light has a sharp edge. When the **Falloff** radius is much larger than the **Hotspot** radius, the light has a soft edge. The **Falloff** area is indicated in viewports by dark blue lines emanating from the light source.

The **Falloff** and its relationship to the **Hotspot** control the sharpness of the light edge, not the sharpness of shadow edges. The sharpness of shadows is controlled by the **Size** and **Smp Range** values under the Shadow Parameters rollout. |
Show Cone	The light's **Hotspot** and **Falloff** areas are indicated on the screen by light blue and dark blue lines called the *cone*. The cone appears on the screen when the light is selected. When the light is unselected, the cone appears only if **Show Cone** is on.
Overshoot	When **Overshoot** is on, the light shines from the entire light plane but casts shadows only in its cone.
Circle Rectangle	Sets the shape of the light cone.
Asp.	Sets the aspect ratio of a **Rectangle** light.

Bitmap Fit   When **Rectangle** is selected as the light shape, you can click **Bitmap Fit** to set the aspect ratio to match a specified bitmap. Click **Bitmap Fit** and choose a bitmap file from the file selector that appears. Clicking a bitmap with **Bitmap Fit** does not set up the bitmap as a projector.

Projector   Turn on this checkbox to enable the projection map.

Map   Before clicking **Map**, set up a map in the Material Editor to be projected from the light. Click and drag the map to the **Map** button, or click **Map** and choose the projector map from the Material/Map Browser.

Shadow Parameters

See the Shadow Parameters rollout under the **Target Spot** entry.

Usage Notes

Direct lights create parallel shadows, and so are excellent for simulating daylight.

A **Target Direct** light is similar to a **Free Direct** light. The only difference is that **Target Direct** lights have a target that can be moved and animated as a separate object. The direction of a **Free Direct** light is controlled by rotating the light itself, while the direction of a **Target Direct** light is controlled by the position of the target. Deciding which type of light to use depends on how you plan to animate the light, or which type of light you find easier to work with.

Creates a target spotlight.

General Usage

On the **Create** panel ![icon], click **Lights** ![icon]. Click **Target Spot**. Click and drag in a viewport to place the light and target. Move the light or target, or change parameters as desired.

General Parameters

On
When checked, the light affects the objects in the scene. When unchecked, the light remains in the scene but does not affect the scene.

This color swatch determines the color and intensity of the light. Click the color swatch to change the RGB or HSV values of the light. You can also change the color and intensity of the light by changing the **R, G, B, H, S** or **V** values below the color swatch.

Exclude /
Include
Click this button to include or exclude objects from the light's effect. When you click this button, the Exclude/Include dialog appears.

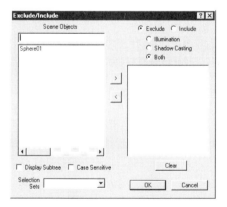

All geometry objects in the scene are listed at the left of the dialog. The controls at the left are the same as those on the Select Objects dialog that appears when you click

Select by Name ![icon]. For information on these options, see **Select by Name** in the *Toolbar* section.

Highlight object names and click the right arrow ![icon] to put the objects on the list at right. To remove objects from the list at right, highlight the object name and click the left arrow button .

At the upper right of the dialog, choose **Exclude** to exclude the objects on the list at right from the effects of the light. This option causes objects on the list at the left to be included. Choose **Include** to include objects at the right and exclude objects at the left.

You can choose to exclude or include the objects at right by choosing **Illumination**, **Shadow Casting** or **Both**. Choosing **Illumination** includes or excludes objects from being illuminated by the light source, while **Shadow Casting** prevents or causes objects from casting shadows. In order to cast shadows with the light, the **Cast Shadows** checkbox under the Shadow Parameters rollout must also be checked.

Clear clears all objects from the list at right and puts them back on the list at left.

Click **OK** to accept settings, or **Cancel** to cancel them.

R G B H S V   These values affect the color and intensity of the light. The **V** value sets the light's intensity along with the **Multiplier** value.

Multiplier   Multiplies the light's intensity (**V** value) to increase or decrease the intensity of the light. The **V** value is limited to 255, but a higher intensity can be created by setting **Multiplier** higher than 1.0. Values between 0 and 1 decrease the intensity. A negative **Multiplier** can be entered to reverse the light effect, removing light from portions of the scene instead of illuminating them.

Contrast   A value from 0 to 100 that affects the amount of contrast between brightly and dimly lit areas. When a light strikes object faces perpendicular to the light direction, light is bright. Faces turned partially away receive dim light. Increase the **Contrast** value to simulate a harsh light.

Soften Diff. Edge   A value from 0 to 100 that affects the softness of the edge between diffuse light (light illuminating the object) and ambient light (areas not affected by the light). Higher values make a softer transition between ambient and diffuse. Increasing this value slightly diminishes the overall brightness of light.

Affect Diffuse	When **Affect Diffuse** is checked, the light affects the diffusely lit areas of objects it strikes, as opposed to highlight areas. This checkbox is checked by default.
Affect Specular	When **Affect Specular** is checked, the light affects the highlight areas of objects. This checkbox is on by default.
Start End	Sets the start and end points for attenuation in units from the light source. Attenuation causes light intensity to increase or decrease as it moves away from the light source. The **Start** and **End** values are expressed in units away from the light source.
Use	When the **Use** checkbox is on under **Near**, the light starts out with zero intensity and gradually increases to full intensity. From the light source to the **Start** point, there is no light. At the **Start** point, the light begins to increase until it reaches full intensity at the **End** point.
	When the **Use** checkbox is on under **Far**, the light dims between the **Start** and **End** points. At the **Start** point, light is at full intensity. The light gradually dims to zero intensity when it reaches the **End** point.
Show	When the **Show** checkbox is checked for **Near**, the **Start** point is indicated by a dark blue circle around the light, and the **End** point by a light blue circle. When the **Show** checkbox is checked for **Far**, the **Start** point is indicated by a yellow circle around the light, and the **End** point by a brown circle.
	The **Use** checkbox can be checked for both **Near** and **Far** at the same time.
Decay	The settings under **Decay** provide an alternate means of fading light as it moves away from the light source. **None** creates no decay, and attenuation takes place only if one or both of the **Use** checkboxes are checked under **Near** or **Far** above.
	The **Inverse** and **Inverse Square** options decay the light as it moves away from the source. With **Inverse**, the intensity of the light at any point is divided by the distance from the **End** point, and continues until it reaches the **Far** point. With **Inverse Square**, the intensity is divided by the squared distance from the **End** point, and continues until it reaches the **Far** point.

Panels

Inverse Square is the real life formula for light moving through air, but your scene might look more realistic with Inverse decay.

If Inverse or Inverse Square decay is used and the Use checkbox for Far is unchecked, the light's intensity is calculated down to very low intensities, making the rendering time longer. Use Far attenuation when using Inverse or Inverse Square decay to minimize rendering time.

Spotlight Parameters

Hotspot   Sets the area where the light shines at full intensity. The Hotspot is expressed as an angle, which is a portion of an imaginary sphere around the light source. The Hotspot angle cannot be larger than the Falloff angle. The Hotspot area is indicated in viewports by light blue lines emanating from the light source.

Falloff   Sets the area where the light intensity fades from full intensity to zero. The Falloff is expressed as an angle, which is a portion of an imaginary sphere around the light source. When the Falloff angle is nearly the same as Hotspot angle, the light has a sharp edge. When the Falloff angle is much larger than Hotspot angle, the light has a soft edge. The Falloff area is indicated in viewports by dark blue lines emanating from the light source.

The Falloff angle and its relationship to the Hotspot angle control the sharpness of the light edge, not the sharpness of shadow edges. The sharpness of shadows is controlled by the Size and Smp Range values under the Shadow Parameters rollout.

Show Cone   The light's Hotspot and Falloff angles are indicated on the screen by an area outlined by light blue and dark blue lines called the *cone*. The cone appears on the screen when the light is selected. When the light is unselected, the cone appears only if Show Cone is on.

Overshoot   When Overshoot is on, the light shines in all directions but casts shadows only in its light cone.

Circle   Sets the shape of the light cone.
Rectangle

Panels

Asp.	Sets the aspect ratio of a **Rectangle** light.
Bitmap Fit	When **Rectangle** is selected as the light shape, you can click **Bitmap Fit** to set the aspect ratio to match a specified bitmap. Click **Bitmap Fit** and choose a bitmap file from the file selector that appears. Clicking a bitmap with **Bitmap Fit** does not set up the bitmap as a projector.
Projector	Turn on this checkbox to enable the projection map.
Map	Before clicking **Map**, set up a map in the Material Editor to be projected from the light. Click and drag the map to the **Map** button, or click **Map** and choose the projector map from the Material/Map Browser.
Target Distance	Displays the distance from the spotlight to the target.

Shadow Parameters

Cast Shadows	Check this checkbox to cause the light to cast shadows.
Use Global Settings	Uses the same **Map Bias**, **Size**, **Smp Range**, **Absolute Map Bias** and **Ray Trace Bias** settings as all other light sources with global settings enabled. By default, the global settings are the default settings for these parameters. If you turn on **Use Global Settings** and change these parameters, the new settings become the global settings, and are used on each light in the scene for which **Use Global Settings** is checked.
Use Shadow Maps	Uses the map method for calculating shadows rather than the ray-trace method. The map method projects a map from the light onto the scene, and calculates shadows from this projection. The shadow map method can produce soft or sharp shadow edges depending on the **Size** and **Smp Range** settings.
Use Ray-Traced Shadows	Uses the ray-trace method for calculating shadows. The ray-trace method projects rays of light into the scene from the light source, producing sharp shadow edges. Ray-traced shadows can produce shadows on wireframe objects. The ray-trace method takes longer to render than the shadow map method.

Panels

Map Bias	Sometimes shadows appear to be shifted slightly toward or away from the object casting the shadow. This is especially noticeable when the object casting the shadow is sitting against the object receiving the shadow, as with a vase sitting on a table. Lower **Map Bias** values pull the shadow closer to the object casting the shadow, while higher values push the shadows away from the object. To find out if the **Map Bias** needs to be adjusted, render the scene with the default value and check the shadows.
Size	The size of the projected map used with the shadow maps calculation method. If shadows are not distinct enough, the **Size** can be increased to improve shadow detail. Higher **Size** values increase rendering time.
Smp Range	Determines the number of times the shadow area is sampled. Lower values produce soft shadows, while higher values produce sharp shadows. Higher values increase rendering time.
Absolute Map Bias	In a rendered animation, shadow edges sometimes appear to flicker due to recalculation of the map bias on each frame. Leave this checkbox on to stabilize calculations of the map bias.
Ray Trace Bias	Shifts a ray-traced shadow toward or away from the object casting the shadow. See **Map Bias** above.

Usage Notes

Target Spot lights are useful for illuminating specific areas of the scene. A **Target Spot** is also ideal for use as a volume light for making a visible light cone generated by an overhead light fixture.

A **Target Spot** light is similar to a **Free Spot** light. The only difference is that **Target Spot** lights have a target that can be moved and animated as a separate object. The direction of a **Free Spot** is controlled by rotating the light itself, while the direction of a **Target Spot** is controlled by the position of the target. Deciding which type of light to use depends on how you plan to animate the light, or which type of light you find easier to work with.

Creates a teapot object.

General Usage

On the **Create** panel ⬚, click **Geometry** ⬚. If necessary, select **Standard Primitives** from the pulldown menu. Click **Teapot**. Click and drag to set the radius of the teapot body. The size of the remaining teapot parts (lid, spout, handle) are fixed in relation to the radius of the teapot.

Keyboard Entry

X, Y, Z   Sets the location of the teapot base center at the X, Y and Z location in the world coordinate system.

Radius   Sets the radius of the largest part of the teapot body.

Create   Creates the object using the parameters entered.

Parameters

Radius   Sets the radius of the largest part of the teapot body.

Segments   Sets the number of segments in each part of the teapot. Segment values for each teapot part cannot be set separately.

Smooth   See this parameter under the **Capsule** entry.

Body
Handle   Includes the specified object as part of the teapot. A teapot can be made up of any combination of parts.
Spout
Lid

Generate   Generates spherical 2x2 mapping coordinates on the
Mapping   body and lid, and cylindrical coordinates on the spout
Coords   and handle.

Usage Notes

The inclusion of a teapot as a primitive object might seem strange, but 3D teapots have a distinguished role in the history of 3D computer graphics. Early 3D pioneers found the teapot shape to be useful for exploring the use of reflection maps. A reflection on a rotating sphere is not much to look at, but reflections on a spinning teapot are constantly changing.

Tessellate

A modifier that adds faces to an object.

General Usage

Select an object. On the **Modify** panel 🔧 , click **More** if necessary, and choose **Tessellate**. Faces are added to the object.

Parameters

Operate On   Choose to tessellate by **Faces** ◁ or **Polygons** ▢ . A face is always triangular, but a polygon may be made up of two or more faces.

Edge   **Edge** tessellation places a vertex in the middle of each face edge and creates three lines connecting these vertices, dividing each face into three faces.

Face-Center   **Face-Center** places a vertex at the center of each face and creates a line from this vertex to each of the three original vertices, dividing each face into four faces.

Tension   Sets the tension for **Edge** tessellation. Each new vertex created at the middle of a face edge is pushed toward or away from the center of the object. Positive values push the vertex away from the center, while negative values push it toward the center of the object. If you want faces to be created without any visible change to the object, set **Tension** to 0.

Iterations   The number of times tessellation is applied. A higher number of **Iterations** creates more faces.

Usage Notes

The **Tessellation** modifier is useful for quickly adding more detail to an object.

You can also tessellate faces with the **Edit Mesh** modifier under the **Face Sub-Object** level.

Tessellation does not make any attempt to smooth out sharp edges on an object. To add faces and smooth sharp edges at the same time, use the **MeshSmooth** modifier.

Creates a text shape.

General Usage

On the **Create** panel ⚞, click **Shapes** ⚟. Click **Text**. Under the Parameters rollout, enter the text in the **Text** window. Choose a font and enter a size for the text with the **Size** value. Click and drag on the screen to create and place the text.

General

See the General rollout under the **Arc** entry.

Parameters

Under Parameters, a font can be chose from a pulldown menu. Available fonts include those installed in Windows NT, and Postscript fonts in the Fonts directory set with the *File/Configure Paths* menu option.

I	Italicizes the text.
U	Underlines the text.
≣ ≣ ≣ ≣	Aligns the text to the left, right or center, or justifies the text. You must have more than one line of text entered to see the effects of these buttons.
Size	Sets the height of the text in units.
Kerning	Adjusts the distance between letters. Decrease this value to move letters closer together, or increase it to move letters farther apart.
Leading	Adjusts the distance between lines of text. You must have more than one line of text to see the effect of adjusting the **Leading** value.
Text	Enter the text. Press **<Enter>** after each line of text to start a new line. The default text is MAX Text. To replace this text, highlight the text and type over it. Text can be pasted from other areas of Windows NT to the **Text** window.
Update	Click to update the text in the viewport. This button is only available when **Manual Update** is checked.
Manual Update	When checked, changes to parameters will not update the text on the screen until **Update** is clicked.

TimeSensor

Causes animation to play at specified times in a VRML scene.

General Usage

On the **Create** panel [icon], click **Helpers** [icon]. From the pulldown menu, choose **VRML 2.0**. Click **TimeSensor**. Click and drag to create the icon. Click **Pick Objects** to choose the objects that will animate.

Time Sensor

Loop	If the animation is shorter than the time between the **Start Time** and the **Stop Time**, turning on this checkbox causes the animation to loop until the **Stop Time** is reached.
Start on World Load	Starts the animation when the VRML file is loaded.
Start Time Stop Time	Sets the starting and ending frames for the animation.
Pick Objects	To pick objects that will animate at the specified time, click **Pick Objects** and pick the objects. As they are picked, the objects appear on the list.
Delete	To delete an object from the list, highlight the object and click **Delete**.
Icon Size	Sets the diameter of the icon. The size of the icon does not affect the operation of the helper.

Usage Notes

The **Loop**, **Start on World Load**, **Start Time** and **Stop Time** parameters apply to all objects picked for this **TimeSensor**. To set different animation times for different objects, use more than one **TimeSensor** in the scene.

Creates a 3D donut.

General Usage

On the **Create** panel ✐, click **Geometry** 🗔. If necessary, select **Standard Primitives** from the pulldown menu. Click **Torus**. Click and drag to set one radius of the torus, then move the cursor and click to set the radius of the torus cross section.

The orientation of the object's local X, Y and Z axes is determined by the viewport in which you begin drawing. The object's local X and Y axes are set on the viewport's drawing plane.

Creation Method

Edge   Draws the torus starting at one end of the torus.

Center   Draws the torus starting at the center.

Keyboard Entry

X, Y, Z   Sets the location of the torus center at the X, Y and Z location in the world coordinate system.

Major Radius   Sets the outer radius of the torus.

Minor Radius   Sets the radius of the cross section circle. You can also think of this value as the difference between the inner radius and outer radius of the torus.

Create   Creates the object using the parameters entered.

Parameters

Radius 1   The radius of the torus center circle. Changing this value changes the overall size of the torus while the radius of the cross section stays the same.

Radius 2   The radius of the cross section circle. You can also think of this value as the difference between the inner radius and outer radius of the torus.

Rotation   Rotates all cross section circles by the specified angle around the entire torus. Entering a value other than 0.0 for **Rotation** will produce visible effects only on toruses with a low **Sides** value, such as 4 or 6. You can also see the effects of rotation when **Twist** is used.

Panels

Twist	Rotates cross section circles progressively around the torus. Each cross section circle is rotated progressively more until the last circle is rotated by the full **Twist** amount.
	Set **Twist** to multiples of 360 for smooth twisting. Twisting at other angles causes a crimp in the torus in the first segment. You can get rid of this crimp by checking **Slice On** and leaving the **Slice From** and **Slice To** values at 0.
Segments	The number of segments around the torus.
Sides	The number of sides on the cross section.
Smooth	Turning on **All** smooths the entire torus, while **None** leaves the entire torus faceted. **Sides** smooths the edges running around the torus. **Segments** smooths the edges running around the cross section.
Slice On	Slices the torus around its local Z axis according to the **Slice From** and **Slice To** angles. The slice function starts at the **Slice From** angle and moves counterclockwise to the **Slice To** angle, cutting out the portion of the torus in this area. The sliced area is capped with two flat surfaces. Slice angles are measured from the object's local Y axis, which is considered to be zero degrees.
Generate Mapping Coords	Generates mapping coordinates on the torus. Sliced areas receive planar mapping.

Creates a knotted torus (3D donut).

General Usage

On the **Create** panel ⟨icon⟩, click **Geometry** ⟨icon⟩. Select **Extended Primitives** from the pulldown menu. Click **Torus Knot**. Click and drag to set one radius of the torus, then move the cursor and click to set the radius of the torus cross section.

The orientation of the object's local X, Y and Z axes is determined by the viewport in which you begin drawing. The object's local X and Y axes are set on the viewport's drawing plane.

Creation Method

Diameter   Creates the torus from one end to the other.

Radius   Creates the torus from the center outward.

Keyboard Entry

X, Y, Z   Sets the location of the torus knot center at the X, Y and Z location in the world coordinate system.

Major Radius   Sets the outer radius of the torus.

Minor Radius   Sets the radius of the cross section.

Create   Creates the object using the parameters entered.

Parameters

The settings under the **Base Curve** section set values along the curve of the knotted torus.

Knot   When this button is on, a knotted torus is created.

Circle   When this button is on, a circular torus is created similar to the standard torus created with the **Torus** option under **Standard Primitives**. However, a circular torus created here can be altered with other settings such as **Warp Height, Warp Count** and **Eccentricity**, while a standard torus cannot.

Radius   The radius of the base curve. Use this value to set the overall size of the torus.

Segments   The number of segments around the torus.

Panels

Panels

P	Sets the number of times the torus turns up and down. This value can be set only when **Knot** is chosen above. To make a torus that curls around itself, make this value greater than half the **Q** value.
Q	The number of times the torus turns around the center. This value can be set only when **Knot** is chosen above. To make a torus that curls around itself, make this value more than twice the **P** value.
Warp Count	**Warp Count** and **Warp Height** produce a series of petals around the center of the torus. **Warp Count** sets the number of petals produced around the torus when **Warp Height** is set to a value other than 0.0. This option is available only when **Circle** is chosen as the base curve.
Warp Height	**Warp Count** and **Warp Height** produce a series of petals around the center of the torus. The height of the petals is **Warp Height** multiplied by the base curve radius. In order to produce petals, you must also set **Warp Count** to a number other than 0.0. This option is available only when **Circle** is chosen as the base curve.

The settings under **Cross Section** set values for the cross section of the torus.

Radius	The radius of the cross section.
Sides	The number of sides around the cross section.
Eccentricity	Changes the relationship between the two axes of the cross section. Values other than 1.0 make the cross section into an elliptical shape. Values between 0.0 and 1.0 make the cross section more like a ribbon, while values between 1.0 and 10.0 stretch the shape in the opposite direction.
Twist	The number of times the cross section twists around the base curve. Note that this setting works differently from the **Twist** parameter for the standard torus. **Twist** for the standard torus works in degrees, creating one full cross section turn with a value of 360. For a knotted torus **Twist**, a value of 1 creates one full cross section turn.
Lumps	Bulges can be created in the torus. Lumps sets the number of bulges in the torus.

Lump Height	**Lump Height** sets the height of the bulges. The bulge height is **Lump Height** times the radius of the base curve.
Lump Offset	Offset for lump placement in degrees along the torus. This parameter is included primarily so lumps can be animated.
Smooth	Turning on **All** smooths the entire torus, while **None** leaves the entire torus faceted. **Sides** smooths the edges running around the torus, leaving the edges running around the torus as faceted.
Generate Mapping Coords	Generates cylindrical mapping coordinates on the torus.
Offset	Offsets the mapping coordinates along the torus U and V. U runs around the cross section while V runs along the base curve.
Tiling	Tiles the mapping coordinates along the U and V. U runs around the cross section while V runs along the base curve.

Usage Notes

The torus is created around a curve called the *base curve*, which may be knotted or circular. A knotted torus is useful for creating abstract 3D objects quickly.

TouchSensor

Creates a sensor that triggers an action or animation when an object is clicked in a VRML scene.

General Usage

On the **Create** panel [icon], click **Helpers** [icon]. From the pulldown menu, choose **VRML 2.0**. Click **TouchSensor**. Click and drag to create the icon. Click **Pick Trigger Object** to pick the object that triggers the animation. Click **Pick Action Objects** to choose the objects that will animate when the trigger is clicked.

Touch Sensor

Panels

Pick Trigger Object	To pick the object that triggers the animation, click **Pick Trigger Object** and pick the object. You can also pick the object by clicking **Select by Name** [icon] or by pressing the <H> key.
Enable	Enables the sensor.
Pick Action Objects	To cause objects to animate when the trigger is clicked, click **Pick Action Objects** and select the objects. As each object is selected, it appears on the list. Animated objects, cameras, lights or **AudioClip** helpers can be selected. When finished, click **Pick Action Objects** again to turn it off.
Delete	To delete an object from the list, highlight the object and click **Delete**.
Icon Size	Sets the diameter of the icon. The size of the icon does not affect the operation of the helper.

Displays object trajectories.

General Usage

Select one or more animated objects. Go to the **Motion** panel ⊕ |.
Click **Trajectories**. The trajectories of selected objects are displayed,
with keys displayed as white boxes on each trajectory.

Sub-Object Level

Trajectories have one **Sub-Object** level, **Keys**. When this sub-object
level is active, keys on the trajectory can be moved, rotated or scaled.
Keys can also be deleted or added with the **Delete Key** and **Add Key**
buttons below.

Trajectories

Delete Key   To delete a key, click on the key and click this button. To
use this button, the **Keys Sub-Object** must be active.

Add Key   To add a key to a trajectory, click this button and click on
the trajectory. To use this button, the **Keys Sub-Object**
must be active.

Start Time   Sets the start and end frames for converting the trajec-
End Time   tory to a spline, or a spline to a trajectory.

Samples   When a spline is converted to a trajectory, **Samples** sets
the number of keys that will be generated on the trajec-
tory. When a trajectory is converted to a spline, **Samples**
sets the number of vertices on the spline.

Convert To   Converts the select objects' trajectory to a spline. The
spline is given the name **Shape##** where **##** is the next
sequential number available for that object name.

Convert From Click **Convert From** and click on a spline to convert the
spline to a trajectory for the selected objects. Each object
now moves along the spline.

Collapse   This option converts animation from a controller to
transformable keys, like those generated by turning on
the **Animate** button and transforming an object. Check
Position, **Rotation** and/or **Scale** to choose the kind of
keys to collapse. Click **Collapse** to convert the selected
objects' animation to transformable keys.

Trim/Extend

A modifier that trims or extends splines.

General Usage

On the **Create** panel ⚡, click **Shapes** 🔲. Uncheck **Start New Shape**. Make one or more intersecting splines as part of the same shape.

On the **Modify** panel 🔧, click **More** to access the list of modifiers. Choose **Trim/Extend**. Click **Pick Locations**. Click on the spline to trim or extend.

Trim/Extend

Pick Locations	Click **Pick Locations** to pick the splines to be trimmed or extended. The **Pick Locations** button remains on so you can pick more than one spline. Right-click on the screen to turn off the **Pick Locations** button.
Auto	Automatically chooses whether to trim or extend a picked spline. 3D Studio MAX first attempts to trim the spline, but if the spline does not intersect other splines in either direction, the spline is extended.
Trim Only	Trims a picked spline. MAX searches along the spline in both directions from the picked location for intersecting splines. If an intersection is encountered at one or both ends, the spline is trimmed to the intersections.
Extend Only	Extends an open spline until it strikes another segment in the shape. A spline is extended in a straight line in the same direction as the end of the spline. The extension becomes a new segment connected to the picked spline.
	If there are no segments for the spline to strike, the spline is not extended. Checking **Infinite Boundaries** can sometimes cause a spline to be extended even when it doesn't strike other segments in the shape.
Infinite Boundaries	When checked, each open spline in the shape is extended infinitely along an imaginary line. These imaginary extensions are not shown on the screen. However, the infinite extensions are used when extending a spline, so it will find segments with which to intersect. Check **Infinite Boundaries** when attempting to extend a spline that does not actually intersect segments in the shape.

View	Projects the splines onto the viewport plane when determining intersections. If splines intersect in the current viewport, then the **Trim/Extend** modifier considers them to be intersecting.
Construction Plane	Projects the splines onto the construction plane when determining intersections.
None (3D)	Only splines that actually intersect in 3D space are considered by the **Trim/Extend** modifier to be intersecting.

Usage Notes

The **Trim/Extend** modifier is available only when a shape is selected.

The **Trim/Extend** modifier works only on intersecting splines within the same shape.

Panels

Tri Patch

Creates a flat rectangular object with deformable patches.

General Usage

On the **Create** panel ⟨icon⟩, click **Geometry** ⟨icon⟩. Choose **Patch Grids** from the pulldown menu. Click **Tri Patch**. Click and drag on the screen to create the patch grid.

Keyboard Entry

X, Y, Z	Sets the location of the patch grid center at the X, Y and Z location in the world coordinate system.
Length **Width**	Sets the dimensions of the patch grid.
Create	Creates the object using the parameters entered.

Parameters

Length **Width**	Sets the dimensions of the patch grid.
Generate Mapping Coords	Generates planar mapping coordinates on the patch grid.

Usage Notes

Patch grids are flat, rectangular objects that can be smoothly deformed. A **Quad Patch** is made with rectangular polygons, while a **Tri Patch** is made with triangular polygons. When a patch vertex is moved, the area around the vertex deforms accordingly to make a smooth transition.

A **Tri Patch** is created with four patch vertices, one at each corner. Do not confuse mesh vertices, which sit at the intersection of each segment, with patch vertices, which may have several segments between them. Patch vertices have movable Bezier handles, and can be used with an **Edit Patch** modifier. See the **Edit Patch** entry for more information on patch vertices.

To increase the number of patch vertices on a **Tri Patch** object, you must use the **Subdivide** option under the **Edge Sub-Object** level of an **Edit Patch** modifier. See the **Edit Patch** entry for more information.

Creates a tube.

General Usage

On the **Create** panel ⟨⟩, click **Geometry** ⟨⟩. If necessary, select **Standard Primitives** from the pulldown menu. Click **Tube**. Click and drag to set one tube radius, then move the cursor and click to set the second radius. Move the cursor and click again to set the tube's height.

Creation Method

Edge	Draws the tube starting at one corner of the tube base.
Center	Draws the tube starting at the center of the tube base.

Keyboard Entry

X, Y, Z	Sets the location of the tube base center along the X, Y and Z axes of the coordinate system.
Inner Radius Outer Radius	Sets the inner radius and outer radius of the tube.
Height	Sets the height of the tube.
Create	Creates the object using the parameters entered.

Parameters

Radius 1 Radius 2	Sets each radius of the tube. The larger of **Radius 1** and **Radius 2** is the outer radius, while the smaller value is the inner radius.
Height	Sets the height of the tube.
Height Segments	Sets the number of segments along the height of the tube.
Cap Segments	Sets the number of circular segments on the tube cap.
Sides	Sets the number of sides on the tube.
Smooth Slice On	See these parameters under the **Capsule** entry.
Generate Mapping Coords	Generates cylindrical mapping coordinates on the tube.

Twist

A modifier or space warp that twists an object.

General Usage

Create an object to twist. This object should have several segments so it can twist smoothly.

To apply **Twist** as a modifier, select the object. Click on the **Modify** panel ![icon] and click **More** if necessary to access the full list of modifiers. Select **Twist**.

To apply **Twist** as a space warp, go to the **Create** panel ![icon]. Click **Space Warps** ![icon] and select **Modifier-Based** from the pulldown menu. Click **Twist**. Click and drag on the screen to create the **Twist** space warp's length and width, then move the cursor and click to specify the space warp's height. Select the object, and click **Bind to Space Warp** ![icon] on the Toolbar. Click and drag from the object to the space warp to bind it to the space warp. To modify the twist, select the twist space warp and go to the **Modify** panel.

Sub-Object Levels

The **Modify** panel for the **Twist** modifier has two sub-object levels.

Gizmo   The controller for the twist. This controller can be transformed and animated to animate the twist.

Center   The center or base point of the twist.

Space Warp Parameters

Length   Sets the dimensions of the space warp object.
Width
Height

Decay   Sets the amount by which the twist will diminish over distance. When **Decay** is 0, the twist angle is the same over the entirety of the object. When **Decay** is higher than 0, the twist decays (becomes less) as the distance between the bound object and space warp increases.

Parameters

Angle   Sets the angle of the twist in degrees.

Bias   A value from -100 to 100 that pushes the twist effect toward or away from the gizmo or space warp center. When **Bias** is 0, the twist is evenly distributed. At -100, the twist is gathered at the gizmo or space warp center. At 100, the twist occurs at the outer edges of the space warp or gizmo only.

Twist Axis   Sets the axis for the twist. The **Twist Axis** uses the gizmo's or space warp's local axis as a reference, not the local axis of the object being bent.

Limit Effect When checked, the twist is limited to parts of the object within the **Upper Limit** and **Lower Limit**. Other parts of the object will rotate in accordance with the twist, but will not twist. When **Limit Effect** is checked, the upper and lower limits appear as two rectangles spanning the length and width of the space warp or gizmo.

Upper Limit Sets the upper limit for the twist. The limit is set in units above the center of the gizmo or space warp. This value must be a positive number. Only vertices that lie between the **Upper Limit** and **Lower Limit** are bent.

Lower Limit Sets the lower limit for the twist. The limit is set in units below the center of the gizmo or space warp. This value must be a negative number. Only vertices that lie between the **Upper Limit** and **Lower Limit** are bent.

UDeflector

A space warp that uses any object as a deflector.

General Usage

Create a particle system object. Create another object to use as a deflector. Move the deflector object so it sits in the path of oncoming particles.

On the **Create** panel ⬚, click **Space Warps** ⬚. Select **Particles & Dynamics** from the pulldown menu. Click **UDeflector**. Click and drag on the screen to create the space warp object.

Click **Pick Object** and pick the deflector object.

Select the particle system object and click **Bind to Space Warp** ⬚ on the Toolbar. Click and drag from the particle system object to the space warp to bind it to the space warp.

Select the space warp object. Go to the **Modify** panel ⬚ to access the parameters. Change the parameters as desired.

Basic Parameters

Pick Object   Picks the object to use as a deflector. After the item is picked, its object name appears next to **Item:** above the **Pick Object** button.

Bounce   See these parameters under the **SDeflector** entry.
Variation
Chaos

Friction   A value from 0 to 100 that determines how much friction affects the particles as they bounce off the deflector. When **Friction** is 0, there is no friction, and objects bounce off at the angle determined by the hit angle and variations set by **Chaos**. When **Friction** is 100, particles bounce in the direction of the face normal at the point of impact. Intermediate **Friction** values pull the bounce angle toward the face normal by a proportionate degree.

Inherit Vel.   See this parameter under the **SDeflector** entry.

Icon Size   Sets the size of the icon. The icon size has no effect on the **UDeflector** effect.

A modifier that allows you to modify UVW mapping.

General Usage

Select an object. On the **Modify** panel 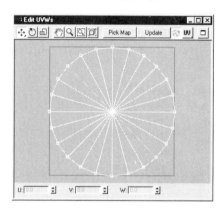, click **More** if necessary, and choose **Unwrap UVW**. Click **Edit** to edit the mapping.

Parameters

Edit
Click **Edit** to display a window where you can edit the current UVW mapping.

In the usual course of work with 3D Studio MAX, mapping is fitted to objects. On this window, the opposite takes place — a representation of the object is fitted to the mapping coordinates. The mapping is represented as a flat rectangle in this window. The object representation might be flattened, stretched or squashed to fit the mapping.

The object is represented as a lattice of UVW faces and vertices. Select vertices and use one of the transforms at the top left of the window to transform the vertices, and thus transform the mapping.

To see a map for reference while working in this window, click **Pick Map** and pick a previously set up map from the **Material Editor** or the **Scene**. You can also choose to view the object representation from **UV**, **VW** or **UW** directions.

Reset UVWs
Resets the object representation to its default configuration.

Channel
Select the mapping channel to edit.

Usage Notes

The **Unwrap UVW** modifier allows you to modify mapping coordinates previously applied to the object. This includes mapping coordinates set by applying a **UVW Map** modifier, or by checking **Generate Mapping Coords** for a primitive or **Apply Mapping** for a loft object.

If an object doesn't have mapping coordinates to pass on to the **Unwrap UVW** modifier, planar mapping is used.

UVW Map

Applies mapping coordinates to an object.

General Usage

Select an object. Go to the **Modify** panel 🔧. Choose **UVW Map**. Adjust mapping coordinates as desired.

Sub-Object Level

Sub-Object The **UVW Map** modifier has one sub-object, **Gizmo**. The gizmo takes the shape of the mapping coordinates. The gizmo can be moved, rotated or scaled to change the mapping.

Parameters

Planar
Cylindrical
Spherical
Shrink Wrap Sets the type of mapping coordinates. For information on mapping coordinates, see the **Mapping Coordinates** entry in the *Mat Editor* section of this book.

Length
Width
Height Sets the size of the mapping coordinates gizmo along its local X, Y and Z axes.

U Tile
V Tile
W Tile Tiles the mapping coordinates in the U, V and/or W directions. **Flip** flips the map in the opposite direction. See the **Mapping Coordinates** entry in the *Mat Editor* section of this book for informationon U, V and W.

Channel Sets the map channel number. Channel numbers allows you to use two different sets of mapping coordinates on an object. When two **UVW Map** modifiers are applied to an object, one set to **Channel 1** and other other to **Channel 2**, maps can be set up to use one channel or the other. In this way, two completely different sets of mapping coordinates can be applied to one object.

A map is assigned to a map channel on the map's Coordinates rollout. For 2D maps, select the **Texture** option and choose **Explicit UVW 1** or **Explicit UVW 2** from the **Mapping** pulldown. For 3D maps, select **UVW 1** or **UVW 2** on the Coordinates rollout. **UVW 1** refers to **Channel 1**, while **UVW 2** refers to **Channel 2**.

X, Y, Z Aligns the mapping coordinates gizmo so it is perpendicular to the object's local X, Y or Z axis.

Panels

Fit	Fits the gizmo to the object.
Bitmap Fit	Fits the gizmo to a bitmap's aspect ratio. Click **Bitmap Fit** and select the bitmap from the file selector.
View Align	Aligns the gizmo with the current viewport.
Reset	Resets the gizmo to its default position and size.
Center	Centers the gizmo on the object.
Normal Align	Aligns the gizmo with a face normal. Click **Normal Align**, then click and drag on the object to select a face normal.
Region Fit	Fits the gizmo to a custom area. Click **Region Fit** and click and drag on the screen to set the gizmo dimensions.
Acquire	Acquires a gizmo from another object. Click **Acquire** and select another object. The gizmo from the selected object's latest modifier is acquired. If an object does not have a modifier with a gizmo, it cannot be selected. You are prompted with two options, **Acquire Relative** and **Acquire Absolute**. **Acquire Relative** positions and orients the gizmo with the object's local axes in the same way it was aligned on the selected object. **Acquire Absolute** does not change the gizmo's location or orientation.

Panels

UVW Xform

Adjusts complex mapping coordinates.

General Usage

Select an object. Go to the **Modify** panel ![icon]. Click **More** if necessary, and choose **UVW XForm**. Adjust mapping coordinates as desired.

Parameters

U Tile **V Tile** **W Tile**	Tiles the mapping coordinates in the U, V and/or W directions. **Flip** flips the map in the opposite direction.
U Offset **V Offset** **W Offset**	Offsets the mapping coordinates along the U, V or W directions.
Channel	Selectts the map channel number to modify. When a map is set up in the Material Editor, you can choose **UVW 1** or **UVW 2** on the map's Coordinates rollout. You can set up some maps in a material as **UVW 1** and others as **UVW 2**. You can also apply two **UVW Map** modifiers to the object with different types of mapping, one with each channel selected. The maps that use **UVW 1** will render with the mapping coordinates set for the **UVW Map** modifier with **Channel 1** selected, and likewise for **UVW 2** and **Channel 2**. This feature allows you to use two sets of mapping coordinates for different maps in the Material Editor.

Usage Notes

The **UVW Xform** modifier is useful for modifying mapping coordinates that cannot be assigned or altered with a **UVW Map** modifier. For example, a loft object receives mapping coordinates along the length of the path when the **Apply Mapping** checkbox is checked. These mapping coordinates don't fall into any of the mapping coordinates categores that can be assigned or altered with the **UVW Map** modifier. However, the **UVW Xform** modifier can be used to shift a loft object's mapping along the path by changing the **V Offset** value, or to rotate the mapping around the path with the **U Offset** value.

Vol. Select

A modifier that defines an animatable selection of faces or vertices.

General Usage

Select an object. Apply a **Mesh Select** modifier to the object and se-

lect faces or vertices. On the **Modify** panel , click **More** if necessary, and choose **Vol. Select**. Choose **Vertex** or **Face** under **Stack Selection Level**. Click **Sub-Object** and choose the **Gizmo** sub-object level.
Animate the gizmo to animate the selection.

Sub-Object Levels

The **Vol. Select** modifier has two sub-object levels.

Gizmo   Bounds the selection set of faces or vertices. The gizmo
        can be animated to animate the selection.

Center The center or base point of any transforms performed
        on the gizmo.

Parameters

Stack Selection Level Choose to select the entire **Object**, or select by
                           Vertex or **Face** with the gizmo.

Selection   Determines how the gizmo will affect the selection
Method   passed up the stack from previous modifiers. **Replace**
            replaces the previous selection. **Add** adds to the previ-
            ous selection, while **Subtract** subtracts from it.

Invert   When checked, the current selection is inverted.

Selection   Determines the type of selection used by the gizmo.
Type   When **Window** is selected, faces or vertices must be com-
            pletely enclosed by the gizmo in order to be selected.
            When **Crossing** is selected, faces or vertices touching the
            gizmo boundary are also selected.

Selection   Determines the shape of the gizmo.
Volume

Fit   Fits the gizmo to the overall shape of the selection
            passed up from a previous modifier. Note that the new
            fitting may not actually encompass the selection when a
            Sphere or **Cylinder** is selected as the **Selection Volume**.
            If no selection has been passed up the stack, the gizmo is
            fitted to the entire object.

Center   Centers the gizmo on the selection passed up from a previous modifier. If no selection has been passed up the stack, the gizmo is centered on the object.

Reset   Resets the gizmo to its default position, size and orientation.

Usage Notes

The **Vol. Select** modifier provides a way to animate a selection set of faces or vertices. When the gizmo is moved or rotated, the selection set changes and is passed up to any modifiers above **Vol. Select** on the stack.

If a vertex or face selection has been made with the **Mesh Select** modifier before the **Vol. Select** modifier is applied, the **Vol. Select** gizmo defaults to a box that surrounds the selection set. If no vertex or face selection was made before applying **Vol. Select**, the gizmo defaults to a box surrounding the entire object.

Wave

A modifier or space warp that deforms an object with an animatable wave.

General Usage

Create an object with several segments along each of its dimensions.

To apply **Wave** as a modifier, first select the object. Go to the **Modify** panel ![icon] and click **More** if necessary to access the full list of modifiers. Select **Wave**. The object deforms according to the wave parameters. Change or animate parameters as desired. You can also access the **Sub-object** level and transform the gizmo to change or animate the wave effect.

To apply **Wave** as a space warp, go to the **Create** panel ![icon]. Click **Space Warps** ![icon] and select **Geometric/Deformable** from the pulldown menu. Click **Wave**. Click and drag on the screen to create the wave's length, then release the mouse and drag to specify the wave's amplitude. Click to set the amplitude. Select the object, and click **Bind to Space Warp** ![icon] on the Toolbar. Click and drag from the object to the wave to bind it to the wave. The object deforms according to the wave parameters. Select the wave. Go to the **Modify** panel to access the wave parameters. Change or animate the parameters as desired. You can also move the wave space warp to change or animate the wave effect.

Parameters

Amplitude 1 The amplitude (height) of the wave in units at the center of the wave space warp object. This value is set to the same value as Amplitude 2 when you create the wave, but can be changed afterward.

Amplitude 2 The amplitude (height) of the wave in units at the edge of the wave space warp object. Beyond the space warp object, this value continues to increase. This value is set to the same value as Amplitude 1 when you create the wave, but can be changed afterward.

Wave Length The length of one wave cycle. By default, the wave space warp displays with two full wave cycles in the wave object. To show more than two wave cycles, increase the **Segments** value below.

Phase A wave is created with a sine wave. **Phase** sets which phase the wave is in, such as whether it is just starting, is peaking or is just ending at any point in time. A **Phase** of 1.0 shifts the wave by an entire wavelength, a **Phase** of 2.0 shifts it by two entire wavelengths, etc. To make the wave look different, **Phase** can be changed to a fractional number such as 0.5. To make the wave undulate over time, animate the **Phase** value.

Decay
Sets the amount by which the amplitude will diminish over distance. When **Decay** is 0, wave height (amplitude) is the same over the entirety of each bound object. When **Decay** is higher than 0, the amplitude decays as the bound object is further away from the center of the space warp. Try fractional values such as 0.2 for best results.

Space Warp Parameters

The **Wave** space warp has three additional parameters. These parameters make the space warp object easier to see and work with, but do not influence the smoothness of the wave on the bound object.

Sides
The number of sides on the space warp object. This value affects the width of the space warp object.

Segments
The number of segments on the space warp object. This value affects the length of the space warp object.

Division
Scales the space warp without altering the wave effect. Higher values make the space warp smaller, while lower values make it larger.

Modify Panel

When a **Wave** space warp is bound to an object, it is placed on the object's stack as an entry called **Wave Binding**. This stack entry can be accessed by selecting the bound object and going to the **Modify** panel.

The **Wave Binding** entry has one parameter, **Flexibility**. This value multiplies the amplitude of the wave, and can be used to animate the amplitude.

Usage Notes

To animate the wave, either the parameters on the wave's Parameters rollout can be animated, or the wave space warp or gizmo itself can be transformed (moved, rotated or scaled) over time. To make a wave undulate, animate the **Phase** value on the wave's Parameters rollout.

Wind

A space warp that pushes particles or objects in a specified direction, simulating the wind blowing the particles or objects.

General Usage

Create a particle object, or create any object. On the **Create** panel ![icon],

click **Space Warps** ![icon]. Select **Particles & Dynamics** from the pulldown menu. Click **Wind**. Click and drag on the screen to create the wind space warp icon. Move or rotate the space warp so the arrow points in the direction in which you wish the wind to move the particles or objects.

To use the space warp with a particle system, select the particle system and click **Bind to Space Warp** ![icon] on the Toolbar. Click and drag from the object to the space warp to bind it to the space warp.

To use the space warp on an object, the space warp must be assigned as part of a dynamics simulation with the **Dynamics** utility. See the **Dynamics** entry for more information.

Parameters

Strength   The strength of the wind. When **Strength** is 0, wind is not in effect. Higher values make the wind stronger. Negative values push objects and particles in the direction opposite the space warp arrow.

Decay   Sets the amount by which the wind strength will diminish over distance. When **Decay** is 0, strength is the same over the entirety of the bound object. When **Decay** is higher than 0, the strength decays as the distance between the bound object and space warp increases.

Planar   Causes the wind effect to always be perpendicular to the plane of the gravity space warp.

Spherical   Causes the wind effect to be spherical, where wind emanates from the center of the space warp in all directions.

Turbulence   When this value is above 0.0, particles and objects randomly change course as they are blown by the wind. Higher values create more turbulence.

Frequency	Sets the rate at which the turbulence is animated. Note that **Frequency** animates the turbulence itself, not the particles. Animating the turbulence is a subtle effect that will only be visible with very large numbers of particles. Leave this value at 0 unless you are animating large numbers of particles and want to vary the turbulence further over time.
Scale	Scales the turbulence effect. When **Scale** is small, turbulence is gentle. When **Scale** is large, the turbulence becomes more erratic. This value does not scale particles or objects, only the turbulence effect.
Icon Size	Sets the overall size of the wind space warp icon in square units. The size of the icon has no relationship to the wind effect. The only purpose for changing the **Icon Size** is to make it easier for you to see and position the icon.

Usage Notes

The **Wind** space warp is similar to the **Gravity** space warp in its effects, but the **Wind** space warp also provides turbulence effects.

XForm

A modifier that transforms an object or selection.

General Usage

Select an object. Go to the **Modify** panel ![icon]. Click **More** if necessary, and choose **XForm**. Click **Sub-Object** and choose the **Gizmo** sub-object level. Use the transforms on the Toolbar to transform the gizmo, and thus transform the object.

Sub-Object Level

The **XForm** modifier has one **Sub-Object** level, **Gizmo**. The gizmo can be transformed with any of the transform buttons on the Toolbar:

![icon] Select and Move

![icon] Select and Rotate

![icon] Select and Scale

![icon] Select and Non-uniform Scale

![icon] Select and Squash

Usage Notes

If a face or vertex selection has been passed up the stack from a Mesh Select modifier applied prior to the **XForm** modifier being applied, the gizmo will transform the selection set only.

At times, using a transform directly from the Toolbar, such as **Select and Non-uniform Scale**, can cause problems later on. For example, if an object is scaled on one axis and linked to a parent object, then the parent is rotated, the linked object might not behave as expected. This is due to the fact that transforms are always applied after modifiers when an object is manipulated, even if the transform was, in actuality, performed before any modifiers were applied to the object.

The **XForm** modifier gets around this problem by giving you a way to perform a transform that is really a modifier. The **XForm** modifier sits on the modifier stack and is applied to the object correctly in relationship to other modifiers.

Material Editor

This section describes the toolbar buttons for the Material Editor, along with each material and map type.

To access the Material Editor, click the **Material Editor** button ⬚ from the toolbar, or choose *Material Editor* from the *Tools* menu. The Material Editor window appears.

The Material Editor window contains its own toolbar buttons, along with several rollouts. The rollouts change depending on the material and map types you're working with.

If you're looking for information on a specific parameter, look under the map or material type that contains the parameter. The parameters that appear when the Material Editor is first accessed appear under the **Standard** entry. **Standard**, the default material type, is the one most often used.

Many maps have common rollouts and parameters. A rollout that appears frequently is described under its own entry, and is referenced in other entries as appropriate. For example, the Noise rollout that appears under many maps is described under its own entry, **Noise Rollout**. There are also entries for overall concepts such as **Mapping Coordinates** and **Environment Maps**. Each of these entries is referenced by other entries as appropriate.

For general information on materials and maps, see the **About the Material Editor** entry that follows this page.

About the Material Editor

The Material Editor is used to set up materials for objects in the scene. Materials set not only the colors and patterns on objects, but also the transparency, shininess, bumpiness and various other attributes of objects.

To access the Material Editor, click the **Material Editor** button 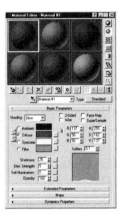 on the toolbar, or choose *Material Editor* from the *Tools* menu. The Material Editor window appears.

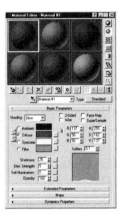

The Material Editor window contains its own toolbar buttons, along with several rollouts. The rollouts change depending on the material or map you're working with.

For information on a specific material, map or toolbar button, see the corresponding entry in this section of this book. For example, to find information on the **Speckle** map, see the **Speckle** entry.

Material Names

Each material is assigned a default name which appears on the Material Editor toolbar. Default material names are **Material #1**, **Material #2**, etc. You can change the material name by highlighting the name on the Material Editor toolbar and typing in a new name. It's a good idea to name your materials with meaningful names as you go along to avoid confusion.

To assign a material to an object, activate the material slot that contains the material, select the object, and click **Assign Material to Selection** on the Material Editor toolbar.

When a material is assigned to an object, it is linked to the object through its name. Material names in the Material Editor do not have to be unique, but names of materials assigned to objects do have to be unique.

What is a Map?

Materials are made up of several attributes, such as shininess and opacity. Many of these attributes can be set with a single number or color for a uniform setting throughout the material. Materials can also be defined partly by or entirely by maps.

A *map* is a pattern of colors. The most commonly used map is **Bitmap**, which is a file containing colored pixels. A **Bitmap** can be any picture file, whether a scanned photograph or a picture drawn in a paint program such as Photoshop. 3D Studio MAX also has other built-in maps that define a pattern of colors, such as **Smoke** or **Dents**. With these maps, you can use two colors to make a pattern, or you can even use more maps to define each color area of the pattern.

Maps can be used to define the overall color of the object, the pattern of shininess, the opacity and the bumpiness of the object, and many other material attributes. Maps are selected from the Material/Map Browser, and are indicated by a green diamond ◈.

The term *map* in 3D Studio MAX can be confusing. There are the maps described above, such as **Bitmap**, **Smoke** and **Dents**, which can be selected from the Material/Map Browser. On the other hand, material attributes defined by maps, such as **Diffuse** and **Bump** under the Maps rollout for **Standard** and **Raytrace** materials, are also called maps. In the 3D Studio MAX documentation, these material attributes are referred to at various times as *maps, map types* and *map channels*. The latter term can be especially confusing as it is also the term used to define the **Channel** parameter for a **UVW Map** modifier.

In this book, the following terminology is used:

- A map selected from the Material/Map Browser is called a *map type*.
- A map name listed on a Maps rollout is called a *map attribute*.

Parent and Child

Materials and maps work with a parent-child relationship. The main material is the *parent* level. By default, this is a **Standard** material, as denoted by the label **Standard** on the button next to **Type** on the Material Editor toolbar.

Under the Maps rollout of the **Standard** and **Raytrace** materials, maps can be chosen to define different attributes of the material. To choose a map, click on the button labeled **None** across from a map attribute name. The Material/Map Browser appears. Choose a map type from the list.

Mat Editor

When you choose a map type, the rollouts on the Material Editor change to display the options for the chosen map type. At this point, you are at a *child* level for the material. To get back to the parent level of the material, click the **Go to Parent** button on the toolbar.

To get back to the child level, go to the Maps rollout and click on the button across from the map attribute name. The button is no longer labeled **None**; it is now labeled with the map type just selected. When you click this button, you are returned to the child level for the map type. These levels of the material are called the *material tree*.

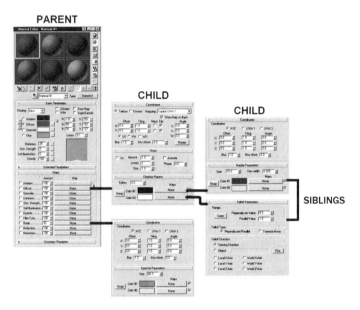

If you have chosen a map type that can be defined further by maps, such as **Smoke**, **Noise** or **Dents**, choosing another map will take you down to yet another child level. Clicking **Go to Parent** will take you back up to the previous child level, and repeatedly clicking **Go to Parent** will eventually take you back to the parent material level.

When two or more maps exist at the same child level, you can click **Go to Sibling** to move from one sibling to the next.

If you can't find the parameters you want on the rollout displayed, it's probably because you're not at the right parent or child level. Click **Go to Parent** to get to the level you want, or go down the child levels until you find it. You can also click **Material/Map Navigator** to display the material and its tree, and click on a child level to go directly to the level.

Materials

There are seven types of materials in 3D Studio MAX.

Standard The standard material type. Most of your materials will be of this default type. Compound materials are made up of other materials, usually **Standard** materials.

Blend A compound material type that blends two materials.

Double Sided A compound material type that uses two materials, placing one on face normals pointing toward the camera and the other on face normals pointing away from the camera.

Multi/ Sub-Object A compound material type that can be made up of any number of submaterials. Each submaterial is assigned to different faces on the object based on the faces' material IDs.

Matte/ Shadow A specialty material type that turns objects into transparent matte objects that can receive shadows.

Raytrace A material type that reflects and refracts the scene around it.

Top/Bottom A compound material type that uses two materials, one for the top of the object and the other for the bottom.

The default material type is **Standard**. To change the material type, click on the button labeled **Standard** next to **Type** on the Material Editor toolbar. A list of material types appears on the Material/Map Browser. Choose a material type. Materials on the Material/Map Browser are preceded by blue sphere, while maps are preceded by a green diamond.

Materials work with a parent-child relationship similar to the one used with maps. The material type is the parent level. Compound materials are made up of other materials, which are children of the parent compound material. A compound material can be made up of other compound materials, but at some point on the chain a non-compound material (**Standard**, **Raytrace**, **Matte/Shadow**) has to be used. A **Standard** or **Raytrace** material may then have maps, which are children of the **Standard** or **Raytrace** material.

Click **Go to Parent** to move up to higher levels of the material tree, or click **Go to Sibling** to move between child materials at the same level.

Adobe Photoshop Plug-In Filter

A map type that applies an Adobe Photoshop Plug-in Filter to a bitmap or animation file.

General Usage

See **Choosing Maps** under the **Material/Map Browser** entry.

Coordinates

See the **Coordinates Rollout (Environment)** entry.

Noise

See the **Noise Rollout** entry.

Adobe Photoshop Plug-In Parameters

Use Plug-in Filters in   Click **Browse** to access the subdirectory where your Adobe Photoshop Plug-ins are stored.

Current Filter   Choose a **Category** and a **Filter** from the pulldown lists.

Input Image   Sets the foreground and background colors for the filter effect. If **Use Alpha Plane** is turned on, then the alpha channel is used to determine which parts of the image the effect should be applied to.

Get Filter Parameters   Displays parameters for the selected filter.

Save as Defaults   Saves the current filter settings as the default settings. If you save these settings as defaults, the next time you choose Adobe Photoshop Plug-in Filters as a map, the default values will be loaded automatically.

Bitmap Parameters

Bitmap parameters load a bitmap or animation file to be used with the filter. For a description of the parameters on this rollout, see the **Bitmap** entry.

Output

See the **Output** entry for information on the parameters on this rollout.

Time Parameters

For a description of these parameters, see the **Bitmap** entry.

A map type that applies an Adobe Premiere Video Filter to another map.

General Usage

See **Choosing Maps** under the **Material/Map Browser** entry.

Coordinates

See the **Coordinates Rollout (Environment)** entry.

Noise

See the **Noise Rollout** entry.

Filter Parameters

Filter Input   Click **Filter Input** to select a map to receive the filter effect.

Filter Selection

>| <|

Filters available in the **Filter Path** appear on the **Filters Available** list at left. To use a filter, highlight the filter and click the right arrow button. The filter appears on the **Filters Selected** list. More than one filter can be applied to the Filter Input at a time. To remove a filter from the **Filters Selected** list, highlight the filter and click the left arrow button.

Filters Parameters   To set up a filter, highlight the filter on the **Filters Selected** list and click **Setup at Start**. A dialog appears with parameters specific to the selected filter.

Filter Path   Click **Add Path** to select a path for filters. The name of the filter path is displayed and the filters in the selected subdirectory are listed on the **Filters Available** list. You can add as many paths as you like, and select them with the **Filter Path** pulldown. In this way, filters from different paths can be added to one filter effect. To remove a path from the **Filter Path** list, select the path from the pulldown list, and click **Remove Path**.

Output

See the **Output** entry for information on the parameters on this rollout.

Time

For a description of these parameters, see the **Bitmap** entry.

# Assign Material to Selection

Assigns the current material to the selected object(s).

General Usage

To assign a material to an object, first select the object with the standard selection tools such as **Select object** . With the desired material slot active, click **Assign Material to Selection** .

Usage Notes

Materials can be assigned to any geometry object, but cannot be assigned to cameras, lights, space warps or other non-geometry objects. The **Assign Material to Selection** button is grayed out when there are no geometry objects selected.

A material assigned to an object in the scene is called a *hot* material. When a material is assigned to an object, small triangles appear at the corners of the material slot, indicating that is has been assigned in the scene. When the object to which the material has been assigned is selected, the triangles are solid. When the object is not selected, the triangles become hollow.

If the material isn't assigned to any objects in the scene, there are no triangles at the corners of the material slot. A material that has not been assigned in the scene is called a *cool* material.

When a material is assigned to an object, it is linked to the object through its name. Material names in the Material Editor do not have to be unique, but names of materials assigned to objects in the scene do have to be unique. If you assign a material to objects, then later create another material with the same name and attempt to assign it to an object, you will be prompted to choose whether to replace the material of the same name assigned earlier. If you choose to do so, the material will be replaced on all objects holding the material of that name.

If you don't want to replace the material, change the material name to a unique name before assigning it to objects in the scene.

Toggles the display of a checkered background in the sample slot.

General Usage

Click the **Background** button ![icon] to turn the sample slot background on and off. The background appears as a series of colored checks in the currently active sample slot.

Usage Notes

The **Background** option is particularly useful with transparent materials. When the sample slot background is black, it may become difficult to discern a transparent material in the sample slot. Turning on **Background** makes it easier to see a transparent material.

The colored check background can be changed to a custom bitmap. To change the background, use the **Custom Background** settings on the Material Editor Options dialog. You can access this dialog by clicking the **Options** button ![icon] on the Material Editor toolbar.

Turn on the **Background** option has no effect on backgrounds in viewports or in rendered images. Only the currently active sample slot is affected.

Mat Editor

 **Backlight**

Toggles the use of a backlight to light the sample object in the sample slot.

General Usage

Click **Backlight**  to turn on the backlight for the active material slot.

Usage Notes

The object in each sample slot is lit with sample lighting to give you an idea of what the material will look like when lit in the scene. By default, the sample object is lit with one light placed forward and to the upper left of the object. Turning on **Backlight** places another light at the back lower right of the sample object, giving you an idea of what the material will look like when lit from the back.

Turning **Backlight** on or off has no effect on the lighting in the scene. It exists solely to help you visualize the material better.

When creating a transparent, two-sided material, turning on **Backlight** and **Background** will give you a more accurate representation of the material.

A map type that uses a bitmap for its definition.

General Usage

See the **Choosing Maps** entry for information on selecting this map for use in a material.

Coordinates

See the **Coordinates Rollout (Environment)** entry.

Noise

See the **Noise Rollout** entry.

Bitmap Parameters

Bitmap parameters control how the bitmap is loaded and resized for the material.

Bitmap   Specifies the bitmap file. Click on the long blank box to access the file selector and pick a bitmap file. When the bitmap has been chosen, the filename appears in the box.

Reload   Reloads the bitmap into memory. When you specify a bitmap, 3D Studio MAX loads the bitmap into memory. If you later change the bitmap in a paint program, the changed bitmap will not be automatically reloaded and displayed in the Material Editor until the *.max* file is closed and reloaded. Click **Reload** to reload a changed bitmap into memory without closing the *.max* file.

Filtering   In most cases, a bitmap is not exactly the same size it will appear in the rendering. When the bitmap must be scaled up or down for use in a material, 3D Studio MAX uses one of two methods to determine how bitmap detail will be treated during scaling. **Pyramidal** is a fast method that works for most scenes, but some of the detail is filtered out in the process. The **Summed Area** method filters out little or no detail and yields superior crispness even when the camera is close up to the object, but takes much longer to calculate. Use **Pyramidal** unless you need high detail for closeups. You can also choose the **None** option to perform no filtering on bitmaps. This may cause the bitmap's pixels to look rough or chunky in renderings.

Mat Editor

Some map attributes use a single value rather than all the RGB information in a bitmap. For example, the height of the bump at any area of a **Bump** map is determined by one value.

For map attributes that use a single value, the options under **Mono Channel Output** are used. For map attributes that use all three RGB values, the options under **RGB Channel Output** are used to determine how the colors in the bitmap are interpreted for use in the material.

Mono Channel Output	Determines where the one value will come from. **RGB Intensity** uses the average of the RGB values. **Alpha** uses the bitmap's alpha channel to determine the single value, where fully opaque areas of the alpha channel yield a high value and transparent areas yield a low value.
	These settings are used only when the bitmap is used to define a map attribute that uses a single value, such as the **Shininess, Shin. Strength, Self-Illumination, Opacity** and **Bump** map attributes.
RGB Channel Output	Determines where RGB colors will come from. **RGB** uses the colors in the bitmap. **Alpha as Gray** uses the bitmap's alpha channel to create a grayscale image. Note that if the bitmap has no alpha channel, using **Alpha As Gray** will yield a completely white output.
	These settings are used only when the bitmap is used to define a map attribute that uses all three RGB values, such as the **Ambient, Diffuse, Specular, Filter Color, Reflection** and **Refraction** map attributes.
Apply	Applies the crop settings below.
View Image	Displays a window where the crop region can be set interactively. Drag the handles at the edges of the marquee to set the crop region. **U**, **V**, **W** and **H** can be set at the top of the window. Click **UV** to change to **XY** coordinates, where **X** and **Y** can be entered to specify the left and top edges of the crop region in pixels. **W** and **H** also change to allow entry of the width and height of the crop region in pixels.
Crop	Crops the bitmap to the crop region.
Place	Scales the bitmap to the crop region, leaving the rest of the bitmap area transparent.
U	Sets the left edge of the crop region as a fraction of the original image width.

Mat Editor

V	Sets the top edge of the crop region as a fraction of the original image height.
W	Sets the width of the crop region as a fraction of the original image width.
H	Sets the height of the crop region as a fraction of the original image height.
Jitter Placement	When this option is checked, the scaled bitmap is placed somewhat randomly. This option creates an effect only when **Place** option is selected, the **W** and/or **H** values are less than 1.0, and the **Jitter Placement** value is above 0.0. The larger the **Jitter Placement** value, the more random the placement.
Alpha Source	The term *alpha* refers to transparency. An alpha channel is information saved with an image that specifies which areas of the bitmap are transparent. If the bitmap does not have an alpha channel, the **Image Alpha** button will be disabled. In this case, you can tell 3D Studio MAX to use the brightness and darkness of the bitmap to determine which parts of the bitmap will show and which won't. When you turn on **RGB Intensity**, black or dark areas of the bitmap will appear transparent while lighter areas will be opaque. If the map is being used as a **Diffuse** map, for example, darker areas of the map will be transparent, showing the **Diffuse** color from the Basic Parameters rollout. Turning on this option is the same as turning on the **Alpha from RGB Intensity** option on the Output rollout.

If the bitmap has an alpha channel, it can also be used to specify which portions of a bitmap will be transparent. Turn on **Image Alpha** to use the image's alpha channel. If you want the bitmap to appear as is with no transparency, turn on **None (opaque).** This is the default setting.

Premultiplied alpha	There are two methods of saving alpha information with a bitmap, *premultiplied alpha* and *non-premultiplied alpha*. Premultiplied alpha means that RGB values have already been adjusted based on the alpha value, and the alpha information is saved along with the RGB information. With non-premultiplied alpha, RGB information has not been adjusted, and the alpha information is saved separate from the RGB information. The premultiplied alpha method is more efficient for compositing.

(side margin) **Mat Editor**

If the bitmap has an alpha channel and you wish to use it with this map, select the type of alpha information saved with the bitmap. If you are not sure, leave **Premultiplied alpha** on. If the bitmap appears much more transparent than you expected in the rendering, try turning **Premultiplied alpha** off.

Output

See the **Output** entry for information on the parameters on this rollout.

Time

When an animation file or files are specified as the bitmap, the options under the Time rollout set parameters for the animated map.

Start Frame	Specifies the frame in the current animation on which the animation file will begin to play.
Playback Rate	Sets the rate of playback for the animated map. A **Playback Rate** of 1 plays one frame for each frame in the current animation. A value of 2 uses every other frame in the animated map, making playback appear to double in speed. A value of 0.5 uses each frame in the animated map twice, making playback appear half as fast.
End Condition	Determines what will happen before the **Start Frame** and after the animated file has reached its end. **Loop** causes the animation to play over again. **Ping Pong** causes the animation to play backward when it reaches the end, then to play forward again when it reaches the beginning. **Hold** causes the first frame to be used until the **Start Frame**, then uses the last frame after the animation has finished playing.

Usage Notes

None of the parameters on these rollouts change the actual bitmap file. They affect only how the bitmap is displayed and used in the material.

The use of a bitmap in a material for any map attribute other than **Reflection** or **Refraction** requires mapping coordinates on the geometry in order for the material to render correctly. To find out more about mapping coordinates and how to assign them, see the **Mapping Coordinates** entry.

Sample Usage

Bitmaps are often assigned to a **Diffuse** map attribute for defining the pattern and colors of a material.

Access the **Material Editor** ![icon]. The first sample slot is selected. Expand the Maps rollout for the material. Across from the **Diffuse** map attribute name, click on the button labeled **None**.

The Material/Map Browser appears. Choose **Bitmap** from the list. The Material Editor rollouts change to show the parameters for the **Bitmap** map type.

Click on the long blank button next to the label **Bitmap** under the Bitmap Parameters rollout. When the file selector appears, highlight the bitmap *abstrwav.jpg*. Click on the **View** button to view the bitmap before you apply it. Close the bitmap display.

Click **OK** on the file selector window to select the bitmap. The bitmap appears on the sphere in the active sample slot.

Experiment with the **Tiling**, **Angle** and other settings and observe the effect on the sample slot material.

When you first accessed the Material Editor, you were at the parent level for the material. Since you chose a map type, you have now moved down to a child level. When you have finished experimenting, click

Go to Parent ![icon]. The rollout for the parent level of a **Standard** type material appears again.

Blend

A material type that blends two submaterials.

General Usage

To create a **Blend** material, click the **Type** button labeled **Standard** on the Material Editor toolbar. The Material/Map Browser appears. Choose **Blend**. Choose whether to keep the **Standard** material as a submaterial or to discard it. The rollout for the **Blend** material appears.

Basic Parameters

Material 1 **Material 2**	Click on the button labeled **(Standard)** to create the first or second submaterial for the blend. The display changes to reflect the settings for a **Standard** material at the child level. You can change the type of material by clicking on the button labeled **Standard** next to **Type** on the Material Editor toolbar. Click **Go to Parent** ⬆ to return to the **Blend** material parent level. To temporarily turn off a material, uncheck the checkbox at right.
Mask	Uses a map as a mask for the blend. Black areas of the mask display the first material, white areas display the second material, and gray areas display a mix of the two. Click on the button labeled **None** to choose a map type. To use the **Mix Amount** instead of the mask, uncheck the checkbox to the right.
Interactive	Determines which material is displayed in shaded viewports. Only one material can be displayed at a time in shaded viewports.
Mix Amount	When **Mask** is unchecked, this value sets the mix amount for the materials. A value of 0 displays only **Material 1**, while 100 displays only **Material 2**. Values between 0 and 100 show a blend of the two materials. These values can be animated to morph between the materials.
Use Curve	Uses the mixing curve to set the sharpness of the transition between the colors of the two materials. When **Upper** and **Lower** are the same, there is a sharp transition between colors. When there is a wide range between **Upper** and **Lower**, the change is more gradual.

Usage Notes

A **Blend** material works similarly to the **Mix** map type. A **Blend** material mixes two materials, while the **Mix** map type mixes two maps.

Mat Editor

Cellular

A map type that generates a cellular pattern.

General Usage

See **Choosing Maps** under the **Material/Map Browser** entry.

Coordinates

See the **Coordinates Rollout (Non-Environment)** entry.

Cellular

Cell Color	Click on the color swatch to set the color of the cells. You can also click on the box labeled **None** to assign a map for the color. When the checkbox next to the map box is on, the map is used for the color. Otherwise, the color swatch determines the cell color.
Division Colors	The color of the divisions between cells is a gradient between two colors or maps. Click on **None** to use a map for either division color, or click on a color swatch to select a color. When the checkbox next to a map box is on, the map is used for the color. Otherwise, the color swatch determines the cell color.
Circular	Creates circular cells.
Chips	Creates cells with straight edges.
Size	Sets the size of cells in relationship to one tile area. See the **Size Parameter** entry.
Spread	Sets the thickness of divisions. Change the value to make the divisions thicker or thinner. With the **Circular** cell type, decreasing **Spread** makes divisions thicker. With the **Chips** type, decreasing **Spread** makes divisions thinner.
Bump Smoothing	If the **Cellular** map is assigned as a **Bump** map attribute on a **Standard** or **Raytrace** material, the **Bump Smoothing** value smooths transitions between colors in the map. Higher values make a smoother transition.
Fractal	Uses a fractal equation to calculate the cell divisions.
Iterations	Specifies the number of times the fractal cell divisions will be computed. Higher values create more detail but take longer to render. This value can only be changed when **Fractal** is selected.

Mat Editor

Adaptive   When checked, the number of fractal iterations changes depending on how close the object is to the camera. When the camera gets close to the object, the number of iterations increases to preserve close-up detail. This checkbox is available only when **Fractal** is selected.

Roughness   Sets the sharpness of the transition between cells and divisions for the **Fractal** option. Higher values such as 1.0 cause a sharp transition from cells to divisions. Iterations must be greater than 1.0 for **Roughness** to have an effect as it sharpens the transition from one iteration to the next. This value is available only when **Fractal** is selected.

The settings under **Thresholds** determine how the division colors will display. These settings are for fine-tuning the cell divisions.

Low   Adjusts the size of the cells. Increasing this value increases the size of the cells. When this value is 1.0, only the cell color shows.

Med   Adjusts the size of the first division color relative to the second. When this value is 1.0, none of the first division color is visible.

High   Adjust the overall size of the divisions.

Output

See the **Output** entry in this section of the book.

Checker

A map type that creates a checker pattern from two colors or maps.

General Usage

See **Choosing Maps** under the **Material/Map Browser** entry.

Coordinates

See the **Coordinates Rollout (Environment)** entry.

Noise

See the **Noise Rollout** entry.

Checker Params

Soften
Softens the edges between checkers. When **Soften** is 0.0, no softening takes place. A very low value such as 0.05 creates a soft edge between checkers. Higher values such as 1.0 or above blur the checkers into a solid color.

Color #1
Color #2
Sets colors or maps that make up the checker pattern. Click on the color swatch to change the color, or click on the box under **Maps** to assign a map for the color. When the checkbox to the right of the box is checked, the map is used. When the checkbox is unchecked, the color in the color swatch is used.

Swap
Swaps the color swatches and/or maps for **Color #1** and **Color #2**.

Mat Editor

Composite

A map type that composites two or more maps using their alpha channels.

General Usage

See **Choosing Maps** under the **Material/Map Browser** entry.

Composite Parameters

Set Number   Click this button to set the total number of maps to composite. The number of maps appears next to **Number of Maps** at right.

Below these values are a number of map entry slots. The number of slots changes according to the number set with **Set Number**. If the number exceeds 10, a scroll bar appears at right where you can scroll through the maps.

To select a map, click on the button next to the map number and select a map from the Material/Map Browser. Each map can be turned on and off with the checkbox to the right of the entry slot.

Usage Notes

An alpha channel is transparency information attached to a file. An alpha channel is saved as a grayscale image. White areas indicate opaque parts of the image, while black areas are transparent. Gray areas are more or less transparent depending on how dark or light they are.

The **Composite** map type uses the alpha channel of each map for compositing. Maps are layered one on top of the other. Dark areas of the map's alpha channel cause the map to become transparent and show the map underneath.

The first map used with **Composite** doesn't have to have an alpha channel, but all subsequent maps should. If a map doesn't have an alpha channel, you can "fake" one with most map types by turning on the **Alpha from RGB Intensity** option under the Output rollout. Dark areas of the map will become transparent, while lighter areas will be opaque.

Coordinates Rollout (Environment)

The Coordinates rollout determines how the map will work with mapping coordinates. Maps that allow environment mapping have the rollout described below. Other maps have a different Coordinates rollout which does not allow environment mapping. The first parameter on the other rollout is **XYZ**. For information on the other rollout, see the **Coordinates Rollout (Non-Environment)** entry.

For further information on environment mapping, see the **Environment Maps** entry.

Texture   Causes the map to be applied to the object per the specified mapping coordinates. Leave this option on if you are using this map as a **Diffuse**, **Bump** or other map type that uses mapping coordinates. This option works in conjunction with the selection from the **Mapping** pulldown.

Environ   Projects the map to the environment (background) using the environment map type specified under **Mapping**. Turn on this option if you are using the map as a background or **Reflection** map. This option works in conjunction with the selection from the **Mapping** pulldown.

Mapping   Specifies how the map is to be applied to the object. If **Texture** is selected, there are three choices.

 Explicit UVW 1 uses the mapping coordinates on mapping channel 1. These are the default mapping coordinates specified under various circumstances:

 - On a primitive, by turning on the **Generate Mapping Coords** checkbox
 - On a loft object, by checking the **Apply Mapping** checkbox
 - When a **UVW Map** modifier is applied to the object and **Channel 1** is turned on under the **UVW Map** modifier's Parameters rollout

 Explicit UVW 2 uses the mapping coordinates on mapping channel 2, which are assigned to the object only if you have applied a **UVW Map** modifier and have turned on **Channel 2** under the **UVW Map** modifier's Parameters rollout.

 Planar from Object XYZ applies the map to the object with planar mapping coordinates. The local plane used is specified by **UV**, **VW** or **WU** below. This type of mapping stays still when the object moves, giving the illusion that the object is moving through the map. This selection does not require mapping coordinates to be applied to the object.

If **Environ** is selected, there are four choices:

Spherical Environment, Cylindrical Environment and **Shrink-wrap Environment** project the map onto a large object surrounding the scene. If the material is being applied to an object, the surrounding object is invisible and the map is projected onto the object from the surrounding object. If the map is being set up as a background, the large surrounding object renders as the background.

Spherical Environment projects the map onto a large sphere surrounding the scene.

Cylindrical Environment projects the map onto a large cylinder surrounding the scene.

Shrink-wrap Environment projects the map onto a large sphere surrounding the scene, but brings the bottom edge of the map to a smooth close. This choice usually works better than **Spherical Environment**.

Screen projects the map perpendicular to the rendered view.

With mapping types other than **Screen**, when the camera moves, the background will change accordingly. With the **Screen** type, the background image will not shift or change when the camera moves. Use **Screen** if you want to see the entire map as a background, but only if you are rendering a still image or if the camera doesn't move during the animation.

Show Map on Back   When the **Planar from Object XYZ** map type is chosen under **Mapping** above, the back of the object will not show the map in a rendering unless this option is checked.

The remaining options work with the **U**, **V** and **W** dimensions of mapping coordinates. For information on how these dimensions relate to mapping coordinates, see the **Mapping Coordinates** entry.

Offset   Shifts the position of the map along the **U** and **V** directions. The **Offset** is expressed as a fraction of the map height or width. For example, an **Offset** value of 0.5 for **U** shifts the map along the U direction by half its width.

Tiling Sets the number of tile areas for the map. Increasing these values causes the map to tile more frequently, making it appear smaller on the object. The number of tile areas is determined by multiplying the **U** and **V** **Tiling** values. For example, if **Tiling** for **U** is 4 and **V** is 3, the material has 12 tile areas. These tile areas are spread over the sample object in the material slot. They affect the number of tile areas both in the material slot and on the object to which the material is applied.

Mirror Mirrors the map as it tiles, flipping the map on every other tile to cause the edges of one tile to match the edges of each adjacent tile. When **Mirror** is on, the map is tiled by twice the **Tiling** value. If **Mirror** is checked, **Tile** cannot be checked for the same coordinate.

Tile Causes the map to repeat, or tile, over the object. If **Tile** is unchecked, the map appears only once on the object.

UV, VW, WU Orients the map to the **UV, VW** or **WU** mapping planes. These options are only available when **Texture** is selected above.

If **Planar from Object XYZ** is selected above, **UV, VW** and **WU** correspond to the object's local XY, YZ and ZX planes respectively.

If **Explicit UVW 1** or **Explicit UVW 2** is selected, the map is oriented to lie on the mapping coordinates' **UV, VW** or **WU** plane. See the **Mapping Coordinates** entry for more information on U, V and W.

Blur Blurs the map based on its distance from the camera. Parts of the object that are further away from the camera will be more blurred than those closer to the camera. The default value is 1.0, which produces a slight blur that works for most scenes.

Blur offset Blurs the map regardless of its distance from the camera.

Angle Rotates the map along the **U, V** or **W** axes.

Rotate Brings up a window where you can rotate the map interactively with the mouse. Move the cursor anywhere on the window, then click and drag to rotate the mapping.

Mat Editor

Coordinates Rollout (Non-Environment)

The Coordinates rollout determines how the map will work with mapping coordinates. Maps that don't allow environment mapping have the rollout described below. Other maps have a different Coordinates rollout which allows environment mapping. The first parameter on the other rollout is **Texture**. For information on the other rollout, see the **Coordinates Rollout (Environment)** entry preceding this entry.

Coordinates

XYZ   Uses the object's local XYZ axis to place the map. The mapping stays still when the object moves, giving the illusion that the object is moving through the map. When **XYZ** is selected, mapping coordinates do not have to be applied to the object.

UVW 1   Uses UVW map channel 1 to place the map. This is the map channel for mapping coordinates generated with the **Generate Mapping Coords** checkbox for a primitive, and with the **Apply Mapping** checkbox for a loft object. It is also the map channel used when a **UVW Map** modifier is applied to the object with **Channel 1** selected.

UVW 2   Uses UVW map channel 2 to place the map. This is the map channel used when a **UVW Map** modifier is applied to the object with **Channel 2** selected.

The remaining options work with the **U**, **V** and **W** dimensions of mapping coordinates. For information on how these dimensions relate to mapping coordinates, see the **Mapping Coordinates** entry.

Offset   Shifts the position of the map. If **XYZ** is chosen above, the shift takes place along the object's local axes. If **UVW 1** or **UVW 2** are chosen, the shift occurs along the along the U, V and W directions of the mapping coordinates.

                     The **Offset** is expressed as a fraction of the map height or width. For example, an **Offset** value of 0.5 for **U** shifts the map along the U direction by half its width.

Tiling   Sets the number of times the map is tiled within the mapping coordinates. Increasing these values causes the map to tile more frequently, making it appear smaller on the object. See the **Size Parameter** entry for information on **Tiling** and its relationship to the **Size** parameter.

Angle   See these parameters under the **Coordinates Rollout**
Blur   **(Environment)** entry.
Blur offset

A map type that creates a random pattern of small dots and shapes.

General Usage

See **Choosing Maps** under the **Material/Map Browser** entry.

Coordinates

See the **Coordinates Rollout (Non-Environment)** entry.

Dent Parameters

Size	Sets the size of dents in relationship to one tile area. See the **Size Parameter** entry.
Strength	A value from 0 to 100 that sets the strength of **Color #2** relative to **Color #1**. When **Strength** is 0, **Color #1** is used the most. When **Strength** is 100, **Color #2** is used the most. Intermediate values use more or less of **Color #2** accordingly.
Iterations	The number of times the dents are calculated. Higher values make detailed dents but take longer to render.
Color #1 Color #2	Sets the colors to be used for dents. Click on the color swatch to change the color, or click on the box under **Maps** to assign a map for the color. When the checkbox to the right of the box is on, the map is used. When the checkbox is off, the color in the color swatch is used.
Swap	Swaps the color swatches and/or maps for **Color #1** and **Color #2**.

Mat Editor

Double Sided

A material type defined by two materials, one for face normals pointing toward the camera, and another for face normals pointing away from the camera.

General Usage

Click the button next to **Type** on the Material Editor toolbar. This button is most likely labeled **Standard**. The Material/Map Browser appears. Select **Double Sided**. The parameters for a **Double Sided** material appear.

Basic Parameters

Translucency Controls the blending of the two materials. When **Translucency** is 0.0, the materials are opaque. When **Translucency** is 50.0, the materials are blended equally. When **Translucency** is 100.0, the materials are effectively swapped. If your object is not transparent, leave **Translucency** at 0.0. For objects which are partially or fully transparent, set **Translucency** to a value between 0.0 and 50.0 to blend the materials.

Facing
Material Click on the button labeled **(Standard)** to create the material for faces with normals facing the camera. The display changes to reflect the settings for a Standard material at the child. You can change the type of material,

or work with a **Standard** material. Click **Go to Parent** ![button] to return to the **Double Sided** parameters. The checkbox to the right of the button can be turned off to temporarily turn off the material.

Back Material Click on the button labeled **(Standard)** to create the material for faces with normals pointing away from the camera.

Usage Notes

Increasing **Translucency** makes the materials look transparent, but not the object itself. To make an object appear transparent, go to the parent level of the **Back Material** or **Facing Material** and lower the **Opacity** value on the Basic Parameters rollout, or assign an **Opacity** map under the Maps rollout.

If **Translucency** is 0.0, only the **Facing Material** appears in the sample slot. To see the **Back Material** in the sample slot, change **Translucency** to 100.0 temporarily. When you have finished working with the **Back Material**, change **Translucency** back to 0.0 or to the desired value.

An environment map is a map used to define some part of the scene. Examples of environment maps are backgrounds, projection maps for lights and fog maps.

An environment map is defined in the Material Editor. An environment map is not part of a material, but is simply a map on its own in the Material Editor. Unlike materials, environment maps don't have shininess, opacity or other settings that materials have. They are simply maps.

To set up an environment map, click **Get Material** [icon]. On the Material/Map Browser, each material is preceded by a blue sphere ●, while maps are preceded by a green diamond ◆. Select a map rather than a material. The map type can then be set up as usual.

A map type in a material slot appears flat with no highlights. When a material slot with a map is active, the **Assign Material to Selection** button [icon] is disabled. This is because a map cannot be assigned directly to an object – only a material can be assigned to an object.

To use the map, access the area of 3D Studio MAX to which you wish to apply the map, and click and drag the map to the appropriate area. For example, to set up a background, choose *Rendering/Environment* from the menu. Click and drag from the map sample slot to the button under the **Environment Map** label on the Environment dialog. Triangles appear around the slot to show that it's currently being used. The map's Coordinates rollout controls how the map will be used as a background. For information on using the map's Coordinates rollout to set up an environment map, see the **Coordinates Rollout (Environment)** entry or **Coordinates Rollout (Non-Environment)** entry as appropriate for the type of map you are using.

If you accidentally choose a map type for a sample slot and wish to change to a material type, click and drag a material from another slot to the map slot.

Mat Editor

Falloff

A map type that generates grayscale values based on the direction of the object's normals.

General Usage

See **Choosing Maps** under the **Material/Map Browser** entry.

Falloff maps are designed for use as **Opacity** maps, to create realistic materials for objects with variable transparency. An example would be a soap bubble, where the edges of the bubble are always more opaque than the center area regardless of the bubble orientation.

Falloff Parameters

Range	These values set the amount of black, white or gray generated by faces depending on the direction in which they point. A value of 0 generates black, 1 generates white, and values between 0 and 1 create a gray color.
	The options in this section change according to the selection under the **Falloff Type** section. When **Perpendicular/Parallel** is selected, the values are labeled **Perpendicular Value** and **Parallel Value**, which set the amount of white generated by faces perpendicular and parallel to the falloff direction. When the **Towards/Away** falloff type is selected, the values are labeled **Towards Value** and **Away Value**, which set the amount of white generated by faces pointing toward and away from the falloff direction. The **Swap** button swaps the values.
Falloff Type	Specifies the method to be used to determine whether faces will be white or black. When **Perpendicular/Parallel** is selected, falloff ranges between faces perpendicular and parallel to the falloff direction. When **Towards/Away** is selected, falloff ranges between faces pointing toward and away from the falloff direction.
View	Sets the falloff direction to the view in the current viewport.
Object	Falloff is directed toward the center of the picked object. Click **Pick** and pick the object.
Local X/Y/Z Axis	Falloff is directed along a local axis of the object to which the material is applied.
World X/Y/Z Axis	Falloff is directed along one of the world coordinate system axes.

Mat Editor

Flat Mirror

A map type that reflects the surrounding scene in a flat surface.

General Usage

See **Choosing Maps** under the **Material/Map Browser** entry.

Flat Mirror Parameters

Apply Blur	Applies blur to the reflection per the **Blur** setting.
Blur	Sets the amount of blur on the reflection. Some blur is desirable to anti-alias the reflection.
First Frame Only	When this option is on, reflection is generated on the first frame only, saving rendering time. This image information is then used throughout the animated sequence. Turn on this checkbox when the reflection is not expected to change over the course of the animation.
Every Nth Frame	When this option is on, reflection is generated at intervals throughout the animation. The frame intervals are specified by the number spinner to the right of the option. When possible, set the number higher than 1 to save rendering time.
Use Environment Map	When checked, the background environment map is taken into account when rendering the reflection. When unchecked, the environment map is ignored during rendering. For more information on environment maps, see the **Environment Map** entry.
Apply to Faces with ID	Applies the flat mirror reflection to faces with this material ID. This checkbox must be turned on and an ID entered when creating a **Standard** material and using **Flat Mirror Reflection** as the **Reflection** map. If you are using this map as part of a **Multi/Sub-Object** material, this checkbox does not have to be turned on. For further information on material IDs and what they are, see the **Multi/Sub-Object** entry.

The settings under the **Distortion** section set up the method and intensity of bumpy variations on the flat mirror surface.

None	Applies no distortion to the flat mirror surface.
Use Bump Map	Uses the currently assigned bump map to distort the flat mirror surface.

Mat Editor

Use Built-in Noise	Uses the noise parameters below to distort the flat mirror surface.
Distortion Amount	This checkbox is only available when **Use Built-in Noise** or **Use Bump Map** is selected.

The settings under the **Noise** section set up the noise effect to be used for distortion. These options are available when **Use Built-in Noise** is selected. There are three noise types, **Regular**, **Fractal** and **Turbulence**. Each uses a slightly different set of equations to calculate the noise effect. **Regular** makes larger noise blobs than **Fractal** and **Turbulence**. Try each type to visually determine the type of noise you want.

See the entry for the **Noise** map type for more information on each parameter on this section of the rollout.

Usage Notes

A **Flat Mirror** map should be used only as a **Reflection** map.

In order to use a **Flat Mirror** map, a unique material ID must be assigned to one face, or to coplanar faces (faces on the same plane). An example of coplanar faces are the faces on a cylinder cap.

To assign a material ID for the reflection, select the object, apply an **Edit Mesh** modifier, and go to the **Faces** sub-object level. Select the flat mirror faces, and change the **Material ID** number under the panel's Edit Surfaces rollout to an unused ID. Many primitives automatically generate material IDs when the primitive is created, but the material IDs are never higher than 6. If you're not sure which ID numbers are already used, enter a higher number such as 8 or 9.

There are two ways to use a flat mirror reflection. If the flat mirror surface uses the same material as the rest of the object except for its mirroring properties, then a **Standard** material can be used. In this case, set up a **Reflection** map with a **Flat Mirror Reflection** map. The **Apply to Faces with ID** checkbox must be turned on, and the flat mirror face's material ID number must be entered as the **Apply to Faces with ID** number.

If the flat mirror surface uses an entirely different material from the rest of the object, then set up a **Multi/Sub-Object** material for the object. For the material number that corresponds to the flat mirror face material ID number, set up a **Reflection** map with a **Flat Mirror** map. In this case, the **Apply to Faces with ID** checkbox does not have to be turned on.

For further information on material IDs and what they are, see the **Multi/Sub-Object** entry.

Selects a new or premade material or map for the current sample slot.

General Usage

Activate the desired sample slot. Click **Get Material** . The Material/Map Browser appears. Under the **Browse From** section, choose the area from which you would like to select a material or map, or click **New** to start a new material or map. Select the material or map from the list. The material or map is placed in the current sample slot.

Usage Notes

Get Material is primarily for browsing for premade materials in the Material Library or Scene. If a map is chosen, the map replaces the entire material, and cannot be assigned to an object. A map alone is useful only for backgrounds and environment effects, not for assignment to scene objects. If you want to assign a map as part of a material, go to the appropriate map attribute on the Maps rollout, such as **Diffuse** or **Bump**.

On the Material/Map Browser, each material is preceded by a blue sphere ●, while maps are preceded by a green diamond ◆. When you choose a map rather than a material, the sample slot will show a flat object with no highlights or other 3D elements. If you accidentally choose a map when you intended to choose a material, drag another material over the material slot to reset it to a material.

Mat Editor

 Go to Parent

Changes the Material Editor window to display the next highest level of the material.

General Usage

While at a child level such as a map, click **Go to Parent**  to go to the next highest level. If you are several levels down, you can click **Go to Parent** repeatedly until you return to the parent level for the material. When you have reached the highest parent level for the material, this button is grayed out.

Usage Notes

For information on parent and child levels in materials, see the entry **About the Material Editor** at the beginning of this section of the book.

 # Go to Sibling

Changes the Material Editor window to display the settings for a child map at the same level as the current map.

General Usage

While at a child level such as a map, click **Go to Sibling**  to go to another child level at the same level on the material tree. You can click this button repeatedly to cycle through all child maps at the current level. If there are no other comparable child levels to go to, this button is grayed out.

Usage Notes

For information on parent and child levels in materials, see the **About the Material Editor** entry at the beginning of this section.

Gradient

A map type that uses a gradient created from three colors.

General Usage

See **Choosing Maps** under the **Material/Map Browser** entry.

Coordinates

See the **Coordinates Rollout (Environment)** entry.

Noise

See the **Noise Rollout** entry.

Gradient Parameters

Color #1 Color #2 Color #3	The colors in the gradient. Click on the color swatch to change the color, or click on the box under **Maps** to assign a map for the color. When the checkbox to the right of the box is on, the map is used. When the checkbox is off, the color in the color swatch is used.
Color #2 Position	A value between 0 and 1 that sets the position of **Color #2** relative to **Color #1** and **Color #3**. Values below 0.5 move **Color #2** closer to **Color #1**, while values above 0.5 move it closer to **Color #3**.
Gradient Type	A **Linear** gradient moves in a straight line from one end of the map to the other. A **Radial** gradient starts at a center point and moves outward in all directions.

The noise parameters on this rollout pertain to noise within the map. The parameters under the Noise rollout pertain to noise over the entire map. Both sets of noise parameters may be used with the gradient at the same time. See the **Noise** entry for information on these parameters.

Output

See the **Output** entry for information on the parameters on this rollout.

Creates a cold copy of a hot material in the active material slot.

General Usage

Activate a material slot for material that has been assigned in the scene.

Click **Make Material Copy** . The material has the same definition and name, but is not assigned to any objects in the scene.

Usage Notes

This button is only available when the active material slot contains a material that has been assigned in the scene.

When a material is assigned to an object, small triangles appear at the corners of the material slot. A material assigned to an object in the scene is called a *hot* material. When you make changes to the material, the material on the object to which it was assigned is automatically updated.

There may be times when you want to make changes to a material without changing the material in the scene immediately. To do this, you can click **Make Material Copy** on the Material Editor toolbar. This creates a copy of the material with the same name which is not connected to the object to which it was assigned. This is a *cool* material. A cool material no longer has triangles at the corners of the material slot. You can also click and drag a hot material to another slot to create a cool material.

You can then edit the cool material without affecting the material on the object. If you later want to apply a cool material to an object, click

Put Material to Scene . This replaces the hot material currently on the object.

If you want to replace a material in a scene, you can also rename a material to the same name as another material in the scene, and click **Put Material to Scene**.

Although there can be several materials with the same name in the Material Editor, each material assigned to the scene must have a unique name.

Creates a preview of an animated material.

General Usage

Animate a material's parameters or colors. You can do this by moving to a frame other than zero, turning on the **Animate** button and changing the material's parameters or colors. Click **Make Preview** . Click **OK** on the Create Material Preview dialog to create the preview of the animated material. The preview created shows just the sample object in the material slot with the animated material.

Create Material Preview

Preview Range	Sets the frame range for the preview. Select **Active Time Segment** to render the entire active time segment, or select **Custom Range** and enter a frame range to preview a specific range of frames.
Frame Rate	Sets the frame rate for preview creation and playback. The **Every Nth Frame** value sets the frame increment for rendering the preview. For example, entering a 5 for **Every Nth Frame** will render only every 5th frame. **Playback FPS** sets the frame-per-second rate of playback.
Image Size	Sets the preview resolution as a percentage of the material slot resolution. **Resolution** displays the resolution of the preview with the current percentage setting.

Usage Notes

The preview is saved as a file named _medit.avi. Click **Play Preview** to see the preview. The file can be saved with a different filename with

Save Preview .

Mapping coordinates tell 3D Studio MAX how a map should lie on an object. Mapping coordinates can be automatically applied to many types of objects. For example, each primitive has a **Generate Mapping Coords** checkbox that automatically applies mapping coordinates to the object.

There are six types of mapping coordinates: **Planar**, **Box**, **Spherical**, **Shrink Wrap**, **Box** and **Face**. Mapping coordinates work with U, V and W coordinates. The diagram below shows U and V mapping coordinates on each type of mapping.

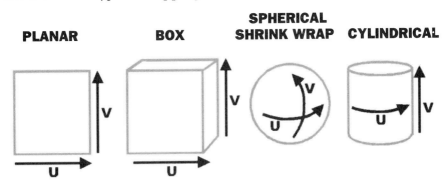

PLANAR **BOX** **SPHERICAL SHRINK WRAP** **CYLINDRICAL**

Face mapping coordinates consist of **Planar** mapping coordinates applied to each face.

The W coordinate is a third coordinate perpendicular to the U/V plane at any point. With Spherical and Cylindrical mapping, the W coordinate points outward from the center of the mapping sphere or cylinder. Most of the time, you will not need to consider the W coordinate.

Instead of using UVW mapping coordinates, you can use XYZ coordinates instead. In this case, when the object moves, the map stays still, giving the illusion that the object is moving through the map. This type of mapping is set up for an individual map on the map's Coordinates rollout. To do this for a 2D map, select the **Texture** option, then choose **Planar from Object XYZ** from the **Mapping** pulldown. For a 3D map, select the **XYZ** option. When XYZ mapping coordinates are used, references to U, V and W on the Coordinates rollout refer to X, Y and Z respectively.

If you attempt to render a scene with objects holding UVW mapped materials and no mapping coordinates, 3D Studio MAX will first assign mapping coordinates to objects if possible. If it cannot, an error message will appear for each object without mapping coordinates.

Every primitive, both standard and extended, has a **Generate Mapping Coords** checkbox on its **Create** and **Modify** panels. Checking this option automatically assigns mapping coordinates to the primitive. Different types of mapping coordinates are applied to different primitives depending on their shapes.

Mapping coordinates can be applied to loft objects by clicking the **Apply Mapping** checkbox under the Surface Parameters rollout of the loft object's **Create** or **Modify** panels. For mapping coordinates applied in this way, the U dimension wraps around the path while the V dimension extends along the path.

For other objects, the **UVW Map** modifier can be applied to generate mapping coordinates. This modifier can also be used on primitives and loft objects.

The **UVW Xform** and **Unwrap UVW** modifiers can also be applied to an object to alter mapping coordinates. See these entries in the *Panels* section for more information.

Mat Editor

Marble

A map type that creates a marble pattern from two colors or maps.

General Usage

See **Choosing Maps** under the **Material/Map Browser** entry.

Coordinates

See the **Coordinates Rollout (Non-Environment)** entry.

Marble Parameters

Size   Sets the size of marble in relationship to one tile area. See the **Size Parameter** entry.

Vein Width   An arbitrary value from 0 to 100 that controls the width of the veins and the background color between them. Values from 0 to 1 create veins that approximate marble. Values above 1 create patterns that don't look much like marble, but can be used for random patterns. A value of 0 removes the veins altogether and leaves a mottled pattern created from varying shades of **Color #2**.

Color #1   The color of the marble veins. Click on the color swatch to change the color, or click on the box under Maps to assign a map for the color. When the checkbox to the right of the box is on, the map is used. When the checkbox is off, the color in the color swatch is used.

Swap   Swaps the color swatches and/or maps for **Color #1** and **Color #2**.

Color #2   The background color for the marble.

Mat Editor

Mask

A map type defined by a color or map, plus a mask to define how much of the color or map can be seen.

General Usage

See **Choosing Maps** under the **Material/Map Browser** entry.

Mask Parameters

Map   Specifies the map to be viewed through the mask.

Mask   Specifies the map to be used as a mask. Where white areas of the mask fall, the map will be displayed. Where black areas of the mask fall, the map will be transparent. In gray areas of the mask, the map will be partially visible.

Invert Mask   Inverts the mask so transparent and opaque areas are reversed.

Usage Notes

The mask determines how much of the map will show, and how much of it will be transparent to show the colors underneath. "Underneath" can mean any map defined at a higher level. For example, the **Diffuse** color on the Basic Parameters rollout is "underneath" the **Diffuse** map on the Maps rollout. When the **Mask** map type is used as a **Diffuse** map, dark areas of the mask will cause the **Diffuse** color on the Basic Parameters rollout to show through.

Mat Editor

Material Effects Channel

Sets the material's effects channel number for use in Video Post processing.

General Usage

To change the **Material Effects Channel** for a material, click and hold on the **Material Effects Channel** button ![icon]. Drag the cursor to the desired number and release the mouse. The **Material Effects Channel** button changes to reflect the chosen number.

Usage Notes

When processing images with Video Post, you can apply special processes called *filters* to rendered objects. These filters are similar to those found in paint programs. For example, you can alter the contrast and brightness of particular objects or materials during rendering.

If you want a filter to be applied only to objects with specific materials, you can specify these materials with the **Material Effects Channel**. When a **Material Effects Channel** other than 0 is set for a material and a filter is set up in Video Post, you can specify that the filter only affect materials with that particular **Material Effects Channel** number. Materials with the default **Material Effects Channel** of 0 are not affected by filters.

If you don't plan to use Video Post filters, you need never change the **Material Effects Channel** number. Changing the **Material Effects Channel** will have no effect on renderings performed without Video Post filters.

For an example of how to use the **Material Effects Channel**, see the **Video Post** entry in the *Panels* section of this book.

Mat Editor

Material/Map Browser

Displays materials and maps for selection.

General Usage

The Material/Map Browser can be accessed in a variety of ways:

- By clicking **Get Material** .
- By clicking a map button labeled **None.**
- By clicking the button next to **Type** on the Material Editor toolbar.

Under **Browse From**, choose the area you wish to browse. Materials are preceded by a blue ball, while maps are preceded by a green diamond. Choose the desired map or material from the list.

The Material/Map Browser is a modeless dialog, which means you can leave it on the screen for as long as you like as you work with other aspects of the Material Editor.

Material/Map Browser Controls

≣	**View List**	Lists materials and maps by name, with maps indented below their parent levels.
≣	**View List + Icons**	Lists materials and maps with names and small thumbnails. Maps are indented below their parent levels.
•	**View Small Icons**	Displays each material and map with a small thumbnail image for each one.
●	**View Large Icons**	Displays each material and map with a large thumbnail image for each one.
°?	**Update Scene Materials From Library**	Updates the materials in the scene with materials of the same name in the current material library. This feature is useful for replacing proxy materials with premade materials from another library. This option will only have an effect if a new material library has been loaded.

| ✗ | **Delete from Library** | Deletes the currently selected material from the library. This option is available only when the **Mtl Library** option is selected under **Browse From**, and a material is highlighted. |

| 🗐 | **Clear Material Library** | Clears the material library of all materials. This option is available only when the **Mtl Library** option is selected under **Browse From**. |

The options under **Browse From** determine what will be displayed.

Mtl Library   Displays materials and maps in the material library.

Mtl Editor   Displays materials and maps in Material Editor slots.

Active Slot   Displays the material in the active slot and/or any maps used to define the material.

Selected   Displays materials and maps on selected objects in the scene.

Scene   Displays materials and maps assigned to objects in the scene.

New   Displays material and map types.

The remaining options allow you to further customize the display of materials and maps.

Show   Select **Materials** to display materials and/or **Maps** to display maps. **Root Only** shows only the parent level of materials. **By Object** displays each object name with its material name below it.

When **New** is selected and **Maps** is on, map categories appear at the lower left of the window. See **Choosing Maps** below for information on these options.

File   These options work with material libraries. Depending on the option chosen under **Browse From**, different options appear under **File**. When **Mtl Library** is selected, **Open** opens an existing material library, **Merge** merges an existing material library with the current library, and **Save** saves the material library with its current name. If **Mtl Library**, **Mtl Editor, Selected** or **Scene** is selected, the **Save As** option allows you to save the currently displayed materials as a material library with a new name.

Choosing Maps

In most places in the Material Editor where a map can be assigned, a button labeled **None** appears. For example, on the **Standard** material's Maps rollout, a button labeled **None** appears across from each map attribute name. To choose a map, click on the button labeled **None**. The Material/Map Browser appears. Choose a map type.

On the Material/Map Browser, when the **New** option is selected under **Browse From**, map categories are listed at the lower left of the Material/Map Browser. Select one of these categories to view maps in the selected category only.

2D Maps   Displays map types that are defined by a 2D pattern.

3D Maps   Displays map types that are defined in three dimensions. The effect of a 3D map is to make the object appear to be carved out of the material.

Compositors Displays map types that composite two or more maps.

Color Mods   Displays map types that modify the colors in a map or on an object.

Other   Displays map types that perform reflection and refraction.

All   Displays all map types.

When any option other than **New** is selected under **Browse From**, the list displays maps that have already been defined. Any map listed can be selected, and is loaded with its current definition. This feature is handy when you want to reuse a map that has already been defined.

When **Mtl Editor**, **Active Slot**, **Selected** or **Scene** is chosen under **Browse From** and an existing map is selected from the list, you are prompted to choose whether to use a **Copy** or **Instance** of the selected map. If you choose **Instance**, every time you make a change to the selected map, all instances of the map will change accordingly.

A selected map type can be replaced with another map type later on. To do this, go to the child level of the map you want to replace. Locate **Type** label on the Material Editor toolbar. The name of the current map type appears on the button next to **Type**. Click on the button. The Material/Map Browser appears. Choose another map type.

Displays the material tree for the currently active sample slot.

General Usage

Activate the desired sample slot and click **Material/Map Navigator** . The Material/Map Navigator appears.

The Material/Map Navigator is a modeless dialog, which means you can leave it on the screen for as long as you like as you work with other aspects of the Material Editor. You can apply a material to an object in the scene from the Material/Map Navigator by dragging the material to the object.

Material/Map Navigator Controls

View List Lists materials and maps by name, with maps indented below their parent levels. Materials are preceded by a blue ball, while maps are preceded by a green diamond.

View List + Icons Lists materials and maps with names and small thumbnails. Maps are indented below their parent levels.

View Small Icons Displays each material and map with a small thumbnail image for each one.

View Large Icons Displays each material and map with a large thumbnail image for each one.

Usage Notes

The Material/Map Navigator is primarily for viewing the material tree. You can also click on a child level to go directly to that level.

Mat Editor

Matte/Shadow

A material type for matte objects.

General Usage

To create a **Matte/Shadow** material, click the **Type** button labeled **Standard** on the Material Editor toolbar. The Material/Map Browser appears. Choose **Matte/Shadow**. The rollout for the **Matte/Shadow** material appears.

Matte/Shadow Parameters

Opaque Alpha	When this checkbox is on, the matte object is rendered on the alpha channel. When this checkbox is off, the matte object does not appear on the alpha channel.
Apply	Environment fog can partially or completely envelop objects in fog. When the **Apply Atmosphere** checkbox is on, any environment fog set up in the scene will envelop the matte object as with any other object.
At Background Depth	When checked, the renderer fogs the scene first, then renders shadows.
At Object Depth	When checked, shadows are rendered first, then fog is applied to the scene.
Receive Shadows	When checked, the matte object receives shadows.
Affect Alpha	When checked, the shadows on the matte object are rendered on the alpha channel. This checkbox is only available when **Opaque Alpha** is turned off. This causes only the shadow on the matte object, and not the matte object itself, to render on the alpha channel.
Shadow Brightness	A value from 0 to 1 that sets the brightness of shadows on the matte object. At 0, shadows are a solid mass of color as set with **Color** below. At 1, no shadows appear. Intermediate values mix the shadow color set below and the colors of the background or environment map.
Color	Sets the color for shadows.

Usage Notes

A **Matte/Shadow** material causes the object and all portions of other objects behind it to be completely see-through, showing only the background (if any) behind the objects. Matte objects are useful for compositing effects. If you have never had to composite images, the reasons for using a **Matte/Shadow** material might be difficult to imagine.

You can use a **Matte/Shadow** material when you are rendering several animated sequences for later compositing. Suppose you have rendered an image to be used as a background. If an object in the scene is supposed to pass behind an object in the background image, you can place a proxy of the background object in the foreground of the scene and apply a **Matte/Shadow** material to it. When the scene object moves behind the **Matte/Shadow** object, the scene object will "disappear" as it moves behind the **Matte/Shadow** object, and the background will show through, giving the illusion that the scene object is actually passing behind the background object.

Mat Editor

Mix

A material type that mixes two colors or maps by a specified amount, or with masking.

General Usage

See **Choosing Maps** under the **Material/Map Browser** entry.

Mix Parameters

Swap
:   Swaps the color swatches and/or maps for **Color #1** and **Color #2**.

Color #1
:   The first color or map to be mixed. Click on the color swatch to change the color, or click on the box under **Maps** to assign a map for the color. When the checkbox to the right of the box is on, the map is used. When the checkbox is off, the color in the color swatch is used.

Color #2
:   The second color or map to be mixed.

Mix Amount
:   When the checkbox next to the far right of **Mix Amount** is turned off, this value sets the mix amount for the two materials. A value of 0 displays only **Color #1**, while a value of 100 displays only **Color #2**. Values between 0 and 100 show a blend of the two colors or maps, with lower values showing more of **Color #1** and higher values showing more of **Color #2**. This value can be animated to gradually shift between the two maps or colors.

Use Curve
:   Uses the mixing curve to set the sharpness of the transition between the two colors or maps. When **Upper** and **Lower** are the same, there is a sharp transition between colors. When there is a wide range between **Upper** and **Lower**, the change is more gradual.

Output

See the **Output** entry.

Multi/Sub-Object

A material type made up of two or more materials.

General Usage

To create a **Multi/Sub-Object** material, click the **Type** button labeled **Standard** on the Material Editor toolbar. The Material/Map Browser appears. Choose **Multi/Sub-Object**. Choose whether to keep the **Standard** material as a submaterial or to discard it. The rollout for the **Multi/Sub-Object** material appears.

Basic Parameters

Set Number   Click this button to set the total number of materials. The number of materials appears next to **Number of Materials** at right.

Below these values are a number of material entry slots. The number of slots changes according to the number set with **Set Number**. If the number exceeds 10, a scroll bar appears at right where you can scroll through materials.

To set up a material, click on the button next to the material number. By default, a **Standard** material is used, but this material type can be changed if necessary. Set up the material as you would any other material. From the **Standard** material child level, you can click **Go to Parent** ![icon] to return to the **Multi/Sub-Object** parent level.

Usage Notes

Every face on every object has a material ID. A material ID is a number that tells 3D Studio MAX how to apply materials to the object.

If an entire object is to receive the same material, you need not concern yourself with material IDs. Material IDs are primarily for assigning more than one material to an object.

To apply multiple materials to an object, use the **Multi/Sub-Object** material. This material type is made up of two or more material types. Each material has a number. When the **Multi/Sub-Object** material is assigned to an object, each numbered material is applied to faces with the corresponding material ID. For example, Material #3 is assigned to faces with material ID 3.

For many objects, material ID 1 is assigned to all faces. For some objects, particularly primitives, two or more material IDs are assigned to the object automatically. For information on which material IDs are assigned to which parts of primitives or other objects, consult your 3D Studio MAX documentation.

Noise

A map type that creates a random pattern from two colors or maps.

General Usage

See **Choosing Maps** under the **Material/Map Browser** entry.

Coordinates

See the **Coordinates Rollout (Non-Environment)** entry.

Noise Parameters

Noise Type There are three noise types, **Regular**, **Fractal** and **Turbu-lence**. Each uses a slightly different set of equations to calculate the noise effect. **Regular** makes larger noise blobs than **Fractal** and **Turbulence**. Try each type to visually determine the type of noise you want.

Size Sets the size of noise effect in relationship to one tile area. See the **Size Parameter** entry.

High Controls the amount of **Color #1** and **Color #2** used in
Low the noise effect. When **High** is 1.0 and **Low** is 0.0, an equal amount of colors 1 and 2 are used. Reduce the **High** value and leave **Low** at 0.0 when you want less of **Color #1** and more of **Color #2** in the noise effect. Increase the **Low** value and leave **High** at 1.0 when you want less of **Color #2** and more of **Color #1** in the noise effect.

Levels The number of times the noise effect is calculated. Higher values yield more detail but take longer to render.

Phase The noise effect appears to be random, but in fact is calculated based on the **Phase** value. When **Phase** is changed, the noise effect changes. Animating the **Phase** will animate the noise function, changing the shape and location of the noise blobs over time.

Color #1 The first color to be used in the noise effect. Click on the color swatch to change the color, or click on the box under **Maps** to assign a map for the color. When the checkbox to the right of the box is on, the map is used. When the checkbox is off, the color in the color swatch is used.

Color #2	The second color to be used in the noise effect.
Swap	Swaps the color swatches and/or maps for **Color #1** and **Color #2**.

Output

See the **Output** entry for information on the parameters on this rollout.

Usage Notes

Some maps have a Noise rollout for creating random patterns in the map. The parameters on the Noise rollout appear on the next page.

Noise Rollout

Some maps have a Noise rollout for creating random variations in the map. The Noise map has its own set of parameters, described under the **Noise** entry.

On   When checked, Noise parameters are enabled.

Amount   The strength of the noise function. Higher numbers make large sweeping variations, while lower numbers make subtle random ripples.

Levels   The number of times the function is applied. With higher **Amount** values such as 20 or 50, the effect of increasing the number of levels is much more noticeable.

Size   The scale of variations in relationship to the size of the geometry.

Animate   Noise is created in a map with a sine wave. When this checkbox is on, noise is animated by shifting the sine wave. This checkbox must be on in order for noise to animate.

Phase   Shifts the starting point of the sine wave. This value can be animated by turning on the **Animate** button at the lower right of the 3D Studio MAX screen, changing the current frame to a frame other than zero and changing the **Phase** value. Changing the **Phase** value alone will not animate the noise; the **Animate** checkbox must also be on. An animated **Phase** value further changes the shifting sine wave animation that is produced when the **Animate** checkbox is on.

Mat Editor

Sets various options for the Material Editor.

Material Editor Options

Manual Update
When this checkbox is on, a sample slot is not updated with current material changes until it is selected.

Don't Animate
When this checkbox is on, animated maps are not updated as you move the time slider or play animation. However, they are updated when you release the mouse after moving the time slider, or stop the animation.

Anti-alias
When this checkbox is on, materials in sample slots are antialiased. Antialiased materials look better but take much more time to display. Turning on this checkbox may slow down operation of the Material Editor considerably.

Progressive Refinement
When this checkbox is on, sample materials are displayed first as chunky pixels, then the chunks are refined. While this display is occurring, you can continue working with the Material Editor. Whenever it has a few idle moments, the Material Editor will continue to refine materials until they are displayed in full detail. If materials are displaying too slowly, turn on this checkbox to speed up your access to the Material Editor functions.

Simple Multi Display Below Top Level
When this checkbox is on, all materials in a **Multi/Sub-Object** material are displayed on the sample only when you are at the parent level for the entire material. When you are at a submaterial level, only the submaterial is displayed on the sample.

Custom Background
Sets up a custom background for sample slots. Click the blank button and select a bitmap name. When you click the **Background** button on the Material Editor toolbar, this bitmap will be displayed as the background. Only a bitmap can be chosen as the custom background.

Renderer
Allows you to choose an alternate renderer for rendering sample slots. In order to choose another renderer, it must first be chosen as the Production renderer. To do this, choose *File/Preferences* from the menu, click the **Rendering** tab, and click **Assign** across from **Production** under the **Current Renderer** section.

Light #1 Color Click the color swatches to change the colors of the two
Light #2 Color default lights used in sample slots. You can also set a Multiplier for each light. Click **Default** to return to the default settings.

Ambient A value from 0 to 1 that sets the intensity of ambient light
Light in sample slots. Click **Default** to return to the default
Intensity value.

Background A value from 0 to 1 that sets the lightness of the plain
Intensity background in sample slots. A value of 0 makes a black
background, while 1 makes a white background and intermediate values make a gray background. This value
does not affect the intensity of the background pattern

that displays when **Background** 🏁 is on. Click **Default**
to return to the default value.

3D Map Sets the virtual size of the sample object in all sample
Sample Scale slots, in units. For example, setting this value to 20 ap-
proximates a sample object that is about 20 units across.
This value is useful when working with 3D maps, where
the size of the effect is related to the size of the object in
the scene. Changing this value does not change the display size of the object in the sample s.ot, but it may affect
how various maps appear on it.

Custom Sets up any 3D object as the sample object. To prepare a
Sample custom object for use as a sample object, set up a scene
Object containing just that object with a camera showing the
view of the object you would like to have in the sample
slot. Optionally, you can also include lights. Save the file
as a *.max* file.

On the Material Editor Options dialog, click the **File
Name** button and select the file. Turn on **Load Camera
and/or Lights**. Click **Apply**, then click **OK**. Activate a

sample slot, then click and hold **Sample Type** 🔵.

Choose the button at the end of the flyout 🔳. The
custom object appears in the slot as it appears in the
file's camera view.

Any lights from the scene are used to illuminate the custom sample object. If no lights are in the chosen file, default lighting is used. If the custom object appears larger or smaller than you would like, return to the saved *.max* file and adjust the camera view.

The custom sample object is treated as any other sample object. For example, you can right-click on the sample slot, choose **Drag/Rotate**, and rotate the custom object in the slot as you would any other sample object.

UVW 2 Mapping Determines how maps on mapping channel 2 appear in sample slots. When **Generated by Object** is selected, mapping coordinates for channel 2 are generated by the default mapping coordinates for the currently displayed sample object. For example, if the sample object is a sphere, then spherical mapping coordinates are used for mapping channel 2. If **Planar** is chosen, maps on mapping channel 2 use planar mapping coordinates in the sample slot. These choices have no effect on the mapping coordinates on objects in the scene.

Slots Sets the number and configuration of sample slots displayed at any one time. There are always 24 sample slots available, but 3x2 configuration shows only 6 at a time, while 5x3 shows only 15 at a time. To scroll to the remaining slots, place the cursor over a sample slot boundary until the pan cursor ✋ appears, then click and drag to the right or left to reveal the remaining sample slots.

Apply Applies changes without exiting the dialog. This option is particularly useful when experimenting with different colors and intensities for lights and the background color.

OK Applies changes and exits the dialog.

Cancel Cancels all changes, including those applied with the **Apply** button.

Mat Editor

Output

A map type that affects how a map is displayed and output.

General Usage

See **Choosing Maps** under the **Material/Map Browser** entry.

Output Parameters

Invert
: Inverts the hue and brightness of each pixel in the map. Saturation is not affected. For example, black becomes white, dark blue becomes pale yellow, bright red becomes bright cyan.

Clamp
: When this checkbox is on, the overall limit of color brightness and saturation remains the same regardless of changes to the RGB Level. Turn this checkbox on to prevent the map from becoming too bright while changing the RGB Level.

Alpha from RGB Intensity
: When this checkbox is on, an alpha channel is generated based on the brightness of pixels in the map. Black areas become transparent, white areas remain opaque and any areas in between are more or less transparent depending on their brightness.

Output Amount
: Affects the saturation and transparency of the map against the color underneath. At the default value of 1.0, the map is opaque with its native saturation. At 0.0, the map is completely transparent and does not show at all. At values between 0 and 1, the map is partially transparent and desaturated.

: At values above 1.0, the map's saturation increases up to the highest value of 100.0. At values below 0.0, the map's hues invert and saturation increases as the number decreases.

RGB Offset
: Affects the brightness of the map. At the default value of 0.0, the map has its native brightness. As this value increases above 0.0, the map becomes brighter and more self-illuminated. The highest value of 10.0 produces a self-illuminated white map. As this value decreases below 0.0, the map becomes darker and more self-illuminated. The lowest value of -10.0 produces a self-illuminated black map.

RGB Level	Affects the saturation of the map without affecting the transparency. At the default value of 1.0, the map has its native saturation. At 0.0 or below, the map is completely black, indicating no saturation. At values above 1.0 up to 100.0, the map becomes more saturated.
Bump Amount	If the map is being used to define a bump map, you can change the **Bump Amount** value to alter the effective intensity of the bumps. If this map is the only map being used to define the bump, then you can change the bump amount by returning to the parent level of the material and changing the value next to **Bump** under the Maps rollout. However, changing the **Bump** value at the parent level changes the entire map. Change the **Bump Amount** value here to change the intensity of the bump map for this map only. This value affects the intensity of the map only if it is being used as part of a bump map.

Usage Notes

Many map types have an Output rollout. The parameters described here are the same as those on the Output rollout for other map types.

The **Output** map type is meant to be used with maps that don't have an Output rollout. This map can be assigned as a child map on any map with child maps, such as **Checker**.

Particle Age

A map type for particles that changes their colors according to particle age.

General Usage

See **Choosing Maps** under the **Material/Map Browser** entry.

Particle Age Parameters

Color #1, 2, 3 When a particle reaches the percentage of its life designated by the **Age** %, its color changes to the corresponding color or map. As a particle ages further, its color gradually changes to the color or map specified for the next age interval. Click on the color swatch to change the color, or click on the box under Maps to assign a map for the color. When the checkbox to the right of the box is on, the map is used. When the checkbox is off, the color in the color swatch is used.

Age #1, 2, 3 Sets the age as a percentage of the particle's total life. For best results, set one **Age** % to 0 and set the color or map for particle birth, then set another **Age** % to 100 and set the color or map for the particle just before it expires. Set an intermediate **Age** % and indicate a color or map for the particle in the middle of its life.

Output

See the **Output** entry for information on the parameters on this rollout.

Usage Notes

Particle Age maps should be used only in materials that will be applied to particle systems such as **PArray**, **PCloud**, **Super Spray**, **Spray** and **Snow**.

The **Particle Age** map is useful for changing particle colors or opacity as they age. To change particle colors and/or opacity according to particle speed rather than age, use the **Particle MBlur** map type.

A map type for particles that changes colors according to particle speed.

General Usage

See **Choosing Maps** under the **Material/Map Browser** entry.

Particle Motion Blur Parameters

Color #1   The color of the particle as it decelerates.

Color #2   The color of the particle as it accelerates.

Sharpness   Sets transparency relative to speed to create a motion blur effect. If **Sharpness** is 0.0, particles are blurry no matter how fast they travel. Higher values make the particles appear sharp when moving slowly and more blurry when moving faster.

Usage Notes

The **Particle MBlur** map only works when the material is assigned to **PArray**, **PCloud**, **Super Spray**, **Spray** and **Snow** particle systems.

The **Particle MBlur** map was designed to be used as an **Opacity** map, causing faster particles to appear more transparent and thus creating a motion blur effect. To use a **Particle MBlur** map in this way, set up the map as an **Opacity** map with the default color and **Sharpness** settings, and apply the material to a system of moving particles.

In particular, **Particle MBlur** is very effective used in conjunction with the **Stretch** value on a **PArray** particle system. See the **PArray** entry in the *Panels* section under **Direction of Travel/MBlur** for more information.

Particle MBlur works with all particle types except **Constant**, **Facing**, **Metaparticles** and **PArray Object Fragments**.

The **Particle MBlur** map type will not work correctly if used as part of a **Multi/Sub-Object** material.

To change particle colors and/or opacity according to age rather than speed, use the **Particle Age** map type.

Perlin Marble

A map type that creates a highly variable marble pattern from two colors or maps.

General Usage

See **Choosing Maps** under the **Material/Map Browser** entry.

Coordinates

See the **Coordinates Rollout (Non-Environment)** entry.

Marble Parameters

Size	Sets the size of the marble pattern in relationship to one tile area. See the **Size Parameter** entry.
Levels	Specifies the number of times the marble pattern will be computed. Higher values create more detail, but take longer to render.
Color #1	The color of the marble veins. Click on the color swatch to change the color, or click on the box under Maps to assign a map for the color. When the checkbox to the right of the box is on, the map is used. When the checkbox is off, the color in the color swatch is used. **Saturation** is a value from 1 to 100 that determines how bright the color or map will appear in the material. Larger values make the color or map brighter.
Color #2	The background color for the marble.
Swap	Swaps the color swatches and/or maps for **Color #1** and **Color #2**.

Usage Notes

The **Perlin Marble** map creates a pattern similar to the one created with the **Marble** map, but it uses a different algorithm which results in a more varied marble pattern.

✎ Pick Material from Object

Selects a material from an object and places it in the current sample slot.

General Usage

Activate the desired sample slot. Click **Pick Material from Object** ✎, then click on an object in the scene. The object's material is placed in the active material slot. Triangles appear at the corners of the material slot to indicate that the material is assigned in the scene.

Usage Notes

Triangles at the corners of a material slot indicate a *hot* material, which means the material is currently assigned in the scene.

If the picked material is already displayed in another slot when you use **Pick Material from Object**, that material becomes an unassigned version of the same material, and the triangles at the corners of the slot disappear to indicate that the material is no longer connected to anything in the scene. In other words, the unassigned material has the same name and attributes as the hot material, but changing the material attributes will not affect objects in the scene. This material is called a *cool* material.

Mat Editor

Planet

A map type that creates patterns to simulate land and water shapes like those on a planetary surface.

General Usage

See **Choosing Maps** under the **Material/Map Browser** entry.

Coordinates

See the **Coordinates Rollout (Non-Environment)** entry.

Planet Parameters

Water Colors   Sets the color for water areas. **Color #1** is the color of the center of water areas. **Color #2** is placed around **Color #1**, and **Color #3** is placed around **Color #2**. **Color #3** meets the land.

Land Colors   Sets the colors for land areas. **Color #4** is the shoreline color, meeting water **Color #3**. **Color #5** is placed next to **Color #4**, then **Color #6** next to **Color #5**, and so on until **Color #8** is used at the center of the land mass.

Continent Size   Sets the size of the continent in relationship to one tile area.

Island Factor   A value from 0 to 100 that sets the level of variation in land colors. At 0, there is no variation. As the **Island Factor** is increased, land colors are varied to produce small spots of land in the water areas and random variations within land areas.

Ocean %   Sets the percentage of the planet's surface to be covered with water colors.

Random Seed   Sets the base number for random calculations. Change this number to observe different land and water configurations. This value cannot be animated.

Blend Water/Land   When checked, the boundaries between land and water are blended slightly. When this checkbox is off, the boundaries are not blended, showing a sharp delineation between land and water.

Usage Notes

Planet maps are useful as **Diffuse** or **Bump** maps for creating planetary surface materials.

Play Preview

Plays the most recent material preview.

General Usage

Create an animated material. Click **Make Preview** to make a preview of the material. Click **Play Preview** to view the preview.

Usage Notes

The preview is saved as a file named _*medit.avi*. The file can be saved with a new name with **Save Preview** .

Mat Editor

Pop-up Menu

When a material slot is right-clicked, a pop-menu appears.

Drag/Copy   Select one of these options to determine what will
Drag/Rotate happen when a click-and-drag is performed on a material slot. With **Drag/Copy**, a click-and-drag copies the
material to another slot. With **Drag/Rotate**, the sample
object is rotated in the slot.

Reset   Resets the sample object to its default rotation. This
Rotation   option has an effect only when **Drag/Rotate** has been
selected and the sample object has been rotated.

Render Map   Renders an animated map to a file. A dialog appears
where you can enter a frame range, resolution and
filename. You can also use **Make Preview** on the Material Editor toolbar to render an animated material.

Options   Accesses the Material Editor Options dialog. This dialog
can also be accessed with the **Options** button on the
Material Editor toolbar. See the **Options** entry in the *Mat
Editor* section of this book.

Magnify   Displays the current material slot in a separate window
that can be stretched to any size. When **Auto** is checked,
the magnified window is updated whenever the material in the corresponding slot is changed. When **Auto** is
unchecked, the magnified window is updated only
when the **Update** button is clicked.

3x2   Sets the display of the material slots to the selected ma-
5x3   trix.
6x4
Sample
Windows

Put Material to Scene

Puts the currently selected material to the scene.

General Usage

Create a material and assign it to an object in the scene with **Assign Material to Selection** . Click and drag the material to another slot in the Material Editor, or click **Make Material Copy** . Make changes to the new material. When the changes are complete, click **Put Material to Scene** . The new material replaces the old one in the scene.

Usage Notes

When a material is assigned to an object, small triangles appear at the corners of the material slot. A material assigned to an object in the scene is called a *hot* material. When you make changes to the material, the material on the object to which it was assigned is automatically updated.

There may be times when you want to make changes to a material without changing the material in the scene immediately. To do this, you can click **Make Material Copy** on the Material Editor toolbar. This creates a copy of the material with the same name which is not connected to the object to which it was assigned. This is a *cool* material. A cool material no longer has triangles at the corners of the material slot. You can also click and drag a hot material to another slot to create a cool material.

You can then edit the cool material without affecting the material on the object. If you later want to apply a cool material to an object, click **Put Material to Scene**. This replaces the hot material currently on the object.

If you want to replace a material in a scene, you can also rename a material to the same name as another material in the scene, and click **Put Material to Scene**.

Although there can be several materials with the same name in the Material Editor, each material assigned to the scene must have a unique name.

Put to Library

Puts the currently selected material to the material library.

General Usage

Activate a material slot. Click **Put to Library** ![icon]. The material is placed in the current material library.

Usage Notes

A *material library* is a collection of materials and their definitions. Saving a material to a material library makes it possible to use the material in another scene. You can put together libraries for each type of work you do. For example, a material library can be made for architectural materials, or for favorite abstract materials.

A material library is saved in a file with the extension *.mat*. The default material library *3dsmax.mat* is loaded when you load 3D Studio MAX.

Click **Put to Library** to put the active material to the library. Once materials are put in the library, the library must be saved in order to retain the materials. To save the library, click **Get Material** ![icon] to go to the Material/Map Browser. Under **Browse From**, choose **Mtl Library**. Click **Save As.** Be sure to enter a new library name. If you don't, then the default library *3dsmax.mat* will be overwritten.

You can also save a library containing just the materials in the Material Editor slots, the materials in the scene or materials on selected objects. To do this, choose **Mtl Editor**, **Scene** or **Selected** under **Browse From**, and click **Save As**.

To create a new library, you must first clear the current library. On the Material/Map Browser, choose **Mtl Library** under **Browse From** and click **Clear Material Library** ![icon] on the Material/Map Browser toolbar. Use **Put to Library** to put new materials to the library. If you do this, be very careful to save the new library with **Save As** rather than **Save**. Clicking **Save** will save the new library with the old library name, wiping out the old library.

To load a saved material library, choose **Mtl Library** under **Browse From** and click **Load**. Choose the material library from the list.

A material type that reflects or refracts the scene around it.

General Usage

Click **Get Material** 🔳. Choose the **Raytrace** material from the list. Set parameters as desired.

Basic Parameters

Shading **2-Sided** **Wire** **Face Map** **SuperSample**	See these parameters under the **Standard** entry.
Ambient	The degree to which light is absorbed or reflected in shadow areas. When **Ambient** is black, light is absorbed, and shadow areas of the material are dark. When **Ambient** is white, light is reflected in shadow areas to the same degree as in lit areas, making shadow areas the same color as the **Diffuse** color.
Diffuse	The color of light reflected by the material in non-highlight areas. This corresponds to real life, where the color of an object is the color of light reflected from it. When the **Reflect** color is white, the reflection of the scene around the object takes over the material, and the **Diffuse** color is not visible.
Reflect	The degree and filter color of reflections. The **V** value of the color determines how reflective the material is, while the color tints the reflection. For example, pure white makes the material very reflective, and reflections are uncolored, as with a mirror. Pure red makes the material very reflective, and all reflections are tinted red. Black makes the material unreflective. A gray color makes a somewhat reflective surface with uncolored reflections.
Luminosity	The degree and filter color of self-illumination. The **V** value determines the degree of self-illumination, while the color tints the material to the degree of self-illumination.

Transparency   The degree and filter color of transparency. The **V** value determines the degree of transparency. Black makes the material opaque, while white makes it fully transparent. Gray makes the material partially transparent. Medium red makes the material partially transparent and tints it red at the same time.

Index of Refr   See these parameters under the **Standard** entry.
Specular Color
Shininess
Shininess Strength
Soften

Environment   Specifies an environment map to be reflected and refracted by the material. If no environment map is specified here, the scene environment map is used, if one is defined under the *Rendering/Environment* menu option.

Bump   Specifies a bump map for the material. This bump map works the same as the **Bump** map under the **Standard** material.

Extended Parameters

Extra Lighting   Sets the color of an additional ambient light that affects this material only. Use this color to simulate radiosity, where the material reflects the color of a bright object nearby. Black makes no extra lighting.

Translucency   Sets the color of a translucency effect on the object. Black makes no translucent effect.

Fluorescence   Sets the tint of a fluorescent effect on the material. When **Fluor Bias** is 0.5, this color has no effect.

Fluor Bias   Sets the intensity of a fluorescent effect. When **Fluor Bias** is 0.5, there is no fluorescent effect. When **Fluor Bias** is greater than 0.5, lights in the scene illuminate the material with a "black light" effect, which emphasizes bright colors. When **Fluor Bias** is less than 0.5, bright colors in the material are dimmed.

Size   See these parameters under the **Standard** entry.
In

Mat Editor

The settings in the **Advanced Transparency** section determine the parameters for refraction (transparency). If the **Transparency** color on the Basic Parameters rollout is black, the material is opaque, and the settings under the **Advanced Transparency** section have no effect.

Transp. Environment When checked, the selected map is used in creating the refraction effect. To assign a transparency map, click the button labeled **None** and choose a map from the Material/Map Browser, or click and drag a map from the Material Editor to the button.

Color When **Color** is checked, these settings determine the color within a transparent object based on thickness. **Start** and **End** determine the range over which color gradually appears. **Amount** sets the intensity of the effect, with 0 eliminating the effect and 1 making the strongest effect. Click the color swatch to set the color, or click the small box to the right of the color swatch to choose a map for the color.

Fog When **Fog** is checked, these settings create fog within a transparent object based on thickness. **Start** and **End** determine the range over which fog gradually appears. **Amount** sets the intensity of the effect, with 0 eliminating the effect and 1 making the strongest effect. Click the color swatch to set the color of the fog, or click the small box to the right of the color swatch to choose a map for the fog.

Raytracer Controls

Raytrace Reflections When checked, this material includes raytraced reflections.

Raytrace Refractions When checked, this material includes raytraced refraction.

Reflect Falloff When checked, reflections dim to black at the specified distance from the object.

Refract Falloff When checked, refraction dims to black at the specified distance from the object.

Options Accesses the Raytracer Options dialog, which can be used to set **Global** and **Local** options. **Global** options are set for the entire scene, while **Local** options affect only objects with this material.

Mat Editor

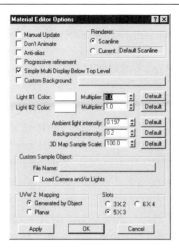

Enable Raytracing turns raytracing on and off. When this option is off, the environment map is reflected and refracted, but objects in the scene are not.

Antialiasing turns antialiasing of refection and refraction on and off. Turning this option off improves rendering time.

When **Self Reflect/Refract** is checked, the object will reflect and refract other parts of itself.

When **Atmosphere** is checked, atmospheric effects such as fog and volume lights are taken into account during raytracing.

When **Reflect/Refract Material ID** is checked, objects affected by a glow or other Video Post filter are reflected or refracted with their Video Post effects.

The **Objects inside raytraced objects** option determines whether objects inside an object assigned a raytrace material are taken into consideration when raytracing is calculated.

The **Atmosphere inside raytraced objects** option determines whether atmospheric effects such as **Combustion** inside an object assigned a raytrace material are taken into consideration when raytracing is calculated.

The **Color Density / Fog** options apply the **Color** and **Fog** settings globally and/or locally, but only if they are checked on the Extended Parameters rollout.

Global Parameters	Accesses the Global Raytracer Settings dialog. See **the Global Raytracer Settings** section later in this entry.
Global Exclude	Displays a dialog where objects can be included or excluded from all reflection and refraction in the scene.
Local Exclude	Displays a dialog where objects can be included or excluded from reflection and refraction from this material.
Override Global Settings	When checked, the settings under **Adaptive Control** and **Blur / Defocus (Distance Blur)** override the global parameters set with **Global Parameters**.
Global->Local	Copies global antialiasing parameters to local settings.
Local->Global	Copies local antialiasing parameters to global settings.
Adaptive Threshold Initial Rays Max. Rays Blur Offset Blur Aspect Defocusing Defocus Aspect	See these parameters under **Global Raytracer Settings** below.
Blur Map	Sets up a map to be used for blur. Lighter areas of the map cause more blurring, while darker areas cause less blurring. The map is used when the checkbox at the far right is checked.
Defocus Map	Sets up a map to be used for defocusing. Lighter areas of the map cause stronger defocusing, while darker areas cause milder defocusing. The map is used when the checkbox at the far right is checked.

Maps

Each map listed on the Maps rollout corresponds to an option of the same name on either the Basic Parameters or Extended Parameters rollout. When the checkbox to the left of the map name is checked, the map is active.

Amount	A value from 0 to 100 that sets the percentage of the option defined by the map. The remainder of the option is set by the option of the same name on the Basic Parameters or Extended Parameters rollout. For ex-

Mat Editor

ample, an amount of 75 for the **Diffuse** map causes the map to define 75% of the **Diffuse** color of the material, while the remaining 25% is set by the **Diffuse** color swatch on the Basic Parameters rollout.

Map   Lists the name of the map.

Ambient   See Ambient under the Basic Parameters rollout.

Diffuse   See **Diffuse** under the Basic Parameters rollout.

Reflect   See **Reflect** under the Basic Parameters rollout.

Transparency See **Transparency** under the Basic Parameters rollout.

Luminosity   See **Luminosity** under the Basic Parameters rollout.

IOR   The lightness or darkness of colors in the map set the index of refraction between 1 and the **Index of Refr.** value.

Spec. Color See **Specular Color** under the Basic Parameters rollout.

Spec. Shininess See **Shininess** under the Basic Parameters rollout.

Shin. Strength See **Shininess Strength** under the Basic Parameters rollout.

Extra Lighting See **Extra Lighting** under the Extended Parameters rollout.

Translucency See **Translucency** under the Extended Parameters rollout.

Fluorescence See **Fluorescence** under the Extended Parameters rollout.

Color Density See **Color** under the Extended Parameters rollout.

Fog Color   See **Fog** under the Extended Parameters rollout.

Dynamics Properties

See these parameters under the **Standard** entry.

About Raytrace Material

This rollout displays information about the authors of the **Raytrace** material.

Global Raytracer Settings

The Global Raytracer Settings dialog is accessed with the **Global Parameters** button under the Raytracer Controls rollout.

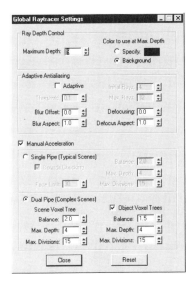

Maximum Depth	Sets the number of times each light ray bounces. Larger values increase rendering time.
Color to use at Max. Depth	This option sets the color used when a ray has bounced the number of times set by **Maximum Depth**. Choose **Specify** to specify a color, or **Background** to use the scene background or the environment background set with the **Environment** option on the Basic Parameters rollout.
Adaptive	When **Adaptive** is unchecked and **Antialiasing** on the Raytracer Options dialog is unchecked, one ray is generated per pixel. When **Adaptive** is unchecked and **Antialiasing** is checked, 4-12 rays are generated per pixel. When **Adaptive** is checked, the number of rays per pixel can range from **Initial Rays** to **Max. Rays**. When **Threshold** is 0, the number of rays per pixel specified by **Max. Rays** are cast. When **Threshold** is 1, the number of rays specified by **Initial Rays** are cast. Intermediate **Threshold** values cast the **Initial Rays** number of rays, then check to see if more rays need to be cast. Higher **Threshold** values are more likely to cast more rays, up to the **Max. Rays** number of rays. The higher the number of rays cast, the more time it will take to render the scene.

Blur Offset   Blurs reflections and refractions. Higher values blur more.

Blur Aspect   Changes the shape of the blur on each pixel. Values above 1 blur more horizontally, while values below 1 blur more vertically.

Defocusing   A value from 0 to 1 that blurs reflections and refractions based on the reflected or refracted objects' distances from the object with the **Raytrace** material.

Defocus Aspect   Changes the shape of the defocus effect. Values above 1 defocus more horizontally, while values below 1 defocus more vertically.

Manual Acceleration   The term *acceleration* refers to the method used to divide up geometry in the scene to prepare for raytrace rendering. When **Manual Acceleration** is unchecked, MAX automatically chooses the calculation method based on the complexity of the scene. The two methods that can be used are called *single pipe* and *dual pipe*. When **Manual Acceleration** is checked, you can manually force MAX to use one of these methods.

A primary function of a raytracer is to figure out which object faces in the scene that each ray will intersect as it bounces around the scene. To accomplish this, raytracing divides the entire scene into a lattice, or framework of cubes. The object faces that sit inside each cube are put in a list. As a ray bounces around the scene and passes through a cube, MAX checks the list of faces in that cube to see if the ray hits any of the faces. This method of subdividing the scene into a lattice makes raytrace calculations more efficient than if the entire scene had to be checked each time a ray was bounced.

The list of faces in each cube are stored in a tree structure. A tree structure is often more efficient for storing and retrieving large amounts of information quickly. Each node (piece of information) on the tree is called a *voxel*.

Single Pipe   Choosing the **Single Pipe** option uses the settings in this section for acceleration. The **Single Pipe** method divides the scene at the face level.

Bounds Checking   When checked, raytracing is optimized for objects with high polygon counts.

Face Limit	**Face Limit** sets the maximum number of faces in each node on the tree.
Balance	Sets the sensitivity of the subdivision calculation. Higher values use more memory but can cause rendering to go faster.
Max. Depth	The maximum number of times a lattice cube can be subdivided to force the number of faces in each subdivision below the **Face Limit**.
Max. Divisions	Dimensions of the lattice. **Max. Divisions** of 15 makes a lattice with 15x15x15 cubes, for a total of 3375 cubes.
Dual Pipe	Uses the **Dual Pipe** method for acceleration. This method divides the scene by objects, rather than by faces.
Scene Voxel Tree	When **Object Voxel Trees** is unchecked, the parameters under **Scene Voxel Tree** are in effect. The **Balance**, **Max. Depth** and **Max. Divisions** parameters are the same as those in the **Single Pipe** section.
Object Voxel Tree	Creates one data tree per object. When **Object Voxel Trees** is checked, the parameters underneath are in effect. The **Balance**, **Max. Depth** and **Max. Divisions** parameters are the same as those in the **Single Pipe** section.
Close	Closes the dialog and keeps the parameters set.
Reset	Resets the parameters on the dialog to their default values.

Mat Editor

Usage Notes

The **Raytrace** material uses *raytracing* to create the colors of the material on the object. Raytracing determines the appearance of the material by sending several rays of light out from the object and bouncing them off other objects in the scene. Raytracing produces accurate reflections, and can also be used to create stunning light effects not possible with the **Standard** material type.

The way an object appears in real life depends on the way it interacts with light. The **Raytrace** material works with absorbed and reflected colors, just as objects do in real life. For example, the **Diffuse** color in a **Raytrace** material is the color of light reflected off the object. Compare this to the **Standard** material, in which the **Diffuse** color is the overall color of the object. The effect is similar, but varies distinctly on very shiny or reflective objects. The **Raytrace** material is excellent for highly reflective surfaces that reflect the scene around them, such as chrome or curved mirrors.

With a **Raytrace** material, the reflected scene must be built around the object. A **Standard** material, however, can be made to appear to reflect a scene by using a bitmap with a picture of the scene assigned as a **Reflection** map. The **Raytrace** material will reflect exactly the scene around it, while the exact reflection with a **Standard** material is more difficult to control. Another option is to use a **Raytrace** map type as a **Reflection** map with a **Standard** material.

Raytrace materials take longer to render. Use a **Raytrace** material only when you need an accurate reflection in your scene. For reflective objects that are not the focal point of the scene, or for images that need to be rendered quickly, use the **Standard** material type.

Mat Editor

Raytrace (map)

A map type that uses raytracing.

General Usage

See **Choosing Maps** under the **Material/Map Browser** entry.

Raytracer Parameters

Trace Mode   Sets the raytrace function that will be performed by the map. Choosing **Reflection** uses the **Raytrace** map for reflection, while choosing **Refraction** uses the **Raytrace** map for refraction. When **Auto Detect** is selected, the map is used for reflection if it is assigned as a **Reflection** map, or refraction if it is assigned as a **Refraction** map.

Options   See the **Raytrace (material)** entry for these parameters.
Global Parameters
Global Exclude
Local Exclude

Background   The background is reflected or refracted as part of the raytrace calculation. Choose to **Use Environment Settings**, to use a color by selecting the color swatch, or to use a custom map for the raytrace background.

Antialiasing

See these parameters under the Global Raytracer Settings rollout of the **Raytrace (Materials)** entry.

Attenuation

The settings under the Attenuation rollout cause objects to be considered by the raytracer only when they fall within a specified distance range from the object with the raytrace map.

Falloff Type   When **Falloff Type** is set to a type other than **Off**, objects are reflected or refracted with varying intensities depending on their distances from the object with the **Raytrace** map. **Falloff Type** uses the **Start** and **End** unit distances from the object. **Linear** causes raytracing to fall off in a straight linear fashion between the **Start** and **End** distances. **Inverse Square** calculates the falloff using the square of the distance to determine the intensity of the reflection or refraction. Only the **Start** value is used, and the end of the attenuation range is reached when the calculated intensity becomes very small. **Exponential** calculated falloff exponentially using the **Exponent** value.

Custom Falloff uses the Start and End ranges along with the settings under the **Custom Falloff** section below.

Color   Specifies how rays appear as they attenuate. **Background** uses the background color, while the **Specify** option can be used to set a custom color for attenuated rays.

Custom
Falloff

Uses the curve displayed to determine attenuation of rays. The left end of the curve sets the ray intensity at the **Start** of the range, while the right end of the curve sets its intensity at the **End**. The **Near** and **Far** values determine the values of the curve at the left and right ends of the curve. A value of 1 indicates full ray intensity, while 0 indicates no intensity. **Control 1** and **Control 2** set values of the curve at intermediate points. Experiment with these parameters to get the desired curve.

Basic Material Extensions

Reflectivity /
Opacity Map

Uses a map to set the degree of reflection or refraction across the object. Lighter areas of the map cause more reflection or refraction, while lower areas cause less reflection or refraction. The number value is the intensity of the raytrace effect, similar to the **Output Amount** value on the Output rollout. See the **Output** entry for information on this value. This map has an effect only if the map **Amount** is not set to 100 at the next higher level of the material.

Basic Tinting

Tints reflected or refracted colors. The **Enable** checkbox turns on the tint effect. A map can be used for tinting, or the color swatch can set a tint color. Increasing **Amount** increasing the tint effect.

Refractive Material Extensions

Color
Density
(Filter)

See the **Color** parameter under the **Raytrace (material)** entry.

Fog

See the **Fog** parameter under the **Raytrace (material)** entry.

About Raytrace

This rollout displays information about the authors of the **Raytrace** map type.

Mat Editor

Reflect/Refract

A map type that reflects the surrounding scene in a curved surface.

General Usage

See **Choosing Maps** under the **Material/Map Browser** entry.

Reflect/Refract Parameters

Source	Sets the source for the six cubic reflection/refraction maps. **Automatic** generates six maps based on the view from the object. When **Automatic** is used, the maps are generated when the image is rendered. The maps are not saved and cannot be accessed or edited. If you want to automatically create cubic environment maps and save them to disk, choose **From File** and see the **Render Cubic Map Files** section below.
	From File uses six files from the disk to create the cubic reflection/refraction map. The files are specified under the **From File** section below.
Size	A value between 1 and 5000 that determines the resolution of the reflection or refraction. Decreasing this value reduces the resolution, which improves rendering speed. To optimize your reflections, start with the default value of 100 and reduce the **Size** parameter to the lowest value possible while still maintaining adequate resolution. When the camera is very close to the reflective/refractive object, you might have to increase this value to keep reflections and refractions sharp.
Use Environment Map	When this checkbox is on, the background environment map is taken into account when rendering the reflection. When this checkbox is off, the environment map is ignored during rendering. For more information on environment maps, see the **Environment Map** entry.

The settings in the **Blur** section determine how much the reflection or refraction will be blurred.

Apply	Applies the **Blur Offset** and **Blur** settings to the reflection/refraction map.
Blur Offset	Blurs the reflection or refraction by the specified amount regardless of its distance from the object. Set this value above 0 when you want the entire reflection or refraction to be uniformly blurred.

Blur   Blurs the reflection or refraction based on its distance from the object holding the material. Parts of the reflection/refraction that are further away from the object will be more blurred than those closer to the object. The default value is 1.0, which produces a slight blur that works for most scenes.

The settings in the **Atmosphere Ranges** section apply to scenes with environmental fog. The **Near** and **Far** ranges specify a fog range relative to the object holding the material. Fog effects that are a distance from the object that falls between the **Near** and **Far** range will be considered in the reflection/refraction rendering. Each distance is computed from the object's pivot point.

Near   Sets the near range from the object, in units.

Far   Sets the far range from the object, in units.

Get from   Gets the **Near** and **Far** range from a camera in the scene.
Camera   To use this option, click **Get from Camera** and select a camera from the scene. The **Near** and **Far** values here are updated to the **Near** and **Far** ranges from the camera's Parameters rollout.

In the **Automatic** section, you can automate the reflection or refraction to be calculated at specified intervals during an animated sequence.

First Frame   When this option is on, reflection/refraction is
Only   generated on the first frame only, saving rendering time. This image information is then used throughout the animated sequence. Turn on this checkbox when the reflection or refraction is not expected to change over the course of the animation.

Every Nth   When this option is on, reflection/refraction is
Frame   generated at intervals throughout the animation. The frame intervals are specified by the number spinner to the right of the option. When possible, set the number higher than 1 to save rendering time.

The settings in the **From File** area specify the six images that define the cubic environment. A cubic environment consists of six bitmaps placed around the scene in a box configuration. The six bitmaps are then reflected in the object to which the material is applied.

The options in the **From File** area are available only when the **Source** above is set to **From File**. During rendering, the maps are placed in the scene in a cubic configuration along each of the world coordinate system's axes, as specified below. The object holding the material reflects or refracts the images, but the cubic maps themselves do not render. If cubic environment files are named in a particular way, you can load one cubic environment map and have the rest automatically load. In order to do this, the map must end with an underscore and a two-letter abbreviation of the map position. For example, loading a file called *view_UP.jpg* for the **Up** map will automatically load *view_DN.jpg* as the **Down** map, *view_LF.jpg* as the **Left** map, etc. The abbreviations are listed below.

Up	Positive Z axis. Abbreviation: _UP
Down	Negative Z axis. Abbreviation: _DN
Left	Positive X axis. Abbreviation: _LF
Right	Negative X axis. Abbreviation: _RT
Front	Positive Y axis. Abbreviation: _FR
Back	Negative Y axis. Abbreviation: _BK

The settings in the **Render Cubic Map Files** section allow you to generate a set of six cubic maps right from this rollout. The six cubic maps are based on the view from a chosen object. The object does not have to be the same one to which the material will be applied. Once the maps are generated, they are automatically placed in the six cubic map entries above.

To File	Click on the box and enter a base filename and extension for the six cubic maps. The underscore and abbreviation for each map listed above will automatically be added to the base filename. Be sure to include the extension, which will tell 3D Studio MAX what file type to use.
Pick Object and Render	To begin rendering the cubic maps, click **Pick Object and Render** and pick the object in the scene. Each map appears briefly as it is rendered. When rendering is complete, the six cubic map files are placed in the entry slots above for the six map types.

Usage Notes

A **Reflect/Refract** map should be asigned only to a **Reflection** or **Refraction** map attribute on a **Standard** material.

Mat Editor

Reset Mat/Mtl to Default Settings

Resets the current map or material to its default settings.

General Usage

Make sure you are at the desired material or map level. Click **Reset Mat/Mtl to Default Settings** ![X]. If you are at a material parent level, the material is reset to its default values. If you are at a map level, the map is reset to its default values.

Usage Notes

When you click **Reset Mat/Mtl to Default Settings** at a material parent level, the material type remains the same. For example, if you reset a **Double Sided** material, all parameters will be reset to default values but the material will still be a **Double Sided** material.

Likewise, when you click this button while at a map level, the map will be reset but will not be removed from the material. To reset the entire material, click **Go to Parent** ![arrow] until you are at the highest parent level for the material, and click **Reset Mat/Mtl to Default Settings**.

RGB Multiply

A map type that multiplies the RGB values from two maps.

General Usage

See **Choosing Maps** under the **Material/Map Browser** entry.

RGB Multiply Parameters

Color #1
Color #2
The two maps or colors to be combined. Click on the box under **Maps** to assign a map, or click on the color swatch to change the color. When the checkbox to the right of the box is on, the map is used. When the checkbox is off, the color in the color swatch is used.

Alpha From
The alpha channel from one or both maps can be used as the alpha channel on the new map that results from the **RGB Multiply** operation. You can choose to use the alpha channel from **Map #1** or **Map #2**. You can also choose **Multiply Alphas** to multiply the alpha channels from both maps. If the chosen map or maps do not have an alpha channel, these settings have no effect.

For information on alpha channels and what they are, see the **Output** entry.

Usage Notes

An **RGB Multiply** map is intended for use as a **Bump** map, where two bump map images are to be combined. **RGB Multiply** will effectively combine the two maps into one bump map.

RGB Tint

A map type that adjusts the red, green and blue values of another map.

General Usage

See **Choosing Maps** under the **Material/Map Browser** entry.

Tint Parameters

R, G, B   Click any one of the **R**, **G** or **B** color swatches to change the intensity of the red, green or blue components of the map.

Map   Click on the button to select a map for tinting.

Sample Type

Sets the type of object to appear in the sample slot.

General Usage

Activate a sample slot. Click and hold on **Sample Type** . Select one of the three sample objects:

 Sphere

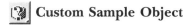 **Cylinder**

Cube

If you have set up a custom sample object, a fourth option will also be available:

Custom Sample Object

Usage Notes

The Sample Type choices are primarily for allowing you to view your material on objects of various shapes so you can better approximate the effect of a material once it is on an object in the scene.

You can also set up a custom object to appear in sample slots. For information on this procedure, see the **Custom Sample Object** description under the **Options** entry.

 # Sample UV Tiling

Sets the tiling for the current material sample slot.

General Usage

Activate a sample slot. Click and hold **Sample UV Tiling** ▣. Select one of the four tiling options:

▣ 1x1

▦ 2x2

▦ 3x3

▦ 4x4

The tiling in the material slot changes to reflect the new selection.

Usage Notes

The tiling set with Sample UV Tiling will not affect the mapping coordinate tiling or any map tiling. It is strictly for viewing the sample.

Sample UV Tiling is set individually for each sample slot.

 **Save Preview**

Saves a material preview under a new name.

General Usage

Animate a material's parameters or colors. You can do this by moving to a frame other than zero, turning on the **Animate** button and chang-

ing the material's parameters or colors. Click **Make Preview**  and make a preview of the animated material. The preview created shows just the sample object in the material slot with the animated material. This preview is saved as a file named _medit.avi_. Click the **Save Pre-**

view button 🖼. Enter a new filename for the preview.

Usage Notes

When you create a material preview, the preview is always saved in the file named _medit.avi_. Only the most recently created preview is saved in this file. To prevent a preview from being overwritten the next time a material preview is made, click **Save Preview** and resave the preview with a new name. This preview can then be viewed at any time by choos-ing _File/View File_ from the menu and choosing the saved preview name.

Mat Editor

Select by Material

Selects all objects with the current material.

General Usage

Activate a material slot with triangles at its corners. Click **Select by Ma-terial** . A selection dialog appears, listing all objects in the scene. Objects with the currently selected material are highlighted. Click **Select** to select these objects, or change the selection set as desired.

Usage Notes

When a material is assigned to an object in the scene, triangles appear at the corners of the material slot. A material in the scene is called a *hot* material. **Select by Material** is only available when a hot material slot is currently active.

Show End Result

Toggles the evaluation of maps and materials up to the current level.

General Usage

Click **Show End Result** to turn it on or off .

When **Show End Result** is off, the material is evaluated only up to the current level. When **Show End Result** is on, all maps and materials that are part of the material are evaluated.

Usage Notes

When creating compound materials, both materials show in the slot, with each material taking up a portion of the sample object. If you want to view just one submaterial while working on it, turn off **Show End Result** while working on the submaterial to display just that submaterial.

Mat Editor

Show Map in Viewport

Displays the current map in all shaded viewports.

General Usage

Create an object and apply mapping coordinates to it. You can do this by turning on the **Generate Mapping Coords** checkbox for primitives, or by applying a **UVW Map** modifier to the object. Turn on shading for at least one viewport. Create a material with a map, such as a bitmap assigned to the **Diffuse** map attribute. At the map's child level, the **Show**

Map in Viewport button is enabled. Click **Show Map in Viewport**. The map is displayed on the object in the shaded viewport.

Usage Notes

Show Map in Viewport is primarily for seeing how a map lines up on an object. This option colors the object in shaded viewports, but will not create other effects such as bumps or reflections. You must render the scene to see these effects.

Only one map can be displayed for a material at a time. If you click **Show Map in Viewport** for a map and another map from the material is already displayed on the object, the new map will replace the old map in the shaded viewport.

The **Show Map in Viewport** button is enabled at a map's child level only. It is available only for 2D maps.

When you select the **Bitmap** map type, you must first choose a bitmap before the map will appear in shaded viewports.

The material does not have to be assigned to an object in the scene in order for the button to be enabled. If you click **Show Map in Viewport** and the material has not been assigned to an object, the map will not appear anywhere in the scene, but will appear as soon as the material is assigned to an object.

You can turn off all maps shown in viewports by choosing *Deactivate All Maps* from the *Views* menu. This action cannot be undone.

Many map types have a **Size** parameter. **Size** sets the size of the map in relationship to one *tile area.*

The number of tile areas in a Material Editor slot is determined by multiplying the **U**, V and **W Tiling** values. For example, if **Tiling** for **U** is 4 and **V** is 3, the material has 12 tile areas. These tile areas are spread over the sample object in the material slot. They affect the number of tile areas both in the material slot and on the object to which the material is applied.

If the **UVW 1**, **UVW 2, Explicit UVW 1** or **Explicit UVW 2** is chosen as the mapping method on the Coordinates rollout, the tile area works in conjunction with the map's **Size** value to determine the size of the effect.

The **Size** value refers to the size of the map relative to one tile area. When **Size** is 1.0, one unit of the map is placed in each tile area. For **Noise**, this is one blob of noise. For **Cellular**, this is one cell with bits of other connecting cells. For **Wood**, the **Grain Thickness** replaces **Size**, but the principle still applies: one chunk of wood grain fits into one tile area. The same is true for the **Planet** map's **Continent Size**, where one continent blob fits into one tile area.

A **Size** of 1.0 places a certain amount of the map in one tile area. A **Size** of 5.0 multiplies the size of the map by 5 in all directions, stretching the map to fill 25 tile areas. A **Size** of 0.5 reduces the map size by ½ in all directions, which reduces the map to ¼ of its size when **Size** is 1.0.

You can always set the size of 2D and 3D maps to your own specifications by keeping in mind the tile area and its relationship to the **Size** parameter. The apparent size of the map in the material slot can be reduced by increasing the **Tiling** values on the Coordinates rollout.

Tile areas on an object can also be affected by the **Tile** values for a **UVW Map** modifier or **UVW Xform** modifier. These tiling values affect only the object to which they are applied, not the tiling in a Material Editor slot. When one of these modifiers is applied to an object, the number of tile areas on the object is determined by multiplying the **Tile** values from the modifier and **Tiling** values from the map's Coordinates rollout. For example, if a map on an object has a **U Tiling** value of 3, and a **UVW Map** modifier is applied to the object with a **U Tile** value of 2, the number of tile areas around the U dimension of the object will be 6.

Mat Editor

Smoke

A map type that generates amorphous smoky patterns.

General Usage

See **Choosing Maps** under the **Material/Map Browser** entry.

Coordinates

See the **Coordinates Rollout (Non-Environment)** entry.

Smoke Parameters

Size	Sets the size of smoke effect in relationship to one tile area. See the **Size Parameter** entry.
Iterations	The number of times the smoke effect is calculated. Higher values yield more detail but take longer to render.
Phase	The smoke effect appears to be random, but in fact is calculated based on the **Phase** value. When **Phase** is changed, the smoke effect changes. Animating the **Phase** will animate the smoke function, changing the shape and location of the smoke tendrils over time.
Exponent	Controls the size and sharpness of the smoke tendrils created with **Color #2**. Increase this value to make the tendrils smaller and sharper.
Color #1	The background color for the smoke effect. Click on the color swatch to change the color, or click on the box under **Maps** to assign a map for the color. When the checkbox to the right of the box is on, the map is used. When the checkbox is off, the color in the color swatch is used.
Color #2	The color of the smoke tendrils. Click on the color swatch to change the color, or click on the box under **Maps** to assign a map for the color. When the checkbox to the right of the box is on, the map is used. When the checkbox is off, the color in the color swatch is used.
Swap	Swaps the color swatches and/or maps for **Color #1** and **Color #2**.

Mat Editor

Speckle

A map type that creates a speckled pattern from two colors or maps.

General Usage

See **Choosing Maps** under the **Material/Map Browser** entry.

Coordinates

See the **Coordinates Rollout (Non-Environment)** entry.

Splatter Parameters

Size	Sets the size of speckles in relationship to one tile area. See the **Size Parameter** entry.
Color #1	The first color to be used in the speckle effect. Click on the color swatch to change the color, or click on the box under **Maps** to assign a map for the color. When the checkbox to the right of the box is on, the map is used. When the checkbox is off, the color in the color swatch is used.
Color #2	The second color to be used in the speckle effect.
Swap	Swaps the color swatches and/or maps for **Color #1** and **Color #2**.

Splat

A map type that creates a pattern like splattered paint from two colors or maps.

General Usage

See **Choosing Maps** under the **Material/Map Browser** entry.

Coordinates

See the **Coordinates Rollout (Non-Environment)** entry.

Splatter Parameters

Size
Sets the size of the splatter dots in relationship to one tile area. See the **Size Parameter** entry.

 # Iterations
The number of times the splatter effect is calculated. Higher values yield more detail but take longer to render.

Threshold
Controls the balance between **Color #1** and **Color #2** in the splatter effect. Decrease this value to see more of **Color #1**, or increase this value to see more of **Color #2** in the splatter effect. When **Threshold** is at about 0.4, the colors are evenly balanced in the effect.

Color #1
The first color to be used in the splatter effect. Click on the color swatch to change the color, or click on the box under **Maps** to assign a map for the color. When the checkbox to the right of the box is on, the map is used. When the checkbox is off, the color in the color swatch is used.

Color #2
The second color to be used in the splatter effect.

Swap
Swaps the color swatches and/or maps for **Color #1** and **Color #2**.

Standard

The standard material type.

General Usage

The Standard material type is the default material type. Unless you specify otherwise, every material you create will be the **Standard** type.

When you first access the Material Editor after starting 3D Studio MAX or performing a *File/Reset* from the menu, the six sample slots hold six default sample materials, each of them the **Standard** material type. The rollouts for the **Standard** material type are displayed by default.

Basic Parameters

The settings under the Basic Parameters rollout determine the general appearance of the material without maps.

Shading
: Choose one of four types of shading for the material. **Constant** shades each face with one color, creating a faceted look on the object. **Phong** shades faces smoothly. **Blinn** is similar to **Phong**, but creates softened highlight edges. This shading type is the default, and will work for most of your materials. **Metal** is designed for use with curved, shiny surfaces. Metal shading uses the **Diffuse** color as the Specular color.

2-Sided
: When checked, the material is applied to both sides of the object's faces, both inside and out. Check this option when some or all of the object's face normals are flipped away from the camera, as when a mesh is imported from a CAD program. To find out more about normals, see the **Normal** entry in the *Panels* section of this book.

Wire
: When this checkbox is turned on, the material renders as a wireframe. The width of the wire is determined by the **Size** setting under the Extended Parameters rollout.

Face Map
: When this checkbox is turned on, maps used to define the material under the Maps rollout are applied to each face of the object.

SuperSample
: When this checkbox is on, the material's highlight is sampled (calculated) repeatedly during rendering. This option also slows down rendering time considerably. Turn on this option only when very high antialiasing is required on bump map highlights, such as when the camera is close to an object with a very detailed bump map.

Maps for each color swatch can be defined by clicking on the small blank button to the right of the color swatch. When a map has been defined, the letter **M** appears in the small box. If the map has been disabled, a small letter **m** appears in the box.

Ambient The color of the object when in shadow. To change the color, click on the color swatch to access the Color Selector. Change the color as desired. When **Lock Color** 🔲 is on, the **Diffuse** color changes when the **Ambient** color is changed.

Diffuse The main color of the object. To change the color, click on the color swatch to access the Color Selector. Change the color as desired. When **Lock Color** 🔲 is on, the **Specular** color changes when the **Diffuse** color is changed.

Specular The color of the highlight. To change the color, click on the color swatch to access the Color Selector. Change the color as desired. A highlight with the **Specular** color appears on the material only if the material is shiny. To see the **Specular** color, set the **Shin. Strength** value on the Basic Parameters rollout to a high amount.

Filter This color is added to the material's color when the material is partially transparent and the **Filter** or **Subtractive** options are selected under the Extended Parameters rollout.

🔲 **Lock Maps** Locks the **Ambient** and **Diffuse** maps. See Maps below.

🔲 **Lock Colors** Locks the **Ambient** and **Diffuse** colors, or the **Diffuse** and **Specular** colors, so they are always the same.

R, G, B The **Red**, **Green** and **Blue** values of the currently selected color. You can change these values here instead of changing them on the Color Selector.

H, S, V The **Hue**, **Saturation** and **Value** settings for the currently selected color. You can change these values here instead of changing them on the Color Selector.

Soften A value between 0 and 100 that softens the highlight. If **Shin. Strength** is much higher than **Shininess**, highlights may have harsh edges. Increase this value to soften highlight edges.

Mat Editor

Shininess	Sets the size of the highlight. When **Shininess** is low, the highlight is large. When **Shininess** is high, the highlight is small.
Shin. Strength	Sets the intensity of the highlight. When **Shin. Strength** is low, the highlight is dim. When **Shin. Strength** is high, the highlight is bright.
Self-Illumination	**Self-Illumination** causes an object to appear to glow with its own light source. This value sets the degree to which the material is self-illuminated. When this value is 100, the material has no dark areas, even when the object is in shadow. This effect can be likened to glow-in-the-dark paint. When this value is 0, there is no self-illumination. Values between 0 and 100 set partial self-illumination.
	Self-Illumination is ideal for materials for light globes and other light sources. However, an object with a self-illuminated material does not actually give off light in the scene.
Opacity	Sets the material's transparency. When **Opacity** is 100, the material is completely opaque. When **Opacity** is 0, the material is completely transparent. Values between 0 and 100 set partial transparency.

Extended Parameters

The **Opacity** section determines how the material will behave when completely or partially transparent.

Falloff	Sets the type of falloff for the transparency effect. When **In** is selected, the object appears more transparent on the inside. When **Out** is selected, the object appears more transparent on the outside. The **Amt** value sets the amount by which to vary the opacity at the extreme inside or outside. For example, if **Opacity** is set to 90, **In** is on and **Amt** is set to 20, the object will render as 70% opaque at its center.
Type	Sets the type of transparency to use when rendering transparent materials. **Filter** tints the material in semi-transparent areas with the **Filter** color on the Basic Parameters rollout or the **Filter** map under the Maps rollout. **Subtractive** subtracts the colors of semi-transparent areas from the colors behind it, making the semi-transparent areas darker. **Additive** adds the colors of semi-transparent areas to the colors behind it, making

Mat Editor

the semi-transparent areas brighter. To choose a transparency type for your scene, set up the model first, then try each one in your scene. The effects of transparency types vary greatly depending on the brightness of colors behind the transparent object. A scene that looks right with a black background will probably have to be adjusted once an environment map and background objects have been set up.

Refract Map/ RayTrace IOR
Sets the index of refraction (IOR) for a **Refraction** map. When this value is 1.0, no refraction takes place. At values below 1.0, the material refracts as if part of a concave surface (curved inward). At values above 1.0, refraction appears to take place on a convex surface, like that of a sphere. For best results, use values between 0.6 and 1.5.

The **Wire** settings indicate how the wire will appear when the Wire checkbox on the Basic Parameters rollout is on.

Size
Determines the size of each wire in the rendered image.

In
Determines how the **Size** setting will be used to set the width of each wire. When **Pixels** is chosen, the **Size** value refers to the width of the wire in pixels. When **Units** is chosen, **Size** refers to the width of the wire in the current units.

The **Reflection Dimming** settings dim reflection maps when in shadow.

Apply
Applies reflection dimming. If this checkbox is off, reflection maps will reflect as brightly in shadowed areas of the object as they do in bright areas. Turning this checkbox on yields more realistic results.

Dim Level
A value from 0 to 1 that specifies the amount of dimming on the reflection across the entire object. When this value is 0, the reflection is completely dark. When this value is 1, the reflection is bright all over. At 0.5, the reflection will be dimmed to half its original intensity all over the object. Use this value to specify the brightness for the darkest reflection areas, then use the **Refl Level** value to set the brightness of the reflection where the material is not in shadow.

Refl Level
A value from 0.1 to 10.0 that sets the brightness of the reflection for areas that are not in shadow. Use low values from 2 to 5 for the most realistic effect.

Maps

The Maps rollout is where you initiate the assignment of map types to the different aspects of the material.

Maps assigned on this rollout override or work with parameters of the same name on the Basic Parameters rollout. For example, a **Diffuse** map with an **Amount** of 100 overrides the **Diffuse** color setting on the Basic Parameters rollout. The **Diffuse** map defines the overall color of the object, and is the map that is used most often.

To assign a map type, click on the button labeled **None** across from the map name. The Material/Map Browser appears, allowing you to choose a map type. For information on each map type, see the corresponding entry in this section of the book. You can also See **Choosing Maps** under the **Material/Map Browser** entry.

When a map is assigned, the checkbox to the left of the map is automatically turned on, indicating that the map is in effect. To turn off a map's effect without losing the map definition, turn off the checkbox. You can turn a checkbox back on at any time to enable the map. Turning on a checkbox when no map has been defined creates no effect on the material.

Mat Editor

Amount	The strength of the map. All maps except **Bump** are limited to a range of 0 to 100. For these maps, the **Amount** sets a percentage strength of the map. For many of these maps, the **Amount** works with the parameters on the Basic Parameters rollout. For example, setting the **Amount** to 60 for the **Diffuse** map causes the map to define 60% of the material's color while the **Diffuse** color on the Basic Parameters rollout defines the other 40%. For **Bump** maps, the **Amount** can range from -999 to 999. An **Amount** of 0 does not use the map, which is the equivalent of turning off the checkbox to the left of the map.
Map	Maps for each aspect of the material are defined under the **Map** label. To assign a map type, click on the button labeled **None** across from the map name. When a map has been defined, the map name appears on the button. You can click on a previously defined map to return to its rollouts and parameters.
	To remove a map, click and drag on one of the buttons labeled **None** and drag it to the map button. To remove the effect of a map but keep all the settings for later use, uncheck the checkbox to the left of the map.

Mat Editor

To copy a map from another slot, click and drag on the map and move it to the new map button.

Ambient   Defines the color of the object in shadow. When **Lock Maps** 🔒 is on, the **Ambient** map is locked to the **Diffuse** map and cannot be changed directly. To change the **Ambient** map definition, you must change the **Diffuse** map definition. Leave **Lock Maps** on for realistic materials. For unusual effects, turn off **Lock Maps** and assign an **Ambient** map different from the **Diffuse** map.

🔒 **Lock Maps**   When this button is on, the **Diffuse** map definition is automatically assigned to the **Ambient** map definition. This is usually desirable. When **Lock Maps** is on, only the **Diffuse** map definition can be changed.

Diffuse   Defines the overall colors and pattern on the object. This map is the most often used map in the Material Editor.

Specular   Defines the colors of highlights. When **Amount** is 100, this map overrides the **Specular** color in the Basic Parameters rollout. Highlights appear only when **Shin. Strength** on the Basic Parameters rollout is greater than zero.

Shininess   Defines the size of the highlight using a map. Bright areas of the map create small highlights, while darker areas create larger highlights.

Shin. Strength   Defines the brightness of the highlight with a map. Bright areas of the map create bright highlights, while darker areas create dim highlights.

Self-Illumination   Defines the degree of self-illumination using a map. Bright areas of the map create high self-illumination, while darker areas create little or no self-illumination.

Opacity   Defines the transparency of the object with a map. When the **Amount** is 100, white areas cause the material to be opaque, black areas are transparent and gray areas are more or less transparent depending on their brightness. When the **Amount** is reduced, pure white areas become transparent to the degree the amount is reduced. For example, with an **Amount** of 75, pure white areas are 25% transparent. This map completely overrides the **Opacity** parameter in the Basic Parameters rollout regardless of the **Amount**.

Filter Color   Defines the filter color used with the **Filter** and **Subtractive** transparent types. This map is only used when the material is partially transparent, and **Filter** or **Subtractive** is chosen under **Type** on the Extended Parameters rollout.

Bump   Creates bumps in the material based on the brightness of different areas of the maps. The **Amount** can range from -999 to 999. A positive **Amount** causes the lighter areas of the map to raise the object in those areas, while darker areas push in the object. A negative **Amount** reverses this effect, with darker areas making raised bumps while lighter areas push the object in. A **Bump Amount** from 30 to 300 produces satisfactory results for most materials.

Reflection   Causes the material to appear as if it is reflecting the map. When the **Reflection/Refraction** map type is used to define this map, the material reflects the scene around it. When another type of map is used, the material appears to reflect the map. For example, using a bitmap of a landscape image as the **Reflection** map makes the material appear to be reflecting the landscape.

An **Amount** of 100 for **Reflection** causes the reflection to cover up the **Diffuse** color or map. Use an **Amount** between 10 and 40 for realistic results.

When the **Amount** is 100, there might not seem to be much difference between the effect of a **Diffuse** map and a **Reflection** map. However, when the scene is animated, the reflection on the object will change as the object or camera moves, while the **Diffuse** portion of the material will move with the object.

Refraction   Creates a refracted image from the map. When the **Reflection/Refraction** map type is used to define this map, the material refracts the scene behind it. To create a completely transparent material that refracts the scene behind it, use the **Reflection/Refraction** map type and set **Amount** to 100.

When another type of map is used to define the **Refraction** map, the material appears as if it is refracting the map. For example, a sky bitmap as the **Refraction** map makes a material that appears to be refract the sky behind it.

Mat Editor

Refraction works in conjunction with the settings under the **Opacity** section of the Extended Parameters rollout. The **Refract Map/RayTrace IOR** sets the index of refraction for the Refraction map. See Extended Parameters above for information on setting the **Refract Map/ RayTrace IOR** for realistic refraction.

When **Amount** is set to 100, any objects inside the object with the refractive material are not visible. Although the refractive object appears to be transparent, it is actually covered with an opaque refraction map made from the refraction information in the map definition.

Refraction mapping creates a rather broad distortion. It is not suitable for subtle refractions such as the slightly bent appearance of a pencil in a glass of water. Use **Refraction** maps for dispersed refractions such as the image of the sky through a bottle.

Dynamic Properties

The Dynamic Properties rollout sets parameters for motion and collision which are used with the **Dynamics** utility. These attributes are applied to an object when the material is applied to the object. These properties can also be assigned to an object with **Dynamics** on the

Utilities panel 🔧. See the **Dynamics** entry in the *Panels* section for information on using dynamic properties in a scene.

Bounce Coefficient A value from 0 to 1 that sets the bounciness of the object. A value of 0, which creates little bounce, is suitable for a bowling ball or cannonball. A value of 1 creates a high bounce like that of a rubber ball.

Static Friction A value from 0 to 1 that sets the object's friction when sitting still. Higher friction makes it harder to get an object moving. A value of 0 creates little friction, while a value of 1 creates an enormous amount of friction. Use low values such as 0.1 or 0.3 for realistic results.

Sliding Friction A value from 0 to 1 that sets the object's friction when moving. An object's static friction can be quite different from its sliding friction. For example, a hockey puck on ice needs a good-sized thwack to get it moving, but once it's going it moves with little friction. As a start, set **Sliding Friction** to half the **Static Friction** value.

A map type that creates a random spotted pattern.

General Usage

See **Choosing Maps** under the **Material/Map Browser** entry.

Coordinates

See the **Coordinates Rollout (Non-Environment)** entry.

Stucco Parameters

Size	Sets the size of stucco dots in relationship to one tile area. See the **Size Parameter** entry.
Thickness	A value between 0 and 1 that sets the amount of blur between the two colors. When **Thickness** is 0.0, the border between the two colors is sharp. When **Thickness** is 1.0, the border is highly blurred.
Threshold	Controls the balance between **Color #1** and **Color #2**. Increase this value to see more of **Color #1**, or decrease this value to see more of **Color #2**. When **Threshold** is at about 0.4 and **Thickness** is 0.2, the colors are evenly balanced.
Color #1 Color #2	The colors to be used in the stucco effect. Click on the color swatch to change the color, or click on the box under **Maps** to assign a map for the color. When the checkbox to the right of the box is on, the map is used. When the checkbox is off, the color in the color swatch is used.
Swap	Swaps the color swatches and/or maps for **Color #1** and **Color #2**.

Mat Editor

Thin Wall Refraction

A map type that creates a refraction effect for thin, transparent objects.

General Usage

See **Choosing Maps** under the **Material/Map Browser** entry.

A **Thin Wall Refraction** map should be used only as a **Refraction** map on a **Standard** material.

Thin Wall Refraction Parameters

Apply Blur	Applies blur to the refraction per the **Blur** setting.
Blur	Sets the amount of blur on the refraction. Some blur is desirable to anti-alias the refraction.
First Frame Only	When this option is selected, refraction is generated on the first frame only, saving rendering time. This image information is then used throughout the animated sequence. Turn on this checkbox when the reflection or refraction is not expected to change over the course of the animation.
Every Nth Frame	When this option is selected, refraction is generated at intervals throughout the animation. The frame intervals are specified by the number spinner. When possible, set the number higher than 1 to save rendering time.
Use Environment Map	When checked, environment maps are taken into account when rendering the refraction. When unchecked, environment maps are ignored during rendering. Uncheck this option if you're using a **Screen** environment map as the background as it will not render accurately with refraction.
Thickness Offset	When an object is refracted in a thin transparent object, the object appears to be slightly offset. **Offset** sets the amount by which the refracted image is offset when seen through the transparent object.
Bump Map Effect	Affects the amount of refraction due to the currently assigned **Bump** map. If no bump map is currently assigned as part of the material, this setting has no effect. Note that the **Thin Wall Refraction** map should not be assigned as a bump map. The **Bump Map Effect** works with whatever bump map is assigned next to **Bump** on the Maps rollout at the material's parent level.

Mat Editor

A material type created from two materials, one for the top and the other for the bottom of the object.

General Usage

To create a **Top/Bottom** material, click the **Type** button labeled **Standard** on the Material Editor toolbar. The Material/Map Browser appears. Choose **Top/Bottom**. Choose whether to keep the **Standard** material as a submaterial or to discard it. The rollout for the **Top/Bottom** material appears.

If you are unfamiliar with the concept of compound materials, see the entry **About the Material Editor** at the beginning of this section.

Basic Parameters

Top Material	Click on the button labeled **(Standard)** to create the **Top Material**. The display changes to reflect the settings for a **Standard** material at the child level. You can change the type of material, or work with a **Standard** material. Click **Go to Parent** 🔼 to return to the **Top/Bottom** parameters. The checkbox next to the button can be turned off to temporarily turn off the material.
Bottom Material	Click on the button labeled **(Standard)** to create the **Bottom Material**.
Swap	Swaps the **Top Material** and **Bottom Material**.
World	Uses the world coordinate system's Z axis as the reference point. All faces pointing in the positive Z direction will receive the **Top Material**, while faces pointing in the negative Z direction will receive the **Bottom Material**.
Local	Uses the object's local Z axis as the reference point. All faces pointing in the positive Z direction will receive the **Top Material**, while faces pointing in the negative Z direction will receive the **Bottom Material**.
Blend	Blends the two materials where they meet. A value of 0 produces no blend, while a value of 100 blends the entire material. A **Blend** value of 10 or 20 works well for most models.

Position	Changes the angle at which a face normal qualifies for the **Top Material** or **Bottom Material**. A **Position** value of 50 is the default. **Position** values below 50 assign the **Top Material** to more faces, while values over 50 assign the **Bottom Material** to more faces. Watch the sample slot in the Material Editor as you adjust this value to see the effect on your material.

Usage Notes

The **Top/Bottom** material type has a very specialized use. It was developed to solve the problem of material assignment on complex objects requiring two materials, each on opposite sides. You probably won't need to use this type of material very often.

The easiest way to think of a **Top/Bottom** material is to consider an object's orientation in the Top viewport. When the **World** option is selected, faces pointing toward you in the Top viewport receive the **Top Material**, while faces pointing away from you receive the **Bottom Material**. More technically, faces of the object with normals that point in the positive Z direction of the world coordinate system receive the **Top Material** while faces with normals pointing in the negative Z direction receive the **Bottom Material**.

For information on face normals and what they are, see the **Normal** entry in the *Panels* section.

Assigns a map that renders vertex colors.

General Usage

See **Choosing Maps** under the **Material/Map Browser** entry.

Usage Notes

The **Assign Vertex Colors** utility on the **Utilities** panel ⟨T⟩ assigns colors to vertices based on the current material. The **Vertex Color** map, when used as a **Diffuse** map, creates a material that can be applied to such an object so the vertex colors will render.

This type of map is intended primarily for programmers and game developers with special needs. For information on how this map is used, see the **Assign Vertex Colors** entry in the *Panels* section of this book.

Water

A map type that creates watery, wavy effects.

General Usage

See **Choosing Maps** under the **Material/Map Browser** entry.

Coordinates

See the **Coordinates Rollout (Non-Environment)** entry.

Water Parameters

Num Wave Sets In large bodies of water, there are usually several sets of waves moving at once. This value sets the number of wave sets. Set this value high for choppy water or low for calm water.

Wave Len Max Sets the maximum wave length for all waves. Each wave length is randomly chosen as a length between this value and the **Wave Len Min**.

Amplitude The depth of the wave is simulated by increasing the contrast between **Color #1** and **Color #2**. A high **Amplitude** produces high contrast between these colors, simulating a deep wave. A low **Amplitude** produces low contrast between the two colors.

Wave Radius Sets the radius, in units, for the base wave. Large values produce large, circular waves, while smaller values produce small, dense waves.

Wave Len Min Sets the minimum wave length for all waves. Each wave length is randomly chosen as a length between this value and the **Wave Len Max**.

Phase Shifts the waves. Animating the **Phase** will animate the waves, changing the shape and location of the waves over time.

Distribution When **3D** is selected, the wave is generated in all three directions. When **2D** is selected, the wave is generated only on the XY plane. When using **2D**, apply a **UVW Map** modifier to the object with **Planar** mapping coordinates to orient the water as desired.

Random Seed Sets the base number for random calculations. Change this number to observe different water effects. This value cannot be animated.

Color #1 The etched color for the wave effect. Click on the color swatch to change the color, or click on the box under **Maps** to assign a map for the color. When the checkbox to the right of the box is on, the map is used. When the checkbox is off, the color in the color swatch is used.

Color #2 The background color for waves. Click on the color swatch to change the color, or click on the box under **Maps** to assign a map for the color. When the checkbox to the right of the box is on, the map is used. When the checkbox is off, the color in the color swatch is used. This color should be darker than **Color #1** to create water effects.

Swap Swaps the **Color #1** and **Color #2** maps or swatches.

Usage Notes

To animate moving water, you can animate either the **Offset** values or the **Phase** value. Animating the **Offset** values moves the water effect but does not change the shape of the waves. Animating the **Phase** alone changes the shape of the waves but does not move them, simulating calm water. To animate moving, changing waves, animate both the **Offset** values and the **Phase** value.

Mat Editor

Wood

A map type that uses two colors or maps to define a wood pattern.

General Usage

See **Choosing Maps** under the **Material/Map Browser** entry.

Coordinates

See the **Coordinates Rollout (Non-Environment)** entry.

Wood Parameters

Grain
Thickness

Sets the thickness of the grain. When **XYZ** is selected as the mapping method on the Coordinates rollout, the thickness is set in units. When **UVW 1** or **UVW 2** is selected as the mapping method, a **Grain Thickness** of 1 takes up the size of one tile area. If **Grain Thickness** is 20, the grain is increased in size by 20 times the size of one tile area. If **Grain Thickness** is 0.5, the grain is reduced to half the size of one tile area.

Radial Noise

Creates noise in the grain on the plane perpendicular to the grain.

Axial Noise

Creates noise in the grain on the plane parallel to the grain.

Color #1

The first color to be used for the wood. Click on the color swatch to change the color, or click on the box under **Maps** to assign a map for the color. When the checkbox to the right of the box is on, the map is used. When the checkbox is off, the color in the color swatch is used.

Color #2

The second color to be used for the wood. Click on the color swatch to change the color, or click on the box under **Maps** to assign a map for the color. When the checkbox to the right of the box is on, the map is used. When the checkbox is off, the color in the color swatch is used.

Swap

Swaps the color swatches and/or maps for **Color #1** and **Color #2**.

Appendix

One of 3D Studio MAX's finest features is its ability to render high quality images. Of course, the quality of the image is determined largely by the artist's ability to assemble the scene with appropriate model detail, lighting and materials. Although 3D Studio MAX is capable of broadcast quality rendering, the bulk of responsibility for image quality falls on the artist and how he or she uses the tools. In other words, you can't expect MAX to make a beautiful rendering out of a sloppy scene.

3D Studio MAX is also capable of rendering broadcast quality images for video output. "Broadcast quality" is a subjective term used by the video industry to mean "good enough to be broadcast on television." Broadcast quality refers to the software's resolution and color depth capabilities as well as its ability to render with realistic color and lighting.

This appendix is devoted to MAX's various tools for fine-tuning a rendering and getting the most of out of its rendering capabilities, including the tools required to create broadcast-quality renderings.

Rendering Concepts

VFB

The VFB (virtual frame buffer) is a viewing window for files. The VFB appears when the scene is rendered, when files are viewed, and anytime a bitmap is to be displayed.

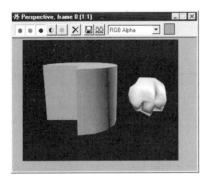

The buttons at the upper left display – **Enable Red Channel** ⬛, **Enable Green Channel** ⬛ and **Enable Blue Channel** ⬛ – display the image's red, green and blue channels respectively. You can also view the image's alpha channel with the **Display Alpha Channel** button ⬛, or see it as a monochrome image with the **Monochrome** button ⬛. The image can also be cleared from the VFB with **Clear** ✕, or saved to a bitmap with **Save Bitmap** ⬛. The bitmap is saved in full RGB color regardless of whether all color buttons were depressed at the time the file was saved.

Click **Clone Virtual Frame Buffer** ⬛ to make a copy of the VFB.

For information on the file and its colors, right-click and drag across the image. A window appears with information about the file, and also about each color over which the cursor moves. When you release the mouse, the last color accessed appears in a color swatch at the upper right of the VFB window. You can copy this color to another color swatch on another window or panel, or you can click the color to see the Color Selector and get the color's values.

You can zoom into the VFB by holding down the **<Ctrl>** key and clicking on the display. To zoom out, hold down **<Ctrl>** and right-click on the display. To pan the display, hold down **<Shift>** and click and drag the cursor.

Although you can change the display to view the file in many different ways, you cannot save the changes on this window. In addition,any changes to the color on the Color Selector are not recorded on the image.

Fields

In video and television, frames are displayed by drawing a series of horizontal lines. As each frame is displayed, first the odd numbered lines of the frame are drawn, then the even numbered lines. This drawing of lines is performed so quickly that the process is not discernible to the naked eye.

A group of odd or even numbered lines is referred to as a *field*. Each frame has two fields, one for odd numbered lines and one for even.

When **Render to Fields** is checked on the Render Scene dialog, each frame is rendered in two passes, one for each field. At the frame time, the odd numbered lines are rendered. Then MAX moves to a time between the current frame and the next frame, and renders the even numbered lines at that point. When a series of frames are rendered in this way and played back on a video monitor, the motion is extra smooth, and simulates playback at double the frame rate. For example, when fields are rendered at 30fps, the playback simulates 60fps, making the motion appear to be very smooth.

For AVI files, film or any other kind of output other than video, a field-rendered animation looks odd when played. Use field rendering only when rendering for video playback.

You can also choose whether to render odd or even numbered lines first when rendering fields by choosing **Odd** or **Even** under the **Rendering** tab of the Preference Settings dialog. This dialog can be accessed with the *File/Preferences* menu option. Most of the time, odd numbered fields are displayed and rendered first, but some equipment may require even numbered fields to be rendered first. For example, if you plan to play the video sequence backward on a device that plays odd fields first, you should render with the **Even** option selected.

Appendix

Super Black

Super black is a sub-black color used in compositing. On some video editing systems, super-black is used as a "transparent" color, so other video elements can be composited behind the rendered image where black appears. Unless you are creating a video animation for a production company that specifically requests super black, you need never concern yourself with it.

3D Studio MAX cannot produce a sub-black color accurately, so when super black is required, MAX treats RGB 0,0,0 as super black and uses a threshold value to ensure that all other black colors in the scene are lighter than this color. Under the *File/Preferences* menu option, under the **Rendering** tab, the **Threshold** value sets the minimum intensity (**V** value) for black colors on rendered objects. For example, if **Threshold** is 15, the **V** value of black objects and dark areas of the screen will never drop below 15. The background area will render as pure black with a **V** value of 0. Editing equipment can then be adjusted to consider this pure black area a super black area.

The **Threshold** value is used in renderings only when the **Super Black** checkbox on the Render Scene dialog is checked before rendering.

Time Code

Time code is a numerical reference for time in video. Time code is represented as **Hours:Minutes:Seconds:Frames**. For example, a time code of 00:01:12:06 is one minute, 12 seconds and 6 frames from the starting point of the time code. Time code is used by video players, and by editing programs such as Speed Razor and Adobe Premiere.

Ticks

Animation time can be counted in many different ways. Sometimes it is necessary to switch from one time-counting method to another. For example, an animation created at 30fps for the American market must be reworked if it is to be played at 25fps for the European market.

To make these types of conversions easier, the *tick* was invented. A tick is 1/4800 of a second. The number 4800 can be evenly divided by 24, 25 and 30, the three most common frame rates.

When the tick is used for counting time, you can switch between different frame playback rates without causing time problems.

Appendix

Video Color Checking

When preparing images for video or broadcast, it is vital that you watch out for colors that don't display well on video. Very bright, highly saturated colors, such as pure red or pure blue, tend to smear or fuzz when played back from videotape. This is particularly important if you are new to video output. A rendering with hot, bright unsafe video colors looks fine on the computer screen, and may even seem stable on the video monitor or first generation of tape. But when VHS copies are produced from the original or the images are broadcast on television, the unsafe colors will be smeared and fuzzy and may even "buzz" (a term for colors that bounce and dance quickly, like a swarm of buzzing bees).

One only has to watch a local cable channel to see smeary, buzzy, fuzzy video in action. Inexperienced video producers often make commercials with extremely bright colors in an attempt to get your attention. This is particularly evident in titling, where the name of the company or product appears on the screen in screaming red or blue.

Smeary, buzzy, fuzzy video is the hallmark of the amateur. Just about anyone can spot the difference between a prime-time commercial and a low-budget cable commercial, but not everyone can say why. Lighting, staging, script, acting and video format play a part in the quality of a commercial, but the most visible difference is the color. Professional looking commercials contain mostly subdued colors with discreet splashes of bright color. During the post production phase of these professional productions, the saturation of bright colors are carefully monitored to ensure they will not fuzz when they appear on your television.

The time to discover that the fruits of your labor are infested with buzzy, smeary colors is not after the final rendering is complete. All that rendering time will be wasted if you have to change the colors and render again. Always check video colors before the final rendering.

Color Checking with 3D Studio MAX

3D Studio MAX has built-in tools to check for unsafe colors. The detected colors can then be adjusted, either manually by yourself or automatically by 3D Studio MAX, to produce more acceptable color ranges in the final rendering.

MAX's tools are not completely foolproof. The only surefire way to tell if colors will smear is to try them out in a variety of output situations, but this would take an inordinate amount of time. MAX's tools provide a quick way to detect and fix obvious video color problems along with subtle violations that might escape the untrained eye. If you're new to video output, these features are particularly useful in training

Appendix

you to find problem areas; after a while, you'll be able to spot and fix them yourself before rendering. Even veteran video artists use color checking before the final render just to be on the safe side.

The brightness of a color on video corresponds roughly to the Saturation setting on the 3D Studio MAX Color Selector. You can avoid color problems by lowering this setting on the colors you use in the Material Editor. However, there will be times when you need a bright color and want to use as high a **Saturation** setting as possible. You may also have lighting requirements that make materials appear more saturated when in the rendering.

In 3D Studio MAX, color checking can be performed in two areas, the Material Editor and the Render Scene dialog. On the Material Editor, the **Video Color Check** button at the right of the sample slots checks the material itself for unsafe colors. Any unacceptable colors are flagged as black in the sample slot. You can then alter the material settings until the sample slot shows no black, indicating that it is free of unsafe colors. This method provides a quick check for non-safe colors, but since color in the rendering relies strongly on lighting and other factors, this method is not foolproof.

To enable automatic color checking when rendering, check the **Video Color Check** checkbox on the Render Scene dialog. This dialog can be accessed prior to rendering by clicking **Render Scene** or choosing *Rendering/Render* from the menu.

Video colors are flagged or altered according to the current preferences. To see or change the preferences, choose *File/Preferences* from the menu and click the **Rendering** tab. Under the **Video Color Check** section, there are three flag/alter options.

Flag with black changes any unsafe color pixels to black in the rendered image.

Scale Luma darkens the pixel colors until they are in the safe video color range. This is the equivalent of reducing the color's **Value** on the Color Selector.

Scale Saturation brings down the saturation of the pixel colors until they are in the safe video color range. This is the equivalent of reducing the color's **Saturation** setting on the Color Selector.

Sample Usage

1. Reset 3D Studio MAX.

2. Access the Material Editor ![icon]. Select any sample slot. Under Basic Parameters, click on the **Diffuse** color swatch. On the Color Selector, change the material to the brightest red with these values:

R 255
G 0
B 0

Note that the **Saturation** on the Color Selector is also at 255, its highest possible value.

3. Create a sphere of any size. Click **Zoom Extents All** ![icon].

4. Apply the bright red material to the sphere.

5. Activate the Perspective viewport. Click **Quick Render** ![icon] to render the scene.

The scene renders as a bright red ball.

6. With the Perspective viewport still active, click **Render Scene** ![icon]. On the Render Scene dialog under **Options**, check **Video Color Check**. Click **Render**.

An area at the upper left of the sphere renders as black to indicate the presence of unsafe colors. Although the shades of red at that part of the sphere might not look as though they are much brighter than the rest of the sphere, the saturation of those colors peak over the top of the acceptable video color range.

7. Choose *File/Preferences* from the menu. Click the **Rendering** tab.

Under Video Color Check, note that **Flag with black** is on. This caused the unsafe pixels to render as black.

8. Turn on **Scale Luma**. Click **OK** to exit the Preferences dialog.

9. Render the Perspective view with **Video Color Check** on.

This time, the offending pixels are rendered as red, but with a shade slightly darker than the original red color on that part of the sphere.

Appendix

The luma values (L) of the offending colors were automatically reduced until they fit in the acceptable color range. Note that the actual material in the Material Editor or on the object has not been changed; only the colors in the rendering were changed.

Although the colors are now in the video color range, you can clearly see the patch where the change occured. When only a few scattered pixels need to be corrected, scaling the luma values does not visibly alter the image. But when large, contiguous areas need correction, such as the portion changed on the sphere, it is better to change the material itself than to scale the luma value.

10. On the Material Editor, click the **Video Color Check** button . This button is located to the right of the sample slots.

In the sample slot, a small portion of the sample sphere turns black. This indicates that the color of the sphere at that spot exceeds the safe video color range.

Note that colors in sample slots are largely determined by the Material Editor's sample lighting setup, which may or may not correspond to the lighting in your scene. As you gain experience with 3D Studio MAX and video output, you will know when you can do without the Material Editor's video color checking feature. If you are new to video output, you should always use this feature to ensure your colors fall within the safe saturation range.

11. Click the **Diffuse** color swatch on the Basic Parameters rollout. On the Color Selector, lower the **Saturation** setting to 240.

The black area of the sphere has now disappeared, indicating that the colors in the material are safe for video. Even though the **Saturation** has been brought down a few notches, the sphere is still a very bright red.

12. Choose *File/Preferences* from the menu. Turn on **Flag with black**. Click **OK** to exit the dialog.

13. Render the Perspective view again.

The sphere renders without a black area on it.

There may be times when bright lighting causes a material to pass the video color checking test in the Material Editor, but not in the rendering. Always check colors in the rendering even if you have already checked them in the Material Editor.

Adjusting Materials for Video

Once color problem areas are found, it's up to you to alter your materials so colors in the final rendering conform to the safe video color range. There are a number of ways to do this.

When creating bright materials, keep each color's **Saturation** well below 255. A **Saturation** of 210 or 220 is sufficient for most bright colors. For dark colors, higher **Saturation** is acceptable but should be used only when necessary. The rule of thumb is that you can mix high **Saturation** with a low **Value** setting, or low **Saturation** with a high **Value**, but you cannot have both high **Saturation** and high **Value** settings.

Bitmaps can also cause problems with saturation. For a very bright bitmap, reduce the overall saturation in a paint program before using it in your scene. Image saturation can be reduced by decreasing both the contrast and brightness of the image. If the bitmap is to be used as a **Diffuse** map in 3D Studio MAX, you can also effectively tone down its saturation right in the Material Editor by lowering the **Amount** of the **Diffuse** map to a lesser value such as 80 or 90. In this case, the additional 20 or 10 percent of the material color will be made up by the **Diffuse** color under Basic Parameters. Set the **Diffuse** color to a medium gray to mute the **Diffuse** component's saturation.

Once you have set up your scene, render a few representative frames of the animation and use 3D Studio MAX's tools to check for unsafe colors. Fix any troublesome areas by altering materials as described above. Once your images pass the test, view them on a video monitor and check again for buzzing areas and bright colors. Repeat this test until your video images are picture-perfect. Then you're ready to render the final animation.

Fine Detail in Video Materials

Very fine, deep bumps on an object may also cause buzzing in the video image. To minimize this effect, reduce the depth of the bumps by decreasing the **Bump** map **Amount**. You can also. increase the width of the bumps by decreasing the **Tiling** values for the bump map. This technique will cause loss of definition on the bumps, but by the time your image reaches the video screen, this level of detail won't be discernible anyway.

In fact, many fine details are lost in the process of going from the computer to video. A detailed object in a dimly lit corner of the scene can often be replaced by a low resolution stand-in. Replacing high-face-count objects with less detailed substitutes can substantially reduce your rendering time.